About the Author

Robert MacNeil Wilson graduated in Mining Engineering at Nottingham University. Aged only 24, he was in charge of a coalface, half a mile underground in one of Warwickshire's mines. During the miners' strike, he lived in a pit village and was in sole charge, on several occasions, when his pit was besieged by massed pickets. Robert is a Chartered Engineer and rock musician.

THE ENEMY WITHIN

Robert MacNeil Wilson

The English Civil War Part II (Personal Accounts of the 1984/85 miners' strike)
by Jeremy Deller, published by Artangel, 2001.

Matador
9 Priory Business Park,
Wistow Road, Kibworth Beauchamp,
Leicestershire. LE8 0RX
Tel: 0116 279 2299
Email: books@troubador.co.uk
Web: www.troubador.co.uk/matador
Twitter: @matadorbooks

ISBN 978 1785893 544

British Library Cataloguing in Publication Data.
A catalogue record for this book is available from the British Library.

Printed and bound by CPI Group (UK) Ltd, Croydon, CR0 4YY
Typeset in 11pt Aldine401 BT by Troubador Publishing Ltd, Leicester, UK

Matador is an imprint of Troubador Publishing Ltd

To my pit dad, Mick Cooper

And to Les Baldry,
a big man who stood by a young 'non-stat', when the chips were down,
on two, seriously interesting occasions.

With my boundless gratitude to Jo Faulkner for her support,
without which 'The Enemy Within' would have remained unwritten.

Special thanks to Moira Clinch
for creating the brilliant, characterful
cover that adorns 'The Enemy Within'.

Contents Page

Peril

Nothing could stop it now. They'd tried everything.

Round the corner, out of sight, he broke into a run. Stooping under the twisted, steel arches, he scurried along between the snaking rails of the narrow track.

He hadn't expected to survive this long; after hours in mortal danger, he revelled in the relief of getting away from where it would burst in on them.

Rounding a bend in the tunnel, he pulled up, his pretence of flight over. He was going nowhere.

He just needed a few seconds alone then he'd be able to see it through.

The suspense had been tortuous but it wouldn't be long now, most likely only a matter of minutes. There was no prospect of escape but at least they wouldn't know anything about it. When it came, it would be instant; oblivion.

They'd be wondering what he was doing.

He cast a glance over his shoulder; the only light was from the lamp on his helmet. No one had followed him.

He shook his head.

As he got down to his knees on the dirt floor between the wooden sleepers, the wet cloth of his trouser legs bunched up, pinching at the skin on the backs of his knees.

'How did I end up here?' He whispered.

People on the surface would be starting their evenings; his mates getting ready to go out for a couple of pints, his family at home, his girl.

His girl.

All of them oblivious to his peril, taking comfort, safety, life itself for granted; in their ignorance, confident of seeing him again.

But he knew his destiny. He would never see their world again. His fate was to die; down here, like this.

He stemmed the flow of thoughts. He hadn't come here for that.

He closed his eyes and cleared his head.

When all else had failed he'd felt the urge to do it, one last time.

He'd tried doing it, back in the heading, without the others noticing. But praying silently, eyes open, had felt inadequate, too likely to be missed, to go unheard at this depth. He had to do it properly, out loud.

He bowed his head, clamped his hands together and took a deep breath. It tasted earthy, redolent of the grave.

He rid himself of it, letting it out in a rush.

'Help me find a way to beat this. Let us live.'

He paused.

'Just let me live to see my wedding day,' he murmured.

Knowing he was asking the impossible, he screwed his eyes tighter, willing his plea on its way.

That would have to do.

In the quiet, his breathing calmed. He opened his eyes and looked around. After that brief moment of hope, it was a bitter blow to find himself still down there, still condemned.

Getting back up to his feet, he noticed the dust stuck to the knees of his orange overalls. Grabbing each leg, in turn, he beat them with his free hand to knock some of it away. He couldn't have the rest of them guessing what he'd been doing.

Straightening up a little, he turned, took another deep breath then stepped out to head back, to face it with them, to see it through, to the end.

PART ONE

Whitacre Heath

Together

Eighteen months earlier.

The humped-backed bridge was barely wide enough for a single vehicle. Knowing it well, he let his car's speed fall away, tensing as the parapets closed in on him from both sides; alert, as always, to the possibility of some idiot appearing, at speed, from the opposite direction and ploughing, head-on, into him.

Cresting the brow, he saw the road ahead was clear. He breathed out then snatched a glance to the left, before accelerating away.

In the middle of the field of tired, autumn grass, an old man was picking his way back along the winding path from the river. A small boy hurried towards him with urgent, bouncing steps, a child-sized fishing-rod over one shoulder.

In the lay-bye, further down the road, a young woman leant on the stile into the meadow. Dashing away the strand of brown hair fluttering over her eyes, she called after the boy. He turned, shouted something in reply then gave her a quick, little wave as he pressed on to where the old man now stood waiting.

As the car window dropped below the hedge, depriving him of anymore of the scene, the woman's slim figure caught his eye as she strode round to the back of her beige Allegro to wait for his car to pass.

He waved in acknowledgement and to share a little in the moment. As he drove past, he looked into her eyes and was rewarded with a slight smile.

Horace had an open face, his eyebrows always slightly raised, his lips always ready to form the gentle smile he bestowed, now, upon his grandson.

'Hiya Grandad! How are they bitin'?'

'Ooh, I've copped a few, but we'll fare better,' Horace said, patting the boy on the shoulder as he frisked past. 'Now you're 'ere.'

They made for the river bank, where the willows draped into the water,

3

Horace's stiff-hipped, rolling gait an awkward contrast to the light, skipping steps of the child in front of him.

Although Horace was diminutive, one consolation of Martin's shortness was that it made their height difference appropriate for that of an eight-year old and his grandfather.

'You made good time, lad,' Horace said.

'I know. Me mum picked me up straight from school so I didn't have to walk back home,' the boy called back, over his shoulder. 'I got all me stuff ready last night and got changed in the car.'

'Good lad, it'll gi' us a good couple of hours, toppin' up that keep-net.'

Settled back at his peg, on his canvas-seated stool, Horace drew satisfaction from how much his grandson had absorbed of the angler's art. Snatching a glance at the little boy to his side, he mused on their similarities.

Facially, the resemblance was unmistakable. They both had the same, gentle manner, both tended to wear expressions of innocent contentment, indicative of a keenness to please, a shared aversion to conflict or upset. Both, so different to Martin's father.

Paul was more like his mother; unlike Horace or Martin, he could never just let things go. But there was nothing that Horace would wish to change about his son.

Horace had been lost in his thoughts as Martin chattered away, quietly, to him, until a question from the boy demanded a response.

'I bet you're excited aren't you, Grandad?'

'What's that, lad?

'I bet you're excited. 'Bout tomorrow.'

'Ar, yes. Tomorrer.'

'You have to smile all day, on your birthday, y'know.'

'I know. And you have to celebrate 'em when they keep a-comin' round when you get to my age.' Horace said, gazing out at where his float bobbed on the river.

'Fifty-seven candles to blow out.' Martin said.

'There ain't that many candles in the world. And, besides, I ain't got that much puff.'

And that was no lie, the years in the thick dust of the coalface had wrought their inevitable toll.

As Martin tried to tantalise his grandfather with the surprise they had in store for him, Horace pondered on his birthday and its implications.

After another nightshift down the pit, he'd emerge, a year older, a year closer to retirement. As usual he'd get home, make a pot of tea and take up a cup each for Pat and himself before taking Sally, their Highland Terrier, over the Heath.

When they got back, he'd get an hour or two in bed, until the squeals and laughter from the school playground roused him.

Three years on from tomorrow and, God willing, he'd still be around to work his last shift underground. With his fortnightly cycle of days and nights ended, the only time he'd go to the pit would be on Fridays, to draw his pension from the pay office windows before meeting up with the other old colliers in the clean side of the colliery canteen for a coffee, a weekly catch up and the inevitable bout of reminisces.

He'd have more time to spend fishing then than he could possibly want.

Would the River Anker still have the same draw for him when he no longer needed its open spaces, fresh air and tranquillity as a contrast to the noise, dust and confines of the face?

Too often, miners' retirements were short-lived. His old mate, Bill, Paul's father-in-law, had lasted only nine months, when he'd finished a couple of years earlier. It seemed as though their bodies just couldn't hold themselves together anymore, once their years of toil down the pit were done.

He felt the world owed him nothing but he did hope that he and Martin would be granted a few more seasons on the river bank and he and Pat would be able to enjoy some time together, on his Coal Board pension.

A couple of hours later, as the sun sank behind the pit headgear on the horizon, Horace looked up at the gathering clouds.

'I think we'd better be goin', lad,' he said. 'It's gettin' proper chilly.'

Martin gave him a sad glance then busied himself, helping to weigh their catch, return the fish to the river's flow, fold their stools and pack away their tackle.

He shared the load with his grandad, carrying their rods and stools, deep in thought as they made their way back along the footpath to the lay-bye where Horace's Reliant three-wheeler waited.

'It must be a bit strange isn't it, going down the pit at night?' Martin said, struggling to imagine his dignified, old grandfather crawling around in the filth to be found, deep underground, in the frightening, alien world of a mechanised coalface.

Horace shook his head, smiling.

'It don't make much difference, not once yer down there, lad.'

'It seems strange, though, to be going to work when everyone else is asleep, in bed. I don't think I could do it.'

'Well, you ain't got to worry about that yet a-while, that's years away. Anyway, work 'ard at school and you might be able to get to university. We've got a young manager who went. It 'elps you get on.'

As they walked on, Horace recalled an incident involving his new boss, who had arrived at the pit two months earlier.

It was the start of Mr Greave's third week at the pit and the young mining engineer had been spending time on 85's face, as he did every day. Horace had been on days that week and, that morning, he had been given a new face-trainee to supervise.

85's was a typical, advancing face. Two hundred yards long, the gate road at each end had extended to over a mile in length, as the face had advanced. The maingate, the intake airway for the face, transported the coal away on its belt conveyor. The tailgate, the return airway, was used to carry supplies to the face, along its narrow-gauge railway.

That morning, thirty metres down the face, the larger of the face's two coal-cutting machines was grinding its way down the four-foot high face towards the maingate. The ferocious roar of the shearer's two discs, as they tore away another strip of virgin coal, carried up to where Horace knelt.

Horace's old mate, Tommy, the face-team's snaker, had shoved the face conveyor over to the newly-cut coalface behind the shearer's cut, using the snaking rams in the bases of the chocks, the powered roof supports that held up the roof, taking care to keep the face straight.

As one of the face team's three chockers, it was Horace's job to crawl through the face behind the snaker, lowering the chocks and drawing them in. [Note 1]

Four feet wide and extending back nine feet, each chock weighed about four-tons. Each had three pairs of hydraulic legs, mounted on a heavy, sledge-like steel base, supporting a canopy; a big, rectangular, steel beam which stopped the roof from caving in on them.

Horace was coaching his trainee in the use of the sturdy, rotary handle, mounted on the middle pair of legs, behind the crawling route on each chock, that controlled all of the hydraulic operations of the adjacent chock. Working each chock in turn, Brett would drop its roof beam, showering Horace and himself with dust and sharp lumps and particles of coal, then pull it in under the power of the snaking ram before pumping it back up, with a deep, reassuring crunch as it took the weight of the newly-exposed roof.

Every few minutes there was a great, rumbling roar as a section of roof broke away and caved in behind the advancing chocks and a cloud of white dust would billow out to add to the thick fog of coal-dust generated by the shearer.

Looking back up the face towards the tailgate, Horace saw the beam of a cap-lamp, carving its way through the thick, airborne dust. By the sharp focus of the light and the flame of the 'silver' Davy lamp hanging from the belt of the man crawling towards them, Horace knew that its owner would be a deputy, overman or manager. Within a few seconds he was able to make out the lean, broad-shouldered figure of their new undermanager.

'Good morning, Horace,' Mr Greaves called out as he closed with them, his voice unimpeded by a dust mask. Horace knew that his boss would have dispensed with that protection in order to be able to shout without impediment and communicate clearly with the facemen with all of his face visible.

Horace pulled down his own rubber mask, with its plastic-housed filter. Its straps dropped from around the back of his head to rub the gritty paste of sweat and coal-dust into the skin on the back of his neck.

"Mornin' Boss,' he shouted over the clattering of the face conveyor.

'How're you doing?' Mr Greaves shouted as reached them.

'Champion, Boss. 'Ow about yerself?'

'Marvellous, thank you Horace. You can't beat it when the coal's pouring off the face. Here, let me climb over you and get past, I don't want to hold you up.'

The undermanager grasped Horace's shoulder as he struggled past then he paused to look back at him.

'You've got a new face-trainee, then.'

'Yes Boss. A young apprentice fitter. Started wi' me today.'

The lengthy waiting list for face-training meant that face-trainees tended to be experienced miners in their late-thirties but Horace's latest companion was only eighteen. Being a fitting apprentice, Brett needed to be face-trained in order to receive practical training to work on all of the equipment on the face.

'Chosen the same route as your Paul, then.' Mr Greaves said.

'That's right, Boss.'

Mr Greaves turned to the trainee and shouted, 'And what's your name?'

'Brett. Brett Redfern, Boss.'

'Right, Brett. And what does that say on your hat?'

The young apprentice cringed.

'Sorry Boss, I was goin' to take it off.'

'What does it say?'

'It says, "Fuck the pit", Boss.' Brett called.

'And is that how you feel about our pit?'

The apprentice hung his head. The undermanager waited.

'No, Boss, it was me mate, fuckin' about. He got hold of me helmet in the fittin' shop and wrote it on in marker pen.'

'Well you can get it off. I'm not having any face-trainee who isn't loyal and committed to the pit on any face of mine. If I catch you with anything like that on your hat again I won't have you on the face for training. And that'll be the end of your career as a fitter. Is that clear?'

'Yes Boss. Sorry Boss.'

'Now I've got an initiative test for you: Get off the face, into the tailgate and find a way to get every scrap of those words off that helmet. And don't come back on the face until you've done it.'

'Okay Boss. But I only started me face-training today, I'm not supposed to get out of arms reach from my supervisor.'

'I know that – but you'll be safe if you crawl straight up the crawling route in the chocks. Off you go.' Mr Greaves said with a jerk of his head in the direction of the tailgate end.

'Okay, Boss.'

The undermanager and Horace squeezed against the chock legs to let the youth scrabble past to crawl away with all the ungainliness of a new starter on the face.

'Sorry Boss,' Horace said, hurriedly. 'I told 'im 'e needed to get it off 'is 'elmet and 'e did say as it was 'is mate as had done it. I shoulda done what you done and got 'im to do it straight away.'

'Never mind, Horace, there's no harm done and, hopefully, it'll have given him some food for thought. He'll be alright. But we can't put up with nonsense like that.'

'No, you're right. I suppose we all did daft things when we was young, Boss.'

'I'm still doing 'em, Horace,' Mr Greaves said, finding it difficult to imagine the steady, old faceman ever having behaved foolishly.

Horace struggled his hand into his trouser pocket and pulled out a crumpled, blackened, paper bag.

'Here y'are, Boss, have an aniseed ball.'

'Thanks Horace, just the job,' Mr Greaves said, reaching into the bag.

'They make yer mouth water,' Horace said. 'Stop the dust dryin' yer mouth.'

Mr Greaves smiled at him, turned and crawled off down the face towards the maingate.

Horace kicked himself, wishing he had made his trainee scrape the offending graffiti off earlier. He might have known the lad would get into trouble, especially given the adhesive Union Jack stuck to the side of his young boss's own helmet.

Horace and Martin reached the stile.

Horace clambered over it, into the lay-bye, and put his fishing bag down. He straightened up then reached over and took the fishing rods and stools from the boy.

As he waited for Martin to climb over the stile to join him, Horace was forced to admit to himself that, despite his use of Mr Greaves as an example, he and his grandson possessed very different characters to that of his driven, young manager.

Stopping the Wheels

Whitacre Heath Colliery. Thursday,
29 September, 1983: 06:46

As the last of the dayshift officials rushed off to the shaft to get down the pit in time for the start of their shift, their few, remaining colleagues on nights drifted away to the pithead baths. With their handovers completed, all the inter-shift banter and horseplay done, the dirty offices fell quiet.

Les Parker leant back in his chair, ready for the pit in freshly-laundered, orange, Sketchley overalls. His left-hand, wrapped around a mug of tea, brushed the surface of the dusty, dry table-top as he drew pensively on the cigarette in his other.

His eyes ranged over the previous day's update of the Colliery Plan, hanging on the other side of the room, next to the door, still bright white against the grimy, coal-etched brickwork of the blue, gloss-painted walls. His ear picked up a residual murmur of conversation from the office on the other side of the officials' empty, communal room.

Narrowing his eyes to slits against the smoke curling up from the end of his cigarette, he dropped his gaze to regard the 'non-stat" sitting on the other side of the big, square table that filled the office. Anyone who knew the seasoned, fiery, old senior overman would expect him to be critical of the pit's youthful, new undermanager.

It was a big jump for a young man; the step up from official to undermanager. Les knew of plenty who hadn't been able to hack it. Failing to establish the credibility and authority needed to lead a sizeable, demanding workforce in the unforgiving, underground environment, they'd been eaten alive, devoured as casually as a bite of snap by the pitmen in their charge.

Like Les, Jim Greaves was dressed in his pit-black but the grease and coal-dust coating his overalls were evidence of a more hands-on approach taken, whilst underground.

Les embarked upon a quiet update for his young boss on some actions he had taken, on his own initiative. Jim continued reading and signing the afternoon and nightshift deputies' statutory reports, nodding and grunting occasionally in acknowledgement.

After a few minutes, the scraping of chair-legs on the floor tiles of the other office signalled that the colliery's Statutory Undermanager, Ken Goodall, had finished putting his own senior overman, Bernie Priest, right for the day.

On their way through to Jim and Les, Bernie stopped off to top up both his own and his boss's mugs, pouring well-stewed tea from the big, dented, aluminium tea-pot on the battered, old, wooden table before splashing in some milk from the bottle on the window ledge. Ken carried on through to breeze into Jim's office.

'Gentlemen, Gerald's coming round to take our orders,' Ken said. 'The rope-capping on Number 2 Shaft this morning's goin' to make the winding time for supplies tight today, so 'e needs to know from us what we need most urgent."

'Here 'e is now, the old lad,' Bernie called out, ambling in behind Gerald Chipman, the colliery's Materials Supply Officer who came bustling through into Jim's office.

'You gentlemen want to make sharp and get down that pit,' Gerald said, as Bernie handed Ken's mug back to him. 'There's a war party on its way. I just over'card Messrs Flavel and Deeming and their side-kick, McFadden, finishing their war dance in the Lodge office.'

Ken Goodall rolled his eyes and groaned, his tea mug just short of his mouth.

'Oh, no. Who's rattled their bloody cage?'

'It'll be what I told you about, yesterday, Boss,' Les said, addressing Jim with weary confidence.

'What was that?' Ken said.

'Les found our friends Crowley, Hedges and Coton hanging around the pit bottom before the end of dayshift, trying to beat the queues and get up the pit early,' the young undermanager said

'So I took their numbers and stopped 'em half an hour off their time,' Les said.

'That'll be it.' Gerald said. 'I saw those three idle fuckers hanging around the Lodge office, first thing.'

'Bloody hell,' Ken moaned. 'I could do without Flavel stirrin' up a load o' trouble this mornin'.'

'We'll deal with them if they come,' Jim said. 'But we'd better get these materials sorted out. Here Gerald, take a chair.'

Gerald, Ken and Bernie joined Jim and Les at the dusty, old, office table. The two undermanagers and their senior overmen indicated their priorities for materials, based on the notes in their coal-blackened pocket books and from what they had learnt from the previous shifts' reports and their de-briefing of the nightshift officials.

As Gerald left with his order book, Les looked over at Jim.

'Where are you goin' today, Boss? I thought you might want to come and have a look at 83's salvage with me.'

'There was thirty-five chocks left to come off the face-line at the end o' nights so we should have 'em all off in the next week. Then it'll just be a week to get the conveyor pans off and the cables to come out, a bit o' scrap and the haulage motors.'

Before Jim could respond, the door of the communal room flew open and the colliery's NUM Secretary fulfilled Gerald's prediction.

'Good morning, gentlemen,' Ken Goodall said, as Dick Flavel stormed in to Jim's office with Keith Deeming, the Lodge President, in tow.

The Secretary's post was actually the more senior, the President being more of a figure-head, so it was Dick, the bigger of the two, overweight men, who spoke first, confirming Les's suspicion about the nature of their perceived grievance.

'About 13's Headers,' Dick snapped at Les.

'Yes.'

'Is it true you stopped their time, yesterday?'

'Yes, I did. I stopped all three of 'em 'alf an hour each.'

'On what grounds?'

'Leavin' the job early, bein' in the pit bottom before time, illegal manriding on non-manriding belts – and walkin' the Loco Road, an unauthorised walkin' route. Four offences.'

'They were wet, that's why they came out early,' Dick said.

'If they were entitled to come out early the deputy would have given 'em a wet note,' Mr Greaves said. 'And he'd have arranged safe transport for them.'

'They asked the deputy for a wet note but 'e wouldn't give 'em one. Said 'e didn't want to get it in the neck from the Senior Overman.' Dick said to Jim, with a nod in Les's direction.

'I'm delighted to hear it,' Les said, with a mirthless chuckle.

Dick shook his head and said to Jim, 'You see what we're up against?'

'Yes,' Jim said. 'You're up against me and my Senior Overman who expect the men to work the full shift they're paid for, down the pit.'

'Those men shouldn't have to stand around in wet clothes, queuin' in the pit bottom at the end o' the shift,' Keith Deeming whined, getting in on the act.

'The only way they could get their clothes wet is if they lay down and rolled around on the floor in that heading,' Jim said.

Dick smacked the table.

'Are you going to sit there and let him talk to us like this?' Dick growled through gritted teeth, twisting away from Jim to appeal to the more senior undermanager. 'Do you want me to call my men out?'

'Now, no one wants that, Dick,' Ken Goodall said, standing up and making downward, placatory gestures with his hands.

'Hold on a minute,' Jim said, leaning forward in his chair. 'They may be your members but they're not your men, they're ours. You don't pay 'em. You might represent them but you don't lead them or manage 'em, we do.'

'If he carries on like this I'll stop those wheels,' Dick shouted at Ken, waggling his head and flailing his hand at the office window in the general direction of the colliery's headgear.

'Is that what yer want?' Dick bellowed. 'When I get back to the office all it takes is four phone calls and I can have every underground district out.'

'Then,' he shouted, turning back to glare at Jim, 'you can explain that to Everitt.'

'That's your answer to everything, isn't it?' Jim snapped. '"I'll stop those wheels". When everyone else is doing their best to keep this pit open, all you can come up with are reasons to stop production. Well, if that's the best you can do, fuck off and try it. But I don't believe the men'll walk for anything as daft as what you're shouting about.'

'And that's your final word, is it?' Dick shouted.

'Yep.'

'Right! Come on Keith, we're wasting us time 'ere.' Dick snapped. He jerked his head to get Keith to lead the way then flounced out of the office after him, across the communal meeting room and out of the door, slamming it behind them.

'Mr Greaves!' Ken Goodall said, looking aghast at Jim. 'You shouldn't speak to them like that. If they go and call the men out, the Gaffer'll go blue. They'll probably go straight to him and get 'im to overrule you.'

'That's his prerogative, but I stick by what I said,' Jim said. 'We can't have them holding the pit to ransom.'

He stood up, picked up his Davy lamp and hooked it onto his belt before pulling his NCB donkey jacket off the back of his chair and gathering up the stack of grimy, shift report books to return them to their slots in the blackened, plywood shelving on the wall in the officials' communal office.

'Anyway, to get back to what you were saying Les; I agree, we should go and have a look at 83's salvage and check over the preparations for sealing it off,' Jim said. 'We'd better get going to catch the eight o'clock run before the shaftsmen start on that rope.'

They were out of the dirty office building and halfway across the yard to the shaft before Les spoke up, as he lit up a last cigarette.

'What Mr Goodall says is right, if the men walk off the Gaffer's likely to back down an' overrule you.'

'Huh,' Jim snorted.

'But you did the right thing, tryin' to call their bluff. They just keep pushin' it 'ere 'cause management's always given in. It's about time someone took a stand.'

As they walked past the big winding-engine house, Jim reflected on the nature of general managers. He'd worked for only two in his brief career. Both had been in their mid-fifties, with a career's worth of mining engineering knowledge and experience. They both behaved in a way that was consistent with the general managers' fearsome reputation for being a breed apart, conforming to the old coal industry joke: "What's the difference between God and a colliery general manager? God doesn't think he's a colliery general manager."

But Harry Everitt seemed nothing like as resolute in his dealings with the union as the General Manager at Canley, Jim's previous pit, had been. From what Jim had been told, he could count on it being a matter of minutes before he'd have the Control Room attendant phoning round to track him down, underground, so his new gaffer could administer a stiff bollocking over the phone.

When they arrived at the airlock through to the pithead, Jim waited while Les stubbed out his cigarette then grabbed the handle of the small trapdoor within the outer, steel air-door and pulled on it, releasing air into the chamber before using the lever, bolted to the side of the great door, to prise it open. He stepped over the high ledge and into the chamber, holding the door open for Les to follow. Les walked straight across the air-lock and when Jim let the outer door crash closed with the force of the air pressure, he repeated the process on the inner air-door, opening up access to the big, sealed building of Number 2 Pit Top.

The mine's great fan drew the air through the many miles of tunnels and faces between this upcast shaft and Number 1 Shaft, the downcast shaft, only twenty yards away, which was used for winding coal. The airtight building and airlock doors were needed to prevent the fresh air from the surface from simply short-circuiting down to where the fan drift, the inclined tunnel back up to the fan-house, forked out of Number 2 Shaft.

'Come on Boss, you nearly missed it,' the banksman called out, beckoning them on towards the waiting cage. "Morning Les.'

Before boarding the cage, Jim stopped to have a word with Jed Davenport, a short rogue with a subversive smile and an unmistakable air of danger. Jed stood between the two tracks used to run supply vehicles onto the cages, wearing the safety chain and harness of his trade.

As a member of the Colliery's specialist elite, Jed had one of the most dangerous jobs in the pit. He and his team of shaftsmen inspected and maintained the shafts, including their linings, winding ropes and headstocks. Today, they would be capping one of the winding ropes. This involved cutting a six-foot sample off the end of the rope then re-capping it before sending the sample away to be measured, to make sure that the amount of stretch in the rope was within the proscribed, safety limit. [Note 2]

Jed and the other shaftsmen were usually here at this time of day, waiting to undertake their daily examination of the shaft as soon as the morning's runs of managers and senior officials and engineers had been wound down the pit.

Jed would often wink at Jim when he was in the company of Ken Goodall and make some insubordinate joke at the more senior manager's expense, whilst offering them both a pinch of snuff before they boarded the cage. Jim had valued those friendly, little acts of kindness, particularly, when he had been new to management, in his first, strange, few days at Whitacre Heath.

Jim took a pinch from the tin that Jed held out for him and breathed in heavily, rubbing the snuff between his finger and thumb under each nostril.

15

'You take your time, Boss,' Jed said, with laboured irony. 'We've got all day.'

'Thanks, Jed,' Jim said, closing his eyes, inhaling and enjoying the nicotine hit and the sensation of the snuff as it cleared his head and nose, before stepping off and hurrying over to where the banksman stood waiting, by the shaft.

After running his hands all over them in a quick search of their clothing for contraband, anything that might cause a flame or incendive spark underground, the banksman took Jim's and Les' 'silver' tallies then stood aside, his arm outstretched in a gracious gesture to direct them onto the dark, rusty, cage. *[Note 3]*

As Jim stepped into its wet, rust-covered confines he spotted a hunched, skinny figure, standing side-on in the darkness at the back of the narrow cage box. As his eyes adjusted to the gloom, he recognised the bony face; its prominent cheekbones, the bags under a pair of malevolent, glaring eyes, the bulging of those eyes matched by the way their owner's Adam's apple protruded from his scrawny neck.

The sneer on the man's thin lips grew as he regarded Jim.

'Good morning, Shuey,' Les said, as he boarded the cage behind Jim.

'What's good aboot it?' Shuey McFadden snarled in his broad, Scottish accent. 'Anyway, have you put they's men's time right for 'em?'

The banksman smirked as he unhitched the light, steel cross-bars and let them slide down the rails on either side of the mouth of the cage, their spacings maintained by light, rusty chains that connected them to form the only barrier that prevented riders from falling out of the cage as it hurtled through the shaft. With a hiss of compressed air, the sliding, mesh gate clattered shut on the shaft.

"What are you doing, going down the pit at this time of day?' Jim said, 'A've bin on union business.'

'That arrangement only applies to the Union President and Secretary,' Jim said. 'There's no agreement for any other rep' to spend time on the bank during the working shift.'

The banksman gave three rings to alert the onsetter at the pit bottom and the winding-engineman that they were about to wind men, then gave two more rings for their cage to be lowered. The cage eased down to below the level of the floor of the pit top and into the darkness of the shaft then plunged away, towards the pit bottom, half a mile below.

Shuey grunted and scowled at Jim.

They stood in silence for the ninety seconds it took for the cage to hurtle down the shaft.

'You can make this the last time you're late down the pit,' Jim said, as it landed at the pit bottom. "Cause I'll be telling the banksmen to stop you from entering the mine if you're late again and stopping you a day's money.'

In the brightly lit, white-washed, brick-lined pit bottom, Jim led the way through the three airlock doors to pass through to the intake side of the mine, accessing the short road that curved round to the Loco Road. He strode on to catch the manriding train, boarding the carriages that were to be drawn inbye to be stabled in sidings three miles in from the pit bottom, ready to transport the men out at the end of the dayshift.

Les and Jim sat on wooden-planked seats, opposite each other, inside the thin steel roof and sides of the white-painted carriage. Shuey had slipped into one of the carriages further back down the train, having hollered at the loco driver, telling him to stop to let him off when they reached the entrance to the East Side of the pit.

The carriages lurched off, snatched forward by the powerful, battery locomotive.

After a few jerks and corresponding, loud, crashing collisions of its buffers, the train's noise dropped away to a steady clatter.

'If Shuey gets his way this weekend 'e'll be stoppin' on the bank all the time,' Les murmured.

'Why's that?' Jim said.

'It's the NUM's Committee elections this weekend. Sunday mornin', at the Miners' Welfare. Shuey's standing against Dick Flavel for the Secretary's position. Fancies hisself as top dog.'

'Good grief,' Jim said. 'The Welfare'd be the right place to hold it; they'd have to be pissed-up to vote for him. Surely, the men wouldn't be that daft?'

'I wouldn't be so sure. He's bin workin' 'ard, politicking behind the scenes, drummin' up support.'

Jim was silent. There were a number of reasons to respect Les's opinion on this, not least because Len was a union official himself, being the local President of NACODS, the colliery officials' union.

'If you think we've got problems with 'the Chuckle Brothers', Flavel and Deeming, imagine what it would be like with Mr McFadden rabble-rousin', full-time,' Les said. 'He may have a Scottish accent but 'e came up here from Kent. They reckon 'e left there 'cause 'e was too bolshie, even for them.'

The Kent Coalfield was famously militant. Jim had heard of one of its pits striking over the colour of soap issued at the pithead baths, refusing to accept green or white bars of the soap stamped with the letters "PHB", insisting, instead, on pink bars. The management had only resolved the dispute by sending someone, in a van, to the local supermarket to buy loads of boxes of expensive, Palmolive soap.

The loco driver stopped the train briefly at a manriding station, allowing McFadden to squirm out of his carriage and slip away, to disappear through the air-doors to the East Side of the mine without a word.

The journey to the colliery's furthest extent was another two miles. It gave Jim a few, rare minutes, in the dark, for thoughtful reflection. On the face of it, he'd been reckless with the NUM officials but, if they did go raising hell with Mr Everitt he was determined to defend his position.

At the far end of the Loco Road, Jim and Les clambered out of their carriage and went straight on and through the air-doors to the supply route's rope haulage in the return airway of the South End of the pit. Jim spotted a 45-gallon oil drum lying on its side at the side of the road and smiled to himself. Someone had drawn a face on its round end with a marker pen. With a down-turned mouth and furrowed brow and a large bag under each eye, each bearing the word "Co-op", it was clear whose likeness it was meant to be, even before reading the words written underneath:

"VOTE SHUEY BAG-EYE – NUM SEC".

They boarded the manriding carriage on the rope-hauled train behind the train guard and travelled in with the run of supply vehicles the train was drawing. When they got off the train, Jim darted over to the telephone, a white, metal box clamped to an upright, steel stanchion, junction support, grabbed the handset and dialled the Control Room's number.

'Hello Jerry,' he said to the day shift Control Room attendant, 'How are things going on 85's.'

'Hello Boss. The maingate shearer's cut up to the tailgate and it's on its way back down the face. It was at 95 chock last report, ten minutes ago, so they should be down into the eighties by now, Boss. They're goin' well, so far this morning.'

'No problems anywhere?'

'Nope, all the belts are running. Everything's fine at the moment.'

So there had been no calls made to initiate the wildcat action that Dick Flavel had threatened and the Gaffer wasn't after his blood, yet.

'OK, thanks Jerry. Keep the trains of empties going to the South End Loader to keep those belts going.'

Despite the determined activity of the teams removing the equipment, in a district where the salvage was well-advanced, it felt like the heart had been ripped out of the worked-out production face. 83's facemen had all been re-deployed, committed to their new face; their old one history, its vibrancy and comradeship and the noise and dust of coal-chasing all gone.

Still, there was a need for urgency. The faster its salvage was completed, the sooner the cost of deploying the salvage men could cease. And in Warwickshire, the race was always on to get the valuable equipment off the mechanised face and out of the district before the static production unit succumbed to the ever-present threat of spontaneous combustion. Then it could be sent off to be overhauled before being re-used on future faces, helping to keep down the running costs of the industry.

Back on the surface, Ken Goodall had called back over to the main offices to speak to the General Manager, following the union rep's' visitation. He found Mr Everitt in the Control Room.

As usual, the big man looked grim and inscrutable as he received his regular, morning update from Jerry on the nightshift's performance and the state of play, early in the dayshift.

After five minutes, Tim Hope, the tall, relatively youthful Operations Manager, arrived at the pit and breezed in with a cheery, 'Good morning Gaffer, 'morning Mr Goodall, 'morning Jerry.'

'Good morning Mr Hope – sorry Gaffer,' Jerry said, responding quickly to Mr Everitt's second-in-command before carrying on with his briefing.

When he was finished, Mr Everitt said, 'So both faces should be on for some coal-turning this morning?'

'Yes Gaffer, barring anythin' unforeseen,' Jerry said.

'They should be fine, Gaffer,' Ken cut in. 'They did a decent job setting the faces up on overtime. The packs are both up to date and the machines were all in cut.'

'Right, we'll see,' Mr Everitt growled. 'An' 'ow are you this mornin', Mr Hope?'

The colliery's two most senior men spoke cordially for a couple of minutes, the younger man succeeding in lightening the General Manager's apparent mood. Ken was reluctant to detract from this, he knew the Gaffer would have a sore head from the beers and whiskeys he'd have consumed the night before. Ken had had a good few himself at his own local. It was an accepted part of their way of life, a way of dealing with the relentless pressures on all of them. Only Tim Hope was a rarity, seeming to be able to sail through everything with his confident smile and only the occasional, social drink.

The last thing Ken needed at this time of the morning was to precipitate an explosion from the General Manager; it could upset things for days. But he felt the need to minimise the likely impact of his young colleague's earlier impetuosity.

As soon as there was a lull in the conversation, Ken sidled over to Mr Everitt.

'Er, can I have a quiet word with you, Gaffer?'

The big man cocked his head and looked down at Ken with a raised eyebrow then nodded towards the Control Room door. They withdrew to Ken's clean-side office, on the other side of the corridor.

'I just wanted to let you know. You may 'ave the union after you this morning. Flavel and Deeming 'ad a bit of an 'eated exchange with Mr Greaves, earlier."

'Oh?' Mr Everitt grunted, dashing Ken a dark scowl.

Ken explained how Les Parker had caught the problem children in the pit bottom, before time, the previous day, trying to beat the end of shift queues to ride the shaft, and then stopped their time. He described how the NUM's representation of the matter had gone.

'So it finished up with them threatenin' to stop the wheels and young Jim tellin' 'em that if that was the best they could do they should fuck off and try it – but 'e made it clear it was wrong when everyone else was working 'ard to keep the pit open, like.'

Mr Everitt's scowl deepened. Ken waited for the inevitable explosion, alert to any change in the depth or rhythm of the Gaffer's slow, deep breathing.

'Well 'e's right,' Mr Everitt said. 'If they come bleatin' to me, I shall tell 'em to fuck off.'

'Righto Gaffer,' Ken said, his face brightening. 'I just wanted to make sure you were pre-warned.'

Immediately after their earlier altercation with the young undermanager and his senior overman, Keith Deeming was scuttling along the red-tiled floor of the corridor round the outside of the lamp cabin, struggling to keep pace with his colleague's furious stride.

What do you make o' that?' Keith said

'Wait 'till we get to the office,' Dick muttered over his shoulder.

Going straight to the union office, rather than to the canteen for a coffee and fag, indicated the seriousness of the matter.

After Keith had darted into the room, Dick looked back down the corridor to make sure they wouldn't be overheard. As Dick closed the office door securely behind them, Keith sat down and leant forward in his chair.

'So, what d'ye reckon's goin' on?' He said.

Dick continued to brood as he moved round to his chair to sit at the desk.

'I dunno,' he said. 'Remember what 'appened a fortnight ago, when we called the men out?'

Jim had been approaching the inbye end of 85's Maingate when he encountered the face deputy, walking out towards him.

Whenever Jim saw Stan Townsend, he was reminded of the cartoon character, Freddie Flintstone. On this occasion, instead of giving the undermanager an update on the state of the face and the production so far, Stan raised a concern.

'The chargehand on 56's Face has phoned across and got Pete, the panzer driver, to call Des off the face.'

Des Proctor was the shearer driver who acted as chargehand for 85's Face.

'What's that all about?' Jim said.

'I spoke to the Control Room and they say that 56's men have walked out because of the level of dust on the face.'

'They've what?' Jim barked. 'Where do they think they're working, in a fucking factory?'

'They do this every so often, over there. They come up with some bit of a grievance and walk out. Between you and me, the problem is the Gaffer always gives in to 'em and then pays the men a full shift, so they get 'alf a day's paid holiday out of it.'

'I'd sack the fuckers. What are they phoning over here for?'

'They'll be tryin' to get our men to back 'em and walk out as well.'

'Oh will they? Well I'll put a stop to that nonsense.'

'Good luck, Boss. But I doubt you'll have much joy with 'em, they won't be able to resist an early bath, getting' down the Miners' Welfare for a few beers and gerrin' paid for it.'

21

'Anyway, if it's alright with you, Boss, I'll complete my mid-shift inspection. If you can keep Control informed, I'll keep in touch with them to keep tabs on developments on my way round the district.'

'I'll go and sort 'em out. But Stan, if anything like this happens again, when I'm not here, I expect you to stay at the face and hold onto them, not just let themselves be talked into walking out. You can do your inspection anytime.'

'Okay Boss, point taken,' Stan said, even though the timings of his mid-shift inspection were specified in the Mines and Quarries Act.

Jim walked up to the coalgate's electrical panels. Suspended from mono-rail on a platform slung over the belt conveyor, these provided the majority of the electricity used to power the face. He saw Des standing next to them, hanging up the phone handset. [Note 4]

As the face's chargehand, Des had a paradoxical combination of roles. It was, in part, like that of a foreman, acting as Number 2 to the face deputy, who was in charge of the face and its team of men and responsible for safety and production but it was also to act as the face team's local NUM union representative.

'Good morning, Des,' Jim said as he reached him.

'Mornin',' Des mumbled.

'What are you doing out here? You should be on that machine, turning coal.'

'I've bin called off the face to call 56's chargehand. I'm tryin' to get through to 'im. It sounds like they're in dispute. They'll probably be wantin' us to walk out with them.'

The two packers and two of the maingate end's three rippers ambled back out to join them.

'Where are you lot off to?' Jim said.

'We're just comin' to see if we're walkin' out or not,' one of the rippers replied.

'Well I can tell you this. Whatever that lot are moaning about on the other side of the pit has got nothing to do with you. Just make sure that everyone on this face knows, if they start strolling out before the end of the shift, I'll stop their money from the time they leave and make sure it's not refunded. The manriding haulage will keep on running supplies and I won't need the belts running so you won't get a ride out and if I catch anyone walking the Loco Road I'll fine 'em.'

'So 56's men can fuck about all they want but I suggest you get back onto that face and turn some coal and keep this pit open.'

The men on the other side of the pit had started to walk out. When they had reached the Loco Road, the General Manager had capitulated and granted the union some concession and the men had been persuaded to return to work. By the time the face team arrived back on the face it was snap-time, the twenty-minute break for food and the breather to which they were entitled in the middle of the shift. As usual, they would be paid for their full shift.

Meanwhile, two miles away, down the South End of the pit, Jim's threats had caused the potential militants to capitulate.

Jim was surprised to find, when he got to the surface at the end of the shift, that Ken Goodall and Stan and the rest of the officials seemed to think he had achieved a unique and significant victory with his robust stance.

'They say he's twenty-six.' Keith said.

'Well, 'e don't look it,' Dick said. 'But he's bloody sure of himself.'

'Accordin' to the blokes at Canley he was a good un' as face deputy and overman on a couple o' tough faces. Knows his stuff. And the men seem to rate 'im.'

'Yeah, but a young non-stat' undermanager wouldn't keep going out on a limb on things like this.' Dick said.

'What do you mean?'

'I mean 'e couldn't afford to take the risk, not at this point in 'is career. 'E'd be sacked if 'e lost 'alf a shift's production. No,' Dick said, gazing up at the corner of the ceiling with unfocussed eyes, 'someone else has to be behind this. 'E's newly appointed, only just started 'ere. 'E wouldn't take a stand like this off his own bat.'

'Well Goodall wasn't backin' him up.'

'No, 'e was running scared as usual. 'E's not in on it,' Dick said. 'That figures 'cause 'e's only an old 'Second Class ticket' undermanager, 'e ain't got a degree and First Class Certificate of Competence like 'is understudy, Greaves. And Hope's not in on it neither, 'e's just serving time on 'is meteoric rise to the top, waitin' for the next general manager's job to come up.'

'You remember what Arnie said?' Keith said, referring to an NUM area representative. 'When he tipped us the wink that young Greavesie was comin' 'ere, 'e said 'e was comin' from Canley, 'ad a good reputation with the men and the union but modelled hisself on Canley's general manager, that old bastard Charlton. 'E reckoned 'e was Charlton's protégé. Perhaps that's it.'

Dick pondered for a few moments then shook his head.

'Nah, I reckon 'e's got backing from his current superiors. This won't be his old gaffer's influence. Either Everitt's decided to take a stand, which seems unlikely...'

'No, he's never done it before.'

'...or else someone at Area's marked Everitt's card and told 'im to get Greaves to provoke us and take a stand.'

'So what do we do?' Keith said.

Dick's eyes narrowed.

'We make sure we don't fall into their trap. Whatever they do, we don't want to rush into any action for the time being. We don't want to end up playing into their 'ands, whatever it is they're up to.'

Dick would need to give this some thought. It felt like a new regime was emerging at Whitacre Heath.

The Challenge

Thursday, 29 September, 1983: 14:32

Angela Barker sat back in her chair and smirked at the evident discomfort of the young man standing before her, in front of her desk. The Colliery General Manager's Secretary had a fearsome reputation; tall, slim and handsome with short, raven hair, many at Whitacre Heath behaved as though the forty-year old was the person actually in charge of the pit. She had only ever been polite and courteous to Jim Greaves since his arrival but he sensed that it would be prudent to continue to tread carefully and treat her with serious respect.

He had been in the dirty offices, interrogating his dayshift deputies and overmen, when she had summoned him on the phone. All she would tell him was that his presence was required, he was not to waste time bathing and changing into clean clothes and that he should report to her, immediately.

So here he was, hovering on the threshold of her office, still wearing his dust-impregnated pit-black, NCB donkey jacket and knee-pads, his helmet under his arm, his hair, face, neck, chest and hands a deep, matt black.

For once, the Gaffer was not in his office.

'He's in the conference room with Mr Hope and Mr Goodall,' Angela said. 'They were in there for the Consultative Committee Meeting. Mr Everitt went straight in there to join them, when he got back from the Golf Club.'

That was a surprise; Jim had felt sure that he was destined to be on the carpet, in the Gaffer's big, formal office, to face some belated fall-out from his earlier altercation with the NUM.

'Should I wait here for him?'

'No, he wants to see you right away,' she snapped. 'Go on, get down there and join them in the conference room.'

He hesitated then turned away and started dawdling his way down the corridor. He stopped, turned around and took a few steps back.

'Angie?'

When Jim had first arrived at Whitacre Heath, Mr Everitt had used this diminutive when introducing his secretary to him. Jim was yet to realise that he was the only person, apart from the Gaffer, to use this familiar form of her name when addressing her.

She looked up at him, over her glasses, eyebrows raised.

'Did he happen to say what it was about?'

'No. He didn't share this one with me. Now, off you go.' She said, brushing him away with a couple of brisk waves of her hand.

Jim picked his way carefully down the corridor, in a futile attempt at minimising the amount of dust falling from his boots and clothing onto the polished, tiled floor.

At the entrance to the conference room he paused. Glancing back along the length of the building, he saw Angela watching him, intently, from her desk, in her office at the end of the corridor.

Four quick flicks of her hand and a mouthed "go on" spurred him into action. He knocked on the door then eased it open.

Sticking his head round it, he peered into the room. Mr Everitt was still in his overcoat, holding court with Tim Hope and Ken Goodall, all three standing at the end of the big, meeting table nearer the door. Mr Everitt looked over his shoulder and caught sight of Jim's coal-blackened face peering round the door.

'Ah, here 'e is. Took yer bloody time, didn't yer? Don't 'ang around out there, get yourself in 'ere.'

Jim shuffled round the door and closed it carefully behind himself. He stood on the narrow strip of wood flooring to avoid blackening the conference room's huge, thick-pile carpet with his filthy, steel toe-capped pit boots.

'Don't stand there lookin' like a spare-prick at a wedding, come over 'ere so I can talk to yer.'

'Sorry Gaffer,' Jim murmured, the General Manager's scowl boring into him as he tip-toed over to join the select group.

'Right Mr Greaves, this transfer of equipment on 22's.'

'Yes Gaffer?'

'That face is vital to Whitacre Heath. Without it, we only have two faces. And two faces ain't gonna produce enough coal for us to break even, especially when you take into account our fixed costs.'

Mr Everitt's enunciation of "fixed" sounded more like "fished". It was apparent from this, the slight swaying of his imposing figure and his beery breath that the General Manager had taken on a considerable amount of booze again at the nineteenth hole. No doubt, his level of consumption would have been, in part, a response to the stress from the matter he was raising.

'If we don't get the retreat face on that district producing on time this pit'll be finished and I'll be finished. And we can't 'ave either of them things happenin', can we?'

'No, Gaffer,' Jim said.

'I had Rawlinson on to me, first thing this mornin',' Mr Everitt growled.

As a Lancaster bomber pilot, Ralph Rawlinson had been a scourge of Hitler's Germany. As the South Midlands' Area Director, he was the fearsome, top man of their NCB area. His senior managers were terrified of him.

'I could hear the old bastard on the other end o' the phone, hufflin' and a-pufflin' 'n' chewin' on his pipe. He's given me 'till the start of December to get that face into production and if it's not goin' by then then I'm gone and the pit's gone.'

'I must have that face turning coal by the first of December, d'y'understand?'

'Yes Gaffer,' Jim said, wondering whether the Gaffer was so inebriated that he'd started harbouring the misapprehension that Jim was involved, somehow, in the lack-lustre, face-to-face transfer on 22's.

He glanced, desperately, at Mr Hope, who was standing behind Mr Everitt.

Tim Hope winked at him.

'Now, Mr Goodall,' Mr Everitt said, turning back to the older undermanager. "Ow many chocks have yer had off the old face, so far?'

'We'd got twelve off by the end of dayshift.'

'Fuckin' 'opeless.' Mr Everitt slurred. 'You've bin fuckin' about, pullin' 'em off that face, fer a week. You've not even managed one a shift. D'ye expect me to go down there and do it me fuckin' sen?'

'No Gaffer,' Mr Goodall said.

'And how many of those twelve chocks are on the new face-line and installed in position?'

'We've bin givin' 'em a quick bit of an overhaul and the first'll be going onto 22B's on afternoon-shift, today.'

'That's a fuckin', long-winded, roundabout, fuckin' way of saying "none, Gaffer", ain't it?' The General Manager barked. 'Just say "fucking none, Gaffer".'

'Sorry Gaffer; none, Gaffer.'

'Huh,' Mr Everitt said, shaking his head. 'Right Mr Greaves, so now yer see what yer up against. I want you to drop everythin' else yer doin' – leave it to Mr Goodall, 'ere – I need you to get yourself down to 22's and get that transfer movin' and get that new face turnin' coal by the first of December.'

He glared intently at Jim for a few seconds.

'D'ye reckon you can do that for me?'

'Yes, Gaffer,' Jim said, with a firm nod of his head.

Mr Everitt's body jerked.

'NO!' He bellowed.

Jim flinched.

'That's not good enough!' He snapped.

'Yer said it too easily.'

'It'll mean triplin' the rate of salvage and actually getting' the installation 'appenin', getting' all the chocks transferred and the face conveyor stripped out and round to the new face and re-buildin' both shearers. Normally that'd take at least 'til the end o' December, from where we are now. Do you realise how difficult it's gonna be, to be up and runnin' by the first of December?'

'Yes, Gaffer.'

'Right. I'll ask you again: Can yer do it for me?'

Jim waited a couple of seconds, drew breath then spoke.

'Yes, Gaffer.'

'*NO!*' Mr Everitt howled through gritted teeth, fists clenched, eyes screwed closed in frustration.

Behind Mr Everitt's back, Ken Goodall looked worried. Tim Hope smirked.

'You're still making it sound too easy. I don't just want a "yes Gaffer" from you. It's all I get around here, fuckin' "yes Gaffer". I want yer to convince me you can actually do it. So, come on; d'ye reckon yer can do it for me?'

Jim took even longer before answering, this time. He took a deep breath.

'I know it'll be challenging, Gaffer, but I'll get onto it, straight away, and develop a project plan that'll get that face installed and turning coal for you by the first of December. Then I'll make it happen.'

Mr Everitt looked Jim in the eye for several long seconds then nodded, barely perceptibly, accepting that Jim was staking his reputation on his promise. His lunchtime session and his exhortations had left him drained but it was clear that he was satisfied that he had impressed the criticality of the project upon his young undermanager.

'Right, you'd better jump to it. And you report to me directly on this job, it's that important. D'ye understand?'

'Yes Gaffer,' Jim said. 'Thanks Gaffer.'

Jim turned to leave.

'And remember, Mr Greaves,' Mr Everitt called after him.

Jim stopped and looked back.

'This is Warwickshire; you've got a race on to get that equipment transferred and the new face retreatin' before that district catches fire.'

'Yes Gaffer.'

Jim closed the big door quietly behind him, stunned by the sudden development. The gratitude he had expressed had been genuine. Given the criticality of the challenge, he'd just received a serious expression of his Gaffer's faith in him.

'Everything alright?' Angela called down the corridor to him.

He looked up and called back.

'Er, yeah, fine thanks.'

He picked his way back to her office and shared what he had just been told with her.

'Well that's good,' she said with a sharp nod of her head. 'Something to get your teeth into."

'Yes,' he said as he turned away with a thoughtful smile. 'Yes it is.'

But it was also a most burdensome responsibility; the pit needed that face. Being a retreat face, its two gate roads were already formed so there was less work to delay the shearers each time they reached the ends of the face and that would make it more productive. [Note 5]

The pit was the very heart of the Whitacre Heath community, its reason for being. If he failed to hit the deadline and the pit were to close it would be with the loss of nine hundred and twenty-eight jobs, not just those of the men but women as well, like Angie, Veronica, on reception, and all of the canteen women.

Jim was convinced that economic pits producing competitively-priced coal were essential for the nation. To allow a pit to fall below that threshold and be terminated would be a shameful failure for a mining professional.

He was amazed that he had managed to satisfy Mr Everitt without having offered any real substantiation for the confidence he had expressed in his ability to fulfil his Gaffer's expectations. He couldn't fathom why Bill Metcalfe, the Services Manager, had not been in on his conversation with Mr Everitt and, therefore, had probably had no say in Jim being put in charge of the job. It was puzzling because Bill's senior role made him accountable for all of the colliery's development work; the heading out of areas of the mine and new districts and their installation and the salvage of worked out faces.

Hurrying up the corridor, past the Control Room, Jim cast those thoughts aside and kicked his mind into gear. He stepped out and rushed back over to the dirty offices, to catch his senior overman before he left for the day.

Les followed Jim into Ken Goodall's office, where Mike Taylor, the afternoon-shift senior overman was sitting, adjusting the straps on his kneepads before going underground.

'Mike, I need some help from you and Les,' Jim said.

'What can we do for you, Boss?' Mike said.

Jim told them about the General Manager placing him in charge on 22's and the criticality of the deadline for installing and commissioning the new retreat face.

'That's some challenge you've got there, Boss,' Mike conceded.

'That's some challenge <u>we've</u> got, Mike,' Les said. 'The Boss is going to need our support on this.'

'Yes, of course,' Mike said.

'The first thing we need to do is to make sure we've got the right people on the job,' Jim said. 'On the mining side I need a deputy on each shift with the organisational and leadership skills to drive progress and make things happen, I don't have time to develop them. Who've we got on 22's on each shift?'

'It's Gary Wallender on afternoons, he's a good lad,' Mike said. 'He's fairly new as a Grade One deputy but he's keen and knows what he's doing. With a bit of support, I reckon he'll deliver for you alright.'

'Is he planned to be the face deputy for production when it gets going?' Jim said.

'That hasn't been decided yet, as far as I know – but I can have a word with Mr Goodall and try to make sure he is. I can see what you're thinking, you want the deputies to be settin' 'emselves up for success during installation.'

'Dead right,' Jim said. 'Who've we got down there on days, Les?'

'It's Billy Thompson on days this week, Boss. He was on 22's for all of its advance so there's a good chance 'e'd have been on there when it goes into production, anyway. He's keen enough but 'e's not experienced on face-to-face transfers. Not many of our officials are, we've not done as many as we should have, in the past.' Les said. 'On nights this week, it's Davy Slack. He's well-named, if I'm honest, not really face deputy material. I'll shift him and get Kempie for that shift for you, starting on Monday, Joss Kemp, he's a good coal-turner and he knows his stuff on installation work 'un all'. He takes no nonsense, Joss, and he's well-respected. Anyway, I'll get up there tomorrow, meself, help you to get Billy focused and we can get some impetus going down there, Boss.'

'Yeah, okay, but I'm going back down to have a look at it this afternoon.'

Len gave Mike a brisk nod.

'That's keenness for yer,' he chuckled.

'Yeah, well, we've only got forty-four days left and we've got to make every hour of every day count if we're going to hit the target the Gaffer's been set,' Jim said. 'I need to see how it's set up then get some creative thought going overnight, ready for getting things moving faster on dayshift. I need to get a detailed project plan put together by the end of tomorrow. I need to be in a position where I've got a forecast for completion – then I'll know if I need to find more ways to accelerate the rate of progress.'

'If you're going up there, I'll come with you then, Boss.' Mike said. 'I can introduce you to Gary and start getting involved myself.'

'Okay, thanks Mike. If you wait here, I'll be back in a few minutes. I just want to nip over to the fitting shop and have a word with Jack Lloyd about the fitters for this job.'

Les walked out of the building with Jim.

'So you're off to see Mad Jack, then,' Les said. 'You must have known 'im at Canley.'

'Yeah, Jack's alright.' Jim said. 'He's an old mate of mine.'

'He seems to know his stuff,' Les said. 'But 'e's a lot older than you'd expect for someone only just making it to that position.'

'Yeah, apparently he blotted his copybook with Mr Charlton, when he was a youngster,' Jim said, referring to the General Manager at Canley, Jim's and Jack's previous pit.

'And then, earlier this year, he upset our Gaffer there one time too many, so he made sure Jack was promoted to Assistant Mechanical Engineer, to get shot of him.'

Mad Jack

Colliery General Managers weren't known for their patience and a few months earlier Mad Jack had finally tested that of his former gaffer, at Canley Colliery, to beyond its limit.

It had started when Mr Charlton's ride-on mower had broken down on the first cut of the year. The next day, when he got to the pit, he phoned Jack and told him to get himself round to his house to repair the machine.

When Jack reported back, it was to confess that he had failed to isolate the problem and inform his gaffer that he had brought the mower back to continue his investigation of it in the colliery's fitting shop.

The repair dragged on for days, the days became weeks. Tracking down Jack became even more difficult than usual. Whenever Mr Charlton did manage to corner the elusive engineer, Jack offered some discouraging status report or implausible excuse.

Finally, convinced that he was being fobbed off, Mr Charlton summoned Jack to his office.

The tall, dark rogue arrived for his appointment and thrust his head round the door into the office of Mr Charlton's secretary, Sheila.

'Come in, Jack,' she said. 'You're going to have to wait a while. Mr Charlton's over-running. He's got Terry in with him. Take a seat and wait quietly – and behave yourself – until they come out.'

'That's nice, ain't it?' Jack said, sauntering into the office and making a bee-line for Barbie Whittaker, Sheila's gorgeous, young assistant.

'As if I'd misbehave meself,' he said, with a characteristic leer.

For fifteen minutes, Mad Jack kicked his heels, larking around, distracting Sheila and Barbie from their work with his tomfoolery and breaking their stapling machine.

Eventually, the big, office door opened and Terry Bond, Canley's Colliery Mechanical Engineer, emerged from Mr Charlton's office with his two Deputy Mechanical Engineers.

'Hello, Jack. Up before the Headmaster again, are you?' Terry called out, looking into Sheila's office. 'What've you been up to this time?'

'No idea, Terry. I expect I'll find out soon enough.'

'Jack, in you go,' Sheila scolded. 'You don't want to keep him waiting.'

Jack ducked out of the office and knocked on the General Manager's door.

'Come in!' Mr Charlton's ferocious bark penetrated its thick, oak panels.

Mad Jack looked at Terry and his other colleagues, winced theatrically, winked then turned the shining brass handle and pushed slowly on it.

'Aah, the Scarlet Pimpernel himself, Mad Jack Lloyd, unless I'm very much mistaken,' Mr Charlton said, as Jack stole into the room.

'Yes Gaffer, you are very much mistaken,' Jack called out cheerily. 'It is, indeed, I, Mad Jack.'

Mr Charlton scowled, as Jack shambled over to stand in front of his desk.

'Right,' he growled. 'It seems to me, my lad, that you've got no intention whatsoever of repairing that fucking mower of mine, so you'd best get it back to my house and I'll repair the fucker meself.'

'Oh! I'm sorry you feel like that, Gaffer. And I don't blame yer gettin' fed up wi' waitin', I've bin gettin' fed up wi' it meself, as it 'appens.' Jack said, leaning on the edge of the General Manager's desk to confide in him. 'But it's bin a bit of a struggle findin' out what's wrong wi' it. And I've been busy, what with the problems on the shearer on West 9's and that.'

'You can spare me that nonsense. And geddoff me fuckin' desk.' Mr Charlton barked.

Jack leapt back and jumped to attention.

"Mad Jack?" – and "Busy?" Two words that don't sit well together, my lad.'

'It's three words, Gaffer.'

'What?'

'It's three words – "Mad", "Jack" and "busy". 'Old on a minute, that's four now!"

'You're bloody well-named you are,' Mr Charlton growled, shaking his head. 'But it's me that must be mad, letting you loose on that mower. Just get it back to my place and I'll sort the fucking thing out meself. My lawns are like a fucking wilderness.'

'Righto Gaffer but it'd be a shame 'cause I managed to find out what the problem was the other day. I needed a couple of parts for the clutch and a new butterfly spring for the carburettor, I'm just waiting for 'em to be delivered now. It's bin a bit of a struggle sourcin' 'em. That model's not in production anymore.'

'"Not in production anymore"? It's brand fucking new.'

'I know, but these are exciting times, Gaffer,' Jack said, giving Mr Charlton a knowing nod. 'It's the rate of technicological progress.'

'Don't give me any o' that bollocks. You get that mower fixed and back to my place or I'll have Mr Bond demote you. And you won't be messin' about with mowers and shearers, I'll have you in that pit bottom sump, working on the submersible pumps.'

'Righto Gaffer, I'll get on to it right away, as soon as I get the parts – 'opefully sometime this week.'

'Not "hopefully" or "sometime this week", get the fucker fixed and back to my place tomorrow or yer fucking sacked!'

When Jack reached the door he stopped and looked back.

'Oh, and the subnursable pump in the pit bottom sump's workin' fine, Gaffer, I had a look at that meself, last week.'

'Get out!' The Gaffer bellowed.

'Bloody lunatic,' he muttered, when Jack had closed the door behind himself.

A week later, Mr Charlton dropped in at home, on his way back to the pit from a meeting at the Area Headquarters.

He could hear a mower at work in his back-garden and guessed that it would be the man his wife had told him about; a persuasive contractor who had called at their door to tell her that he had started providing a lawn-mowing service in their village. Frustrated by their own mower's interminable absence and the state of their lawns, she had engaged the man and agreed to pay him ten pounds a time.

The need to employ someone to do the work rankled with Mr Charlton and the extortionate fee had outraged him. But, in the circumstances, he had felt unable to take his wife to task for her decisive action.

Having let himself in through the front door he went through to the kitchen to join her for a cup of tea. Looking out of the kitchen window, he saw that the lawn mower careering around his back-garden was identical to his own.

Then he recognised its reckless operator.

Outside, on the mower, Mad Jack saw the wide-eyed stare on the familiar, furious face at the kitchen window and cursed his luck. He'd thought he'd be safe at that time of day, moonlighting on the General Manager's mower, whilst working nights.

He could tell from Mr Charlton's thunderous expression that he was about to be subjected to the full force of one of the Gaffer's legendary rages.

As Mr Charlton stormed out of the back door, Mad Jack continued his reckless progress around the back lawn with an exaggerated air of nonchalant innocence.

Mr Charlton strode out and stood in the path of the mower. Feigning surprise, Jack pulled up.

'Oh, 'ey-up, Gaffer,' Jack called over the noise of the engine. Mr Charlton reached for the ignition key and turned it off.

'"Hey-up" is it Mr Lloyd? Perhaps you'd like to explain yourself.'

'Oh, certainly Gaffer,' Jack said. 'I got them parts yesterday and fixed the mower – as promised. But, you know 'ow I never leave any repair down the pit without makin' sure it's a hundred per cent? Well, I thought I'd better do the same wi' your mower and gi' it a good test-drive.'

'Any road, I think you'll agree, I've done enough of the lawn – it seems to be working fine.' Jack said. 'Perhaps, now yer 'ere, you'd like to finish it off?'

Thursday, 29 September, 1983: 15:33

When Jim walked into the fitting shop he found Mad Jack diverting five afternoon-shift surface fitters from their work. Jack looked over his shoulder, saw Jim and broke off from his performance.

'Ey up, stand by yer beds, lads; it's the Undermanager.'

The mechanical engineers and electrical engineers were service providers to the mining engineers who managed all of the collieries' underground operations. The reality generally appeared different due to the disdain with which the sophisticated, specialist engineers viewed their more basic, mining engineering colleagues.

Jim had never been able to understand the antipathy between mining engineers and the other two disciplines. It was apparent that every pit's success was dependent on all of them playing their part and pulling together. Consequently, Jim adopted a more collaborative approach and treated his electrical and mechanical colleagues with respect and patience, even when they tried to perpetuate the traditional rivalries. This had helped to foster the positive relations he enjoyed with engineers like Mad Jack and his ability to call on them for support.

Despite this, Jack chose this moment to reinforce old prejudices by launching into a joke at the expense of mining engineers.

'Here lads,' Jack said. 'I heard a good un' about this undermanager the other day.'

Before Jim could interrupt him to tell him that he was in a hurry, Jack had regained the fitters' attention and was away.

'This old colliery general manager gets transferred to take over another pit. On 'is first day 'e decides to check out the calibre of 'is team. 'E summons the Undermanager, the Colliery Mechanical Engineer and the Colliery Electrical Engineer. When 'e's introduced hisself and told 'em what 'e expects of 'em, 'e says he's got a test for 'em – to test their competence, like.'

"E gives each of 'em a different pair of ball bearings and says, "Go away and investigate 'em and come back tomorrer and tell me what you've discovered about your particular pair of ball bearin's".'

'They report back the next day and the old General Manager tells the Colliery Mechanical Engineer to go first and tell 'im what 'e's found out about his'n.'

'The Mechanical Engineer says, "The ball bearings you gave me appeared to be identical to the naked eye but I measured them both with callipers and found that their diameters differed by five thou'. So I conclude that they are not the same size".'

'"Good," says the General Manager, looking suitably impressed. Then 'e says to the Colliery Electrical Engineer, "And what did you find out about your ball bearings?"'

'The Electrical Engineer says, "The ball bearings you gave me appeared to be identical to the naked eye. I measured them both with callipers and found that their diameters were the same, so I passed a current through both of them and found that their electrical resistances were different so I concluded that they are made of different material".'

'"Very good," says the General Manager. So 'e says to the Undermanager, "And what did you find out about your ball bearings?"'

'The Undermanager looks embarrassed and says, "I'm sorry, Gaffer, but I can't tell you anything:

– I lost one and broke the other".'

Whilst Jack's audience guffawed, Jim smiled and nodded. Not only was the joke about mining engineers' general abuse of the heavy industrial equipment entrusted to them, Jim knew it was particularly apt because an overman on the mining side at Canley, his and Jack's previous pit, had actually managed to lose a huge machine section underground.

'Any road,' Jack said. 'I bet this ain't a social call. What're y'after?'

'That's a nice way to greet an old mate,' Jim said. 'But, as it happens, you're right, I do need to speak to you about something.'

'Come into the office,' Jack said.

As Jack led the way to a compartment formed by two steel partitions with big glazed windows in the corner of the old building. Jim walked behind him on the workshop's sticky floor of solid, black grease, weaving between the ancient, battered workbenches, relieved to find that he was not to be subjected to more of Jack's ribbing in front of the fitting shop staff.

'I've just been given a new job,' Jim said.

Jack turned and said, "As 'e put y'on nights, then?'

'No, just a change of responsibilities.'

When they were in the office, Jack closed the steel door with uncharacteristic delicacy then went round to sit down behind the office desk.

'I was surprised you'd managed to steer clear of bein' put in charge on nights or afternoons when you first turned up here,' he said. 'Given as 'ow they always stick 'non-stats' on the backshifts. I thought to meself when I 'eard they'd put you in charge of 'alf the pit "that looks like penny-pinchin' to me, getting a Non-Stat' to do a Statutory Undermanager's job". They shoulda given you the Stat' job by rights.' *[Note 6]*

'Yeah, you could be right but I don't care, I'd rather have the authority and responsibility the Gaffer's given me. And, anyway, it's a more civilised way of life on days.' Jim said. 'It was probably all those years on nights regular that sent you off your head.'

'Nah, I was born like this,' Jack said. 'Anyway, what're you after this time?'

Jim explained Mr Everitt's expectations and the criticality of getting 22B's into production by the target date.

'I need to make sure I've got a good senior fitter on each shift,' Jim said. 'Someone who can get the best out of the rest of the fitting team and overcome problems.'

'Got anyone in mind?'

'Well, you've been here longer than me and, anyway, they're your men so you know them,' Jim said. 'But, I could do with Paul Wood from down my end of the pit as one of them.'

Jack gave him a crafty, wry smile.

'What?' Jim said.

'Oh, nothing. It's just you're obviously after the best men I've got.'

'Of course. It's that important. The Gaffer wasn't joking, he's bloody worried about this. If we don't hit that deadline this pit's finished.'

'Righto, now I know what we're up against I'll make sure you're manned up on all three shifts. Woody's stopped over on overtime so I'll have a word wi' 'im when 'e gets up the pit. Let 'im know 'e's on the transfer list.'

'Thanks Jack. Oh, and could we make it from tomorrow? Tell him to come and find me in the dirty offices before the start of dayshift so he can go to 22's with me. We'll be putting a big shopping list together. We're going to need your persuasive powers in getting hold of what we need and getting some stuff fabricated as quick as possible.'

'Righto Jimmy,' Jack said. 'Us old Canley men have to stick together, y'know.'

As Jim reached the door of the fitting shop, Jack appeared at the office door and called after him.

'I knew I'd 'ave you findin' ways to keep me busy as soon as I 'eard you was comin' to Whitacre.'

'Just trying to keep you out of mischief, Jack,' Jim called back.

He gave Jack a thumbs up and a cheery, winning grin then strode off to find Mike Taylor and get back underground.

Back Down

Thursday: 16:14

As Jim headed for the pithead for the second time that day he was conscious of the inevitable, marked contrast between his appearance and that of the afternoons' senior overman striding out alongside him. Whilst Jim's face and hands were completely black from his earlier shift underground, Mike's had the fresh cleanliness of a pitman starting his shift. The men who rotated on days and nights saw the afternoon-shift as a breed apart; they only ever saw them clean-skinned so they tended to refer to them as "the 'Erefords" or "white-faced uns".

Jim and Mike passed through the airlock and into the airtight, surface building of Number 2 Pit Top, rode the shaft and walked through the pit bottom air-doors and round to the Loco Road to catch a lift on a locomotive, drawing empties to the South End of the mine.

The height of the mine cars made it impossible to climb in and out of them so the two men squeezed into the single, driving position at the rear end of the loco, designed to be just big enough for the driver to perch comfortably, on his own, when driving the loco in the opposite direction.

The loco accelerated away, out of the brightness of the pit bottom station, into the dark of the unlit Loco Road. Supported by great, steel arches, it was tall and wide enough to accommodate the two tracks on which trains travelled in both directions of their endless cycles; taking empties inbye to the loaders and drawing full mine cars back out to the pit bottom.

After they had bounced and shaken their way at speed for more than ten minutes, they approached a well-lit stopping point. The driver drew the train to a halt to allow his two passengers to alight, next to the access through to the East Side of the pit.

Jim struggled out, relieved at having escaped with his life, without having been thrown from the cramped, precarious space between the back of the speeding loco and the first of the train of empty mine cars. When he was clear, Mike extricated himself.

Calling out their thanks to the loco driver, they strode over to the big air-doors to enter the East Side of Whitacre Heath Colliery and take the long hike out to 22's, the district whose fortunes would determine whether the pit had a future beyond the first of December.

Party Night

'**M**artin, clear your Lego up, love. It's nearly tea-time.' Susan sang out from the kitchen.

Pat bustled back in from laying the table, an urgent finger pressed to her lips to silence her daughter-in-law.

'What?' Susan whispered.

'It's alright, luvvie, it's just Horace. He's dropped off in his chair.' Pat said. 'I can never get him to sleep-in long when he's on nights, he always wants to enjoy a bit of daylight. He's been down the allotment today. Let's leave him to the last moment before we rouse him.'

Susan carried on getting cutlery and plates out.

'Paul's the same,' she murmured. 'I don't know how he manages it. About three hours is all he gets, less if Jenny wakes him. Is she alright with Horace?'

'Yes, they're fine.'

The baby was in the lounge, cradled in her grandfather's right arm, lulled to sleep, like the old miner, by the drone of the television's regional news and the heat from the anthracite burner in the hearth. Sally, the Highland Terrier, lay on the deep pile carpet, at her master's feet, watching Martin as he started to pack away his plastic bricks, as soundlessly as possible, on the hearth rug.

At the sound of a key in the front door lock, Sally huffed then scampered out into the hall to meet Paul.

Horace's eyes opened and drifted up to the rolling hills of North Warwickshire in the landscape hanging over the hearth and its coal-fired boiler. It was one of the water-colours that had helped Paul to gain his Grade 1 GCE in Art, fifteen years earlier. His teacher had wanted to buy the picture but Paul had refused several, increasing offers, keeping it to give to his parents the following Christmas. The only other work that Paul had retained from those days was his action-packed oil-painting of Giacomo Agostini, racing to another Motorcycling World Championship, that hung over the fireplace in his own house.

Paul hung his coat on the newel post at the foot of the stairs then eased the lounge door open. The little terrier bitch continued to fuss round his feet, demanding his attention. Paul bent down and ruffled her ears.

'Hello, Sal',' he murmured, as she licked at his hand.

'Hello son,' Horace croaked.

'Hello Dad, happy birthday,' Paul said quietly, conscious that his father was only just awake. He leant down towards the sleeping baby in his father's arm and gave one of her tiny, bare toes a little tug.

Martin looked up eagerly at his dad. Paul ruffled his hair.

'Hello, son,' he said.

He went through to the kitchen, kissed Susan and then the cheek his mother offered as she stood at the cooker.

'Hello love,' Pat said.

'Hello Mum. Gammon for tea?'

'Yes, your Dad's favourite, with pineapple and chips – and an egg as well, seeing as it's his birthday.'

'Lovely. Can you do us an egg as well?'

'I was doing you one anyway, love.'

Paul returned from the kitchen and sat down on the armchair nearest to Martin.

'What d'ye reckon to the news?' Horace said.

'The pay claim? I can't work out what they're up to.'

The news on the radio at five p.m. had reported the National Coal Board's rejection of the NUM's demand for a 5.2 percent increase in the miners' basic pay and its determination to see improved efficiency as well as "reductions in productive capacity", which sounded like a veiled way of saying pit closures, before it would consider any pay-rises.

'I heard that,' Susan said. Normally, she would have left the men to their pit-talk but for once she joined them. 'Sounds a bit ominous, if you ask me.'

'Nah, I don't think it's a problem,' Paul said. 'They've been pressin' for more productivity and improved quality for a while now, and that's all to the good. We should be doing that. It's only five years ago the Government published the Plan for Coal. That's 'cause the country needs as much coal as we can produce to fuel growth.'

'But that would have been the last Government,' Susan said.

'It would,' Paul said. 'But all three parties signed up to it.'

'Well, I hope you're right,' Susan said.

She decided it was best to let the matter drop and retreated to the kitchen. She didn't want to risk irritating Paul and impairing Horace's enjoyment of his birthday celebration.

'What worries me is how it might affect the union committee elections on Sunday,' Paul said. 'We don't want the lads feelin' threatened and letting McFadden squeeze 'imself in. There'd be no end of trouble if we ended up with that nutter as Union Secretary.'

Horace sat quietly.

The last thing he wanted was a strike in pursuit of a pay claim. With only three years left until retirement, he wanted to keep earning and topping-up his life-savings. Losing pension contributions as a result of being forced out on strike would be a bitter pill to swallow.

Horace shook off those thoughts.

'Ave you had a good day, son?'

'Yeah. Didn't 'ave to go far today. I 'ad a couple o' lads with me, working on a compressor in the pit bottom. But I'll be back on the same shift as you, stopping on days next week. The young boss has asked Mad Jack for me, wants me to lead the fitting team on that shift on 22's salvage.'

'What's that?' Susan said, leaning back round the kitchen door. 'You're on days next week?'

'Yeah, I'm goin' over the East Side, on a transfer job.'

Susan smiled and nodded at him; every week of nights avoided was a bonus, another week of normality.

'What's a transfer job, Dad?' Martin asked, kneeling upright on the hearth rug to look up at his father.

'It's where we take the face equipment straight off one worked-out face and onto the next. 22's, which I'm going onto, was an advancing face – you know what I told you about how a face works; it was about two 'undred yards long and it took about four an' a half foot o' coal for a mile 'n' a half in the Top Bench seam. Well, that old face stopped a couple o' weeks ago, when they'd gone far enough, but they've left the lower seam, the Bottom Bench coal, still to get, underneath where they were working.' Paul said, using his flattened hands, held over each other to represent the seams.

'So now they've driven another face-line between the two gate roads, under the thin band of soft dirt that made up the floor of the advancing face, about seventy yards back from the old face, underneath the waste where the roof collapsed as the old face went forward. They've got the two shearer's off, the big machines that cut the coal, now we need to take off the chocks that support the roof and then the panzer – the face conveyor. As each of the chocks that support the roof comes off the face we give it a quick once-over then move it straight onto the new face-line, facin' in the opposite direction, ready to retreat back along the gate roads – that's why we call it a retreat face. It'll come back all the distance that the upper face travelled, takin' another four foot six of coal.'

Paul spoke with enthusiasm, born from the knowledge that new faces meant continued life for his pit. Martin looked with wide-eyes at his father as he tried to picture the intense industrial activity that would be taking place, at that moment, in the confined spaces, deep below ground, not that far from the safety of his grandparent's home.

'We've got to keep the pit open so you can come and work with me,' Paul said. 'Like 'ow I went and worked with yer grandad.'

'I told him he ought to try for university – then he could go down the pit in a job like our young boss.'

'Nah, he wants to be a proper pitman.'

Unusually, Horace spoke out.

'Oh, I dunno, I'd say Mr Greaves is a proper pitman. He can wield a shovel wi' the best of 'em. An' 'e knows what 'e's doin' when 'e's timberin' up a fall. I reckon 'e's 'ad some pit-work, already.'

Paul looked thoughtful and nodded.

'Dad, can we give Grandad his birthday present now?' Martin said.

'Wash your hands everyone, I'm serving up now,' Pat called from the kitchen.

'We'd better do it after dinner or we'll get in trouble with Grandma,' Paul said.

'Here, let me take her,' Susan said, coming in and stepping over to Horace to bend over him and pick up Jenny who was still cradled, asleep, in his arm.

Horace stayed in his armchair, waiting while Paul and Martin slipped out to wash their hands in the downstairs loo before stirring himself. When his son and grandson bustled back in and made their way to the dining table, round the corner of the L-shaped room, next to the kitchen, he heaved himself up stiffly from his armchair

44

Weighing It Up

Thursday: 18:14

Being more familiar than Jim with the East Side of the pit, Mike took the lead as they marched along the East Main Supply Road. During their half-mile hike, they discussed the challenge and potential ways to accelerate the transfer until they saw the lights ahead, illuminating the junction with 22's Tailgate.

Mike had made the necessary courtesy call from the surface, so they didn't need to phone Gary Wallender as they entered his district. They turned right into the tailgate, hung up their donkey jackets, took off their light, orange, workwear jackets then pressed on, in shirtsleeves, towards the face.

A couple of hundred yards into the district, Jim played his cap-light over the rust-coated, steel arches that supported the roof at yard intervals, noting their condition.

When first set, the arches would have been twelve feet wide and ten feet high. Although the floor had heaved up to close the height of the road to little over six feet, the arches weren't too badly deformed. This was good news because the two gate roads were going to take a severe hammering from the massive pressures thrown in front of the new face as it retreated back along the gate roads. A good, arched profile would give the best prospect of maintaining a workable access to the face when those enormous, additional forces bore down on it.

Jim and Mike had to step out to complete the walk to the face in forty minutes. As they approached the new face-line they encountered the equipment that had been dismantled and salvaged from 22A's, the worked out, advancing face, and stored in the side of the road.

First, there was a hundred metre length of road in which some of the face conveyor parts had been stacked. Closer to the face they encountered the major components of the shearers, the two, big coal cutting machines.

Jim stopped and gazed at them, daunted, suddenly, by the volume of work needed to turn 22B's into a producing face before the looming deadline. 22's looked a world away from a production district.

Leaving Jim to his thoughts, Mike drew away then strode on, for another thirty yards, to where a gang of three facemen, in grey, Sketchley vests and y-fronts, were making slow progress at turning a powered roof support down onto the new face which ran away to the left of the gate road. Having found out, from the men, where he could find their district deputy, Mike strolled back to Jim.

'Gary's on the old face with the salvage team, Boss.'

'Okay,' Jim said. 'Let's go and find him, we can take a look at the salvage work first.'

As they made their way along to where the road ended at the old advance face, they passed the trio struggling with the powered roof support.

"Afternoon, men,' Jim called out.

"Ey-up, Boss,' they chorused back.

'Is this the first chock going on?' Jim said.

The oldest of the men, chewing a screw of tobacco, spat a mouthful of black juice onto the floor then wiped his mouth with his wrist.

'Yes, Boss. We're just tryin' to get a system going – find the best way o' doin' it.'

That's what it looked like. This critical step in the process had been neither planned nor even considered.

The last fifty metres of road to the face was filled by the other eleven chocks that had been drawn off the old face and stored there, collapsed to their lowest level.

At the dead-end, at the end of the gate road, Jim and Mike ducked under the lip to access the old, worked out face-line. This was like a low, wide road with straight legs and cross-members installed to support the roof as the chocks were removed. They found the deputy twenty metres down the face-line, supervising six men drawing another chock off the face. Gary saw the two visitors approaching and scurried back up the face, doubled over under the cross beams of the steel square-work.

'Do you know Mr Greaves, Gary?' Mike said. 'He's been put in charge of the transfer work.'

'We 'aven't been introduced but I've seen you around, Boss,' Gary said, shaking Jim's hand.

'Yeah, I recognise you, Gary,' Jim said. 'It looks like your biggest problem in the short-term is where you're going to put the chocks as they come off the face. That gate road's full of 'em already.'

'Yeah, I know. We need to start banging 'em onto the new face.'

'What's your target for transferring chocks?'

'How do you mean, Boss?' Gary said.

'How many chocks are you targeted to get off the old face and onto the new face each week?'

'We 'aven't got a target, as such. It's just shift 'em and get 'em transferred as fast as we can. But I suppose that means at least we ain't restricting ourselves to any number either.'

Jim smiled.

'Well, at least that's positive thinking. Let's have a look at how they're getting the chocks off.'

Jim started walking, doubled up, down the old face.

22A's armoured face conveyor had been stripped down to just the line of steel pans, each pan being five feet long and thirty inches wide. The flights and the hefty chain that had dragged the coal off the working face had all been stripped off it. The chain had been stacked in bundles, out in the tailgate, each consisting of half a dozen flights and their connecting chains, adding to the overwhelming clutter out there.

The panzer's pans were being used as a track for dragging the chocks off the face. Each chock had to be manoeuvred painstakingly up and into position on the pans and then turned ninety degrees to face away from the face, towards the tailgate.

The gang of salvage men were moving the chock along the panzer using the snaking ram built into the chock's base to pull against a long, heavy chain coupled onto a plate secured to the dirt floor in the tailgate. Each time-consuming cycle of extending the ram, dropping a link of chain into its steel claw then drawing it in moved the chock less than a yard. *[Note 7]*

'That's a slow old job you've got there, lads,' Jim said.

'Yer not wrong there, Boss,' one of the men said. 'It's a mug's game, this.'

'Let's go and have a look at the new face-line,' Jim said, heading back out to the tailgate. When he got out into the road, he stood up and looked at the traffic jam of chocks extending back out along the tailgate towards the new face-line.

'This is a bottleneck,' Jim said. 'We've got to get them moving onto the new face.'

Gary and Mike both nodded then looked at each other, shame-faced. They might be lucky and get the first of the one hundred and sixty-seven chocks onto the face, that shift, but it would still have the majority of the new face's two hundred metre length to travel. Even then, it would take even more time to turn it into position. At this rate they would exceed their deadline by months. It was becoming apparent why Mr Rawlinson had applied such pressure on Jim's gaffer.

22B's would be taking the Bottom Bench, underneath where 22A's, the old, advance face, had extracted the Top Bench, the two seams being separated by little more than a foot of friable shale.

The two officials followed Jim as he eased his way back past the three men manoeuvring the first chock. They ducked and scrambled down the couple of feet from the tailgate and onto the new face-line.

Like 22A's, 22B's face-line was another, low, wide road supported by steel square-work. They made their way, bent over, all the way through its two hundred metre length, with nothing to see in the absence of any installed equipment, to emerge and stand up in the maingate. Jim led the way back inbye, from the new, retreat face-line towards 22A's worked-out face-line.

'I just want to see what state the road's in, in terms of the seal on the waste side,' Jim said.

He knew from experience that it was vital to minimise the air passing through the waste, the area which had caved behind the advancing face, in order to prevent spontaneous combustion and the start of a dreaded heating. *[Note 8]*

For thirty metres, the yard spacings between the steel arch legs had been lined with lagging boards slotted between the legs. These had been roughly plastered with hardstop, a pink, gypsum product. Thereafter, none of the arches had been lagged.

'Have you pumped any hardstop in here to complete the seal?' Jim asked Gary, pointing at the spaces between the lagging boards and the corrugated sheets that lined the road behind the steel arch legs and bows.

'Er, no Boss,' Gary murmured.

Jim had seen enough.

'What do you reckon, Boss?' Mike said.

'I reckon we're making hard work of the transfer, it's a slow old process,' Jim said. 'And we're playing with fire, not getting a good seal on both roads. I'll order the equipment and get men deployed on all that, starting on nights.'

'Right,' he said. 'Thank you, Gary. I'm going to head out now. Are you going or staying Mike?'

'I'll walk out with you then head up to 56's to see how the coal-turning's going this afternoon.'

When they had completed the long walk out of the maingate, Jim parted company with Mike who headed inbye, towards 56's face district. Jim cut back through the air-doors to the tailgate to retrieve his coat and jacket before returning to the intake. Its fast moving, main conveyor belt was loaded with coal from the face that Mike would be visiting. Jim vaulted onto the racing, troughed surface then lay down, pushing some of the bigger lumps away from himself, in front and behind, to make his ride less uncomfortable.

As he sped along on the belt, he thought about the enormous amount of work needed to get the face-to-face transfer going at a respectable rate of progress and then go on to exceed it so they could hit the target of producing coal in seven weeks' time.

He remained amazed that the transfer had been embarked upon with virtually no preparation or planning and then left to happen without any serious interest from senior management. It was not surprising that the Gaffer had nominated Jim to get 22's transfer going because it was clear that Bill Metcalfe had neglected it completely.

In the three months that Jim had been at Whitacre Heath, the Services Manager had only been down the pit for one brief visit. He couldn't imagine how anyone could operate like that. Apart from anything else, he must be bored out of his mind. It was surprising that Mr Everitt hadn't got rid of him, particularly since it would have made a useful contribution to the economies that the pit needed to find.

Jim considered his new challenge; as well as introducing a faster way of getting the chocks off the old face and round onto the new face-line, he was going to have to change the attitudes and expectations of everyone involved on 22's.

A Surprise

They completed Horace's birthday tea with a rendition of 'Happy Birthday' and Martin helping his grandad to blow out the candles on the chocolate cake that Pat had baked.

Horace had received a birthday card and a present of a lamb's wool jumper from Pat when he'd taken her cup of tea up to her in bed, first thing that morning. As they finished their pieces of cake, Paul handed his dad another card.

'And that's got your present in it,' Martin called out, bouncing up and down on his chair.

When Horace opened the envelope a piece of paper dropped out of it. He picked it up and found that it was an advert', cut neatly from the Tamworth Herald, for the coming weekend's game fair at Arbury Hall.

'We thought you, me and Martin could go there on Saturday,' Paul said. 'Then we'll buy you a new fishing tackle bag. You can choose what you want from one of the trade stands there.'

'And that's the big surprise!' Martin cried out.

'Thank you all, that's a very kind thought,' Horace said, beaming at them. 'My old one's like me, seen better days.'

Horace's ancient tackle bag was like a faithful, old friend to him but he couldn't disappoint them.

'When we get my new bag you can have me old one,' he said to Martin. 'So you'll have a bag of your own, when we go fishin' next year.'

'Woa, thanks Grandad.'

His grandad's bag would make him a real fisherman. Its numerous pouches and compartments had always fascinated him. He could see himself caring for it, just like his grandfather had for years, keeping it clean and its contents tidy and business-like.

'Colin said he'd come and join us for a lads' day out. 'E's made sure he's not gonna be working overtime on Saturday.'

Paul and Colin had been mates since primary school. Both were good sportsmen, Colin still playing for the Miners' Welfare's football team, although Paul had retired from it after Jenny's birth. They shared a longstanding passion for motorbikes. When Colin was sixteen he had bought Paul's second-hand, but pristine, Honda 50 moped, providing some of the money Paul had needed to graduate to a Honda 350. As soon as they could, they had both moved onto bigger, British bikes.

Pat got up from the table to step into the kitchen.

'I called round on Doreen today and saw little Emily,' she said. 'Diane was telling me about Emily and Jenny playing, round at your house on Monday.'

Pat and Doreen had been friends even before they started at the pit village's school and Diane, Doreen's daughter who was married to Colin, had formed a similarly enduring bond with Susan when they were tiny children. Pat felt sure that a third generation of female friendship was about to be established between their families.

'Yes,' Susan called back to Pat. 'It's nice how it's worked out, both of us having little girls.'

Colin and Diane were both slight and diminutive. Although Colin was two years younger than Diane, they had been drawn together by their respective friendships with Paul and Susan. With Susan having proved to be good enough for Paul, Colin had not been surprised to find that Susan's best friend was the right girl for him.

Pat returned to the table.

'I've got you a couple of bags of tuffees to take with you tonight, for you and your mates on the face, seeing as how it's your birthday,' she said. 'Here, I'll put them in your snap bag so you don't forget them.'

She waved two family-sized packets of Cadbury's Chocolate Éclairs at Horace before putting them into the surprisingly pristine shopping bag that he used to take his snap box of sandwiches and small, metal, vacuum flask of tea down the pit for the mid-shift break of twenty minutes.

The smart, leather bag was a deep, elegant maroon, so thin that it was almost flat, with a neat, gilt catch. Horace had bought it for his mother, when she was in her last years. She had hardly ever used it. When she died and they had been sorting through her things, Horace had agreed that it was too much of an old lady's bag for Pat but had decided it would serve as a snap bag for him. It meant that he was always reminded of his old mum when they broke for snap on the face.

'And before you ask, you don't have to worry, I've put your bag of aniseed balls in as well,' Pat said.

Horace beamed up at her.

'Please can I have a toffee, Grandad?' Martin said.

'No, love, don't be cheeky,' Susan scolded. 'They're for your Grandad and the other men to enjoy, when you're asleep in bed.'

'It's alright, Susan, they're for me and my mates and Martin's my best mate so 'e can 'ave one,' Horace said.

He eased one of the bags of sweets open.

'Here y'are, lad,' Horace said as he handed Martin a crafty couple of toffees. 'Miners love these, it's special pitmen's suck. It gives us the energy to really get them shovels goin'.'

'Thanks, Grandad,' Martin said, with one of the sweets bulging in his cheek. 'Please may I leave the table?'

'Alright, but finish clearing up your Lego,' Susan said.

Paul helped his mother clear the table before flopping down on the settee to join Horace who was watching a programme on the television on the wildlife of the Serengeti.

The majority of the country's coalmines were located in rural areas and, even though they were engaged in one of the heaviest of industries, like many colliers, Paul and his father were true countrymen. Despite the apparent paradox, pitmen tended to enjoy the contrasts between their work in the confined depths of the mine and the fresh air and natural environment of their surrounding countryside. So, it was not surprising that pitmen were fascinated by programmes featuring wide open spaces with animals in the wild, and that they would often talk about them, on their way into the pit or at snap-time, during their next shift underground.

As soon as the programme finished, Paul sprang into action to take Susan, his weary son and grouchy baby home and allow Horace some quiet time in which to ready himself for his shift down the pit.

As they got their things together, the BBC Nine O'clock News came onto the television. Again, the NCB's rejection of the NUM's pay claim was the main story. Paul and Susan paused and listened to the report start after the headlines. It was telling them nothing they hadn't heard already.

'Come on, let's go,' Paul said. 'There's no joy for any of us in that lot.'

After an exchange of kisses and hugs, Horace and Pat stood at the front door to see the young family off, on their way down the road.

'See you on Saturday, Grandad,' Martin called out.

'Yes lad, I'll be lookin' forward to it,' Horace called softly after him. 'Goodnight and God bless.'

'Goodnight Mum,' Paul called. 'G'night Dad, see you in the canteen in the mornin'.'

'I'll put the kettle on for a coffee and fill your flask while you get your shave, love,' Pat said, as Horace closed the door.

'I wish you could've had the night off.'

'Aw, ne'er mind,' Horace said. 'I'll save it for something more important. Besides, it's the last shift o' the week.'

Acceleration

Jim had collected a few, large sheets of graph paper from the Surveying Office, ready to draw up the plan for running the equipment transfer and commissioning as a project. He wouldn't be able to work on that until the end of the day. Before then, he needed to order some of the additional equipment they were going to need, then make the most of the last two shifts of the week; getting the best out of the team on days before setting up the afternoon-shift for success.

Before going underground, he called over to the office in the fitting shop. Mad Jack was already in his pit-black.

'I thought I'd come down the pit, wi' you and Woody,' Jack said. 'Bring a few lads wi' me and give you a hand in settin' 22's up.'

'It's a good thought,' Jim said. 'But it would probably be best if you could get hold of some of the kit we need most urgently before you come down.'

'What're you after?'

'I need a couple of medium-sized Pickrose motors. We've got no haulage system off the old face or onto the new. If we carry on as we are we'll be here 'til next Christmas, never mind this one. I was hoping you could call in a favour or two. Or nick 'em from somewhere.'

As usual, a crafty smile formed on Jack's face at the prospect of a bit of banditry.

'I'm sure I'll be able to find a couple from somewhere for yer. Leave it wi' me.'

'We'll need three low-loaders, as quick as poss',' Jim said. 'They're just dragging the chocks off the face and down the road, steel to steel, on the panzer with rams. And I need you to get someone to knock up a steel ramp and platform so we can pull each chock onto it to haul it up to the same height as the low-loaders so we can just slide the chocks across onto them with a couple of rams. They're having to bugger about lifting and jacking 'em up at the moment. And I'll need one of those for the new face-line as well, for unloading them quickly and safely.'

'An' I suppose you'll be needin' all this by yesterday, as usual,' Jack said, with a wry smile.

'When we start the week on Sunday night we'll only have forty-three days left to get over a hundred and sixty chocks off the old face and into position on the new face then get all the rest of the kit onto the face, re-built and working. It'd be a challenge even if it had all been well-planned – and this hasn't. We have to take that deadline of the first of December seriously; Rawlinson doesn't bugger about.'

'You're not wrong there. And I don't want this place closing round me y'ead and getting sent back to Canley, not the way things were, back there wi' Jackie Charlton, the old bastard.' Jack said. 'Is there owt else you can think of?'

'Yes, could you get the carpenters to make a couple of notice boards and paint them up for me. Here,' Jim said, handing him a scrap of paper. 'I've drawn up what I need.'

Jack glanced at the sheet and smirked.

'Bloody hell, Jimmy.'

'What's up?' Jim said.

'Well, you'd never 'ave made a draughtsman, would yer?'

'What do yer mean? It's good enough to give 'em an idea of what I need'

'It's a typical, bloody mining engineer's effort, that.'

'Bollocks, there's no engineering tolerances involved.'

Jim's sketch was intended to be the design for two progress-monitoring blackboards. Four feet high by three feet wide, under the heading "22's Chock Transfer" the boards would have a table indicating, from top to bottom:

"Total number of chocks = 167";

"Chocks transferred – total" with a space for the number to be chalked in;

"Chock transfer target this week" to enable an increasing target to be chalked in as the systems and process improved; and

"Chock transfer: Actual this week" on the table's bottom row.

'Can you make sure they get them done today?' Jim said. 'I want to get 'em carried in and put up on Sunday night to provide the team with a bit of impetus.'

'Alright Jimmy,' Jack said. 'That little shoppin' list's going to keep me out of trouble for a bit.'

'There you are then; that's a result for everyone!'

Friday: 06:20

Jim darted back across to the dirty offices, met up with Paul Wood and managed to get down the shaft in time to catch the first train along the Loco Road to the East Side of the pit. Billy Thompson, the dayshift deputy for 22's that week, was short and stocky with a keen expression. The three of them met up and sat in the same four-seat bay in the front manriding carriage.

Jim pulled his pocket book and pencil out of his top pocket and drew a sketch showing the way he wanted things setting up on the old and new faces, particularly the layout of the two haulages.

'I'll get Gerald Chipman to starting sending in the rails, sleepers and fishplates we need for laying the track. We need to get track laid all the way through the new face-line – and down the old face to the next chocks to come off.' Jim said. 'Billy, you need to deploy some men to rip out a few sleepers and rails from the tailgate haulage track, next to the new face-line. Les Parker is going to find out where the nearest Jim Crow is for us and get it transported to 22's so we can bend rails to lay a curve out of the old face and another curve into the new face to give us a continuous route for transferring chocks. *[Note 9]*

'We'll need a couple of return wheels so we can install one of the Pickroses as an endless haulage to transport the chocks through the new face-line. Then for salvaging the chocks off the old face-line, the slight gradient up to the tailgate means we can set up a simple, direct haulage.' Jim said. *[Note 10]*

'That should knock hours off each chock transfer,' Billy said, nodding.

When the train reached the East Side's manriding station, they clambered out with the crowd of men bound for that part of the mine and marched inbye to the meeting station for 22's, on the junction of the East Main Return and 22's Tailgate.

'We'll wait here while Billy deploys all the men,' Jim told Paul.

From the way Billy spoke to his men and their responses, it was apparent that he exerted positive leadership over them. When he had allocated tasks to all of his team he joined Jim and Paul on the long walk along the tailgate to the two faces. One advantage of 22B's being a retreat face was that this distance and travelling time would be reduced progressively as the face extracted its way back along its gate roads, gradually increasing the time available for turning coal each shift throughout the district's remaining life.

Jim led the way, explaining his plan in more detail, as they marched along in single file.

When they arrived at the inbye end of the tailgate, Paul said, 'I think that all makes sense, Boss. I can't think of anythin' you've missed.'

'Yeah,' Billy said. 'I suppose it just makes you wonder why we didn't set it up like that from the start.'

They walked up to the group of chocks that cluttered the length of tailgate between the two faces.

'Paul, I want your opinion on the value of what's going on here,' Jim said. 'I know load-bearing tests are recommended these days on face-to-face transfers but we haven't got one of those underground test rigs they use for it. It seems to me that the fitters might just be tampering with things that ain't broken. I reckon we could be as well-off saving some time and just transferring 'em straight onto the new face. [Note 11]

Ten minutes later, having conferred with the fitters working on the chocks, Paul went back to Jim.

'I reckon you're right about that maintenance, Boss, they ain't doing anythin' we can't do wi'out. I'll get 'alf a dozen pressure gauges and get Jack to deploy a couple o' fitters to test their yield pressures when they're in position on the new face.'

Just before snap time, the supply lads on the endless haulage delivered the Jim Crow device for bending rails on a trolley to the inbye end of 22's Tailgate. As soon as Jim was confident that the man operating it was competent and the curved rails they needed would be fabricated by the end of the shift he reinforced his instructions to Billy then headed out of the pit, taking Paul with him.

On the surface, Jim told Paul to go and shower and change so he would be ready to join him in drafting the project plan without covering it in coal dust. Jim called in at the dirty offices and briefed Mike Taylor, the afternoon-shift senior overman and Gary Wallender, 22's afternoon's deputy, on what they needed to do to continue establishing an efficient transfer system.

Friday: 13:50

Over in the clean offices, Jim and Paul sat at a corner of the desk in Ken Goodall's office with mugs of tea, listing the activities that would make up the content of the project plan.

First, there was all the remaining work needed to establish an effective haulage system for the chocks. Allocating estimated times against each of the activities involved in this showed they would not be ready to start transferring chocks from the old to the new face until the following Thursday; the essential preparations for the big push to hit the deadline would use up ten percent of their remaining time. Transferring the chocks would take up another twenty-eight days.

Paul recommended allowing a couple of shifts for getting the shearers' three discs moved into the position on the face (one for the tailgate shearer and two for the maingate shearer). They agreed they would need five days to strip the track off the new face and install the panzer's pans.

Jim knew, from previous experience, that they should be able to install the conveyor chain and flights onto the panzer in under two days, if they took an approach he had used before.

The last big jobs would be getting the shearers components onto the face and assembled and getting the panzer's tailgate drive installed and commissioned. Jim agreed with Paul's estimate that it was likely to take another eight days to get the machines ready for turning coal.

They added up the number of days they had allocated. If everything went smoothly, it would take them until the fourteenth of December to get the face into production; they would miss Mr Rawlinson's deadline by a fortnight.

And that would be the end of the pit.

They checked every element. The limited access to the new face meant that everything was sequential; virtually nothing could be done in parallel.

'What about overtime?' Paul said.

Jim did a couple of quick calculations.

'The only way we could get close to a plan that achieves the deadline would be by manning the job up with overtime throughout the week and on all six shifts, every weekend,' Jim said. 'I'd struggle to get approval for that. It would probably blow the pit's entire overtime budget and we need to stay within that, otherwise they'll be closing us for being unprofitable.'

Friday: 16:45

Mr Everitt had maintained an inscrutable silence while Jim talked him through the slightly neater version of the project plan that he had managed to redraft, hurriedly, for their meeting in the General Manager's office. Jim finished by telling him the bottom-line, in terms of the number of shifts required.

'The only way I can think of us being ready for turning coal by the first of December would be with the heavy use of overtime,' he said.

The stark news was delivered, the subsequent silence excruciating.

'We could quibble about this,' Mr Everitt grunted. 'You'll 'ave got too much or too little against some of the activities. But you're gonna be in the right ballpark. Which means, whatever we do, what you say is likely to be right, we ain't gonna get that face turning by the first of December, without findin' some more time.'

'So, you can 'ave your overtime. But I want the number of men working on each weekend shift to be less than sixty percent of what's been deployed on each shift during this week. And just use the bare minimum in your extra half-shifts during the week to keep the critical jobs moving.' *[Note 12]*

'Thanks Gaffer,' Jim said. 'The other thing I thought we might try, given how tight it is, was to introduce a special incentive bonus for 22's men. I was thinking we could set a target for each activity, like number of chocks transferred per week.'

The General Manager nodded.

'Write up 'ow you see that workin' and talk it through with Mr Hope.'

''Okay,' Jim said, gathering up his papers. 'Thanks, Gaffer.'

Jim left the General Manager's office, found Tim Hope, the Operations Manager, and arranged to meet him after the weekend, on the Monday afternoon, to go through a more developed proposal for the bonus.

The Game Fair

The arch was flanked by two round towers in the same reddish-brown stone as the wall and barely wide enough for one wagon. He stayed low, deep in the hay, holding his breath. Hopefully the Sheriff's men, big, fierce but stupid, would not stop and search the cart that was carrying him through the main gate, into the heart of the stronghold of his arch-enemy, the Sheriff of Nottingham. If they did, he was ready for them, sword in hand.

Saturday, 1 October, 1983; 09:30

As soon as they cleared the archway, Martin switched back to reality. Peering over his grandad's shoulder, he spotted his dad's mate waiting for them, as arranged, arms folded, leather gauntlets in one hand. It was only ten miles from Whitacre Heath to Arbury Hall, on Nuneaton's south eastern outskirts, so Colin was dressed, for their day at the game fair, in jeans, walking boots and his heavy biker's jacket, rather than his full leathers.

His motorbike was up on its stand, its red paint and chrome gleaming in the autumn sun, his full-face helmet, on the bike's seat, still bearing its adhesive Union Jack. Like Paul, he had stuck this on, a few years earlier, when they had toured France and Belgium, visiting the Flanders field where two of Paul's great-uncles had fallen, whilst serving with the Northumberland Fusiliers in the Great War.

Paul swung the Allegro over to the left and pulled up, leaving the way into the park clear. Its back-door flew open and Martin scrambled out and darted over to his dad's best friend, turning his head sideways and wrapping his arms around his waist on impact.

'Oof!' Colin grunted, smiling.

"Mornin',' he called over to Paul and his dad, as Horace wound his window down.

'You lead the way 'n' I'll follow yer,' Colin called, pulling on his gloves.

'Can I ride on the back of Uncle Colin's motorbike?' Martin shouted back to his dad.

Paul gave him a stern, half-smile.

'Alright. But sit tight.'

Colin put his helmet back on. He pulled the bike back off its stand, swung his leg over the seat and kick-started the bike, its warm engine firing first time. Leaning down, he grasped Martin's arm in his gauntleted hand and helped the boy clamber up behind him.

Despite his confidence in Colin as a rider, as Paul drove along the long, straight road, down the middle of the broad avenue of trees, he glanced in his rear view mirror to make sure that Martin was safe. The sight of Colin's stately progress on his slow-moving bike reminded Paul of past arrivals at campsites around the country; Susan on the pillion of his Commando, Diane perched behind Colin on the Triumph.

A dun-coated attendant with a leather satchel stopped them as they reached the gateway into the car park. Paul paid for all four of them; three pounds for each of the men, fifty pence for Martin. He could count on Colin to pay him back later.

As Paul parked up at the end of a line of cars, Colin's bike pulled up alongside, dipping abruptly on its front suspension as it halted by his window. By the time Paul got out of the car, Martin had jumped down and Colin was swinging his leg over, to alight from his bike.

'What's the ground like?' Paul said.

'Seems firm enough but I've got me piece of board to lay under the stand – so it won't topple onto your motor,' Colin said.

Paul took Colin's helmet and gloves from him, stowed them in the Allegro's boot and slammed the lid shut.

As soon as Colin had hauled his bike back up onto its stand, they were off. As Martin started to trot off towards the great, Gothic building, he called out to them.

'It's like a castle or a palace. It looks like the Queen could live here.'

'Arbury Hall's bin the 'ome of the Fitzroy Newdigates for centuries,' Horace called after him. 'They're a big land-ownin' family that used to own pits, like the Dugdales who owned our pit and Baddesley before they were all Nationalised so as the country owns 'em. The Newdigates had 'alf a dozen pits, one of 'em was actually named after them, 'til it closed a year or so ago; Newdigate Pit.'

Martin stopped and stared at the building.

'A famous novelist came from 'ere.' Paul said, as he came up alongside his son. ''Ave you heard of George Eliot?'

'Er, I think so,' Martin said, taking his grandad's hand and starting to walk with him.

'You've probably 'eard the name 'cause it's the same as the 'ospital in Nuneaton where you was born,' Horace said.

'Yeah, probably,' Paul said. 'Anyway, George Eliot wrote some famous books, over a hundred years ago. One o' them was called the 'Mill on the Floss'. The watermill in that story would 've been round 'ere, somewhere. But George Eliot wasn't the writer's actual name. Writers change their names sometimes but she did it 'cause no one would 've taken a woman writer seriously in them days. Anyway, her dad worked here on Arbury Hall land an' she was born 'ere.'

They walked down the south-facing slope from the back of the great house, making their way, with other small groups of visitors, towards the lake and the stalls specialising in fishing tackle. Paul was determined to get his dad to choose his new fishing bag before the best of them had been snapped up.

After visiting three stands, Paul and Martin managed to persuade Horace to allow them to buy an expensive bag, rather than one of the more modest examples in which he had pretended interest.

Paul kept checking the programme of events in the Main Ring to make sure that Martin did not miss out on any of the activities they had ear-marked. As they strolled past the Land Rover stand, with sharp reports carrying across to them from the clay-pigeon shooting competition in progress on the other side of the lake, Martin was amazed to see one of the sturdy vehicles on display, clinging halfway up an impossibly severe, man-made, rocky incline. Beyond this, they found the group of steam-powered traction engines whose tooting whistles had attracted them from across the park. One of them was attached by a long, canvas belt from its drive roller to a huge threshing machine. Several men were feeding sheaves of corn into the terrifying contraption.

As the four of them stood and watched the chopped straw cascading from its back and the hessian sacks attached to its side filling with grain, Horace was transported back to his childhood.

His father had been a harsh man who had thrashed him frequently. It was partly that word that had brought his early years to mind, but there was something else. Seeing the men and machine working reminded him of idyllic times in his boyhood summers, when he had been able to escape to an oasis of happiness; helping to bring in the harvest on his school friend's family farm.

At lunchtime, they called back at the angling stands, collected Horace's new tackle bag and took it back to the car park, where they stood around the boot of Paul's car and tucked into the picnic lunch that Susan and Pat had prepared for them.

Martin skipped alongside Colin as they returned to enjoy the remainder of the fair.

In amongst trade stands dealing in camping equipment, Paul looked admiringly at a gleaming Primus with smart, brass fittings. It was a Rolls Royce job, compared to the basic paraffin stove he and Susan had taken on their camping trips with Colin and Diane, before their children were born. But his stove had always been perfectly adequate. With some juggling, they had even been able to use it to prepare simple, hot meals.

One of their favourite destinations had been a little over two hours away; straight up the A5 and onto the road that wound through the mountains of Snowdonia to Harlech, on the North Wales coast.

But with the brown leaves littering the damp grass and the insipid warmth of the sun's rays, it was a particular camping trip that the two couples had made to the Lake District, in the autumn, nine years earlier, that came to Paul's mind.

On the first day, he and Colin had led the way as they climbed a series of peaks, mostly content to stomp along in their walking boots in companionable silence.

The two girls had always been fit and athletic and, as usual, had kept up easily, with graceful strides, chatting happily together. Occasionally one of them had called forward to her man to get him to substantiate something she had said to her friend.

When they had arrived, late on the Friday evening, Paul had used the Primus stove to prepare a simple meal at their tents, an old favourite on their trips; corned beef with processed peas and Cadbury's Smash, instant mashed potato.

On the Saturday morning, Paul had called on the wife of the farmer who owned field in which they were camped and had bought some freshly-laid eggs for a few pence. They had started the day with boiled eggs cooked on their little stove.

That evening, after a full day's hill-walking, they had walked to a local pub for a bar snack; chicken and chips in the basket for Paul and Colin (preferable to bready, rusk-filled sausages they always thought) and scampi for the girls.

Later, when the two couples were in their hardy, little tents, the damp, cold night air had chilled Paul and Susan as they scrambled to get into their double sleeping-bag and they had cuddled up together, close and tight.

For a moment Paul remembered sliding his hand up the cool, smooth skin of Susan's long, lean thigh and over her goose-pimpled arm, his finger-tips brushing the hardness of her nipples through the soft, thin cotton of her tee-shirt.

With Colin and Diane so close in their own tent, they had needed to be as discreet as they had been when living with Pat and Horace, when first married. The suppression of their sounds and Susan's uncharacteristic quietness had made their love-making illicit, their excitement leaving them flushed with warmth and love.

On the next day's walking, when Paul and Susan had set off together, holding hands, Colin and Diane had held back to follow them, a little way behind.

It had made it even more special when Paul and Susan had found out that she was pregnant to know that their first child had been conceived, in love, on that chill night in Cumbria.

As the four of them made their way back to the car park, Paul and Colin discussed the following day's union elections and the implications of Shuey McFadden's bid for the office of Lodge Secretary.

'If you go round to Paul's place, I'll meet you both there,' Colin said to Horace, back at the vehicles, as they got ready to head back to Whitacre Heath. 'Then we can walk round to the Welfare together and pick up Brian on the way.'

For the first time that any of them could remember, the outcomes of the two key union positions were something other than a foregone conclusion.

The Election

N o one was drinking; the bar wouldn't be open for another forty minutes. Even so, the ballroom at the back of the Miner's Welfare was full of noise and bustle, its air thick with clouds of swirling, cigarette smoke.

Many of the men had thick heads from their Saturday nights and were looking forward to the business being done so they could get to the bar for a quick hair of the dog.

It was crowded but not as full as Keith had expected; two hundred and seventy out of a total workforce of nine hundred and twenty-eight. He'd expected twice as many for the morning's elections.

Dick wasn't surprised. He'd had a more realistic recollection of the attendances at previous Committee elections. Apart from that, with some older members taking voluntary severance or early retirement, to make way for miners transferring from other pits that had closed recently in their Area, they were getting progressively more members living further away and unwilling to make the journey to attend union meetings.

Arnie Campion, an area NUM rep', was chairing the proceedings. Hammering with his gavel, he called the gathering to order with some difficulty.

Dick's heart sank.

'Good morning Comrades,' Arnie shouted, irritating a significant proportion of the assembly visibly with his choice of address.

'Thank you all for coming to take part in these important proceedings to elect the Committee members of the Whitacre Heath Colliery Lodge of the National Union of Mineworkers and your Union health and safety representatives.'

'I'm sure you're all as aware as I am that these elections take place at a time when strong union representation is even more important than usual.'

'Many of you will know the process but for the sake of new members and to refresh memories I'll explain how it works.'

'First, I'll be invitin' proposals and seconders for the two principal positions; the posts of Lodge President and Lodge Secretary. Then the candidate or candidates for the position of Lodge President will have the opportunity to speak on their own behalf or have a comrade proposing them speak out for them. Then they'll be asked to leave the room and the tellers will count the votes on a show of hands.'

'If it's too close to call, we'll have a paper ballot. If a candidate is proposed and seconded unopposed we'll just hold a vote on the basis of all in favour and all against.'

'After the election of your President, we will move onto the selection of Lodge Secretary. When we've completed the elections for those two posts, you'll be called upon to elect your Treasurer and other Committee members.'

Both Dick and Keith looked serious and nervous now, even though, as far as they knew, Keith would be unopposed, as he had been in all of his previous elections.

Dick was really worried. For the first time since he was shoed-in, when his old predecessor had retired from the industry and stood down as Secretary, he faced competition.

People kept assuring him they would be voting for him and that his opponent offered no serious threat but nothing was certain in this game. And McFadden was a dark horse; Dick had heard all sorts of stories about what he'd been getting up to behind the scenes.

Shuey McFadden looked like a changed man; uncharacteristically cheerful and positively enjoying the event.

"That mischievous bastard's got nothing to lose," Dick thought.

Dick just hoped everyone would remember everything he'd done and achieved for them in the past and see McFadden for what he was.

Dick was under no illusions; he was as self-seeking as he considered any politician had to be but he reckoned he was far more motivated by the members' and Union's interests than that snake in the grass.

It seemed like a lifetime ago but had been only two and a half weeks earlier that Shuey had sidled up to Dick.

'Hey, Dickie, I hope you ain't going to take this wrang but I'm thinkin' I'll be throwin' me hat in the ring.'

'Oh yeah?' Dick had said. 'And what d'ye mean, exactly.'

'I wanna stand for the position of Lodge Secretary, at these forthcoming elections, like. No hard feelings, I'm hopin', an' all that, it's just I'd like to give it a pop this time, d'ye ken?'

Dick had nodded. He was not surprised; Keith had already tipped him the wink that the skinny, little weasel was thinking of trying to stitch him up like this.

'Arr, you have to do what yer think's right,' Dick had said, before turning his back on Shuey and walking away.

Keith had tried to reassure him.

'You'll be alright, they vote for us as a pair, like a package – the old dream team. They wouldn't see me an' 'im working for 'em like you and me 'ave.'

Whatever the likelihood, Dick had spent the seventeen days since Shuey's bombshell thinking about the consequences of what would happen if he actually lost to him. Many in management would probably laugh their cocks off. And most of the deputies and overmen. Some, in particular, would be unbearable. You couldn't be Mr Popular when you were fighting for your members' interests; he'd made some enemies.

What would he do if he lost his position? It was who he was, what he was; Union Secretary. To go from being top dog to being nothing was unthinkable. But all too possible.

What could he do? He'd be back down the pit, on shifts. At his age. He had thought he was done with all that, forever.

It was all happening too quickly.

The President's post was announced. Arnie invited proposals. Ted Carter, on the Lodge Committee, proposed Keith, someone else seconded him, Dick didn't see who. There were no other proposals. Ted said a few words about Keith. Arnie asked Keith to leave the room. A show of hands.

'All in favour? .... And against? Keith Deeming is duly re-elected to the post of Lodge President.'

Keith returned to handshakes and congratulations. Dick shook his partner's hand and gave him a sick smile. It had been as easy as that for Dick on previous occasions. It wasn't going to be so easy this time.

Arnie called next for nominations for the post of Lodge Secretary. Keith proposed Dick and Ted seconded him. Ivan Wykes, who had transferred from Newdigate Colliery eighteen months earlier was recorded as Shuey's proposer. Then Dick found that it was Malcolm Kennedy who would be seconding Shuey. Although of Scottish origin, Kennedy was another former Kent miner, like McFadden.

"Another bolshie, Kent bastard," Dick thought.

As Kennedy indicated his support for McFadden, he stared Dick out with a mirthless smile. Meanwhile, Shuey beamed with a pleasure that Dick had never believed the miserable bastard to be capable of.

With no more nominations forthcoming, Arnie invited Dick or his proposer to address the meeting.

Keith had offered to sing Dick's praises, as he had on previous occasions and Dick had accepted. But that had been before McFadden had revealed that he would be throwing his Tam O'Shanter into the ring. Dick had been regretting that casual decision ever since he had known about his emergent opponent. It had been fine having Keith when it was all a foregone conclusion but with an actual contest, Dick was thinking he would have been better advised to have spoken out, more persuasively, for himself.

As in previous elections, Keith's commendation of Dick was expectant of victory to the point of complacency. In essence, it was "we all know Dick Flavel" and a rambling recommendation which Keith summed up by calling out, 'So I recommend to you that we stick with Dick Flavel, the man we all know and trust.'

Then Shuey got up, his shoulders and back hunched, to speak on his own behalf.

'Err, thank you, Mr Chairman. No one's going to catch me speaking ill of my comrade, Dick Flavel. He's got many years' service in for the union and 'e's been Secretary for a good while. But, I suppose it's like every dog has its day etcetera, etcetera and p'r'aps ye might be thinkin', like me, that it's, well, time fer a change, like.'

'Like I say, no one'll catch me speaking ill of Dick Flavel but, basically, like – Keith Deeming's speech was pretty much a recommendation to vote for more of the same.'

'But we all know what that means, don't we? We've all seen how there's bin a cosier relationship developin' between Dick Flavel and the senior management, a case in point being Mr Flavel's capitulation over the recent, unpleasant issue of stoppages of time for some men. Now, as I say, no one's gonna catch me speaking ill of Dick Flavel and Dick's entitled to his own style but, I can tell yuz, Shuey McFadden is nae appeaser, I'm no arse-licker and I never wull be.'

'With the Coal Board's rejection of the NUM's latest, reasonable pay claim and demands for Union members to start swallowing any closures the Board might think fit, you might be excused for feeling that you might need – that the time's right – for a younger Lodge Secretary with a fresh approach and more fire in 'is belly.'

'There's no one got more respect for my dear friend and comrade, Dick Flavel, than me but I can tell all of yuz present here today that I'd be proud to serve you as your Lodge Secretary but that I'll respect and abide by the democratic decision as I ask ye'all tae vote for me.'

Many present read Dick's lips as he muttered to Keith.

'The slimy, little bastard.'

'Aye, vote fer Shuey,' Malcolm Kennedy called out, glaring at Dick.

Arnie asked the two candidates to leave the ballroom until recalled.

Exiled in the closed and empty bar with McFadden, Dick resisted the temptation to have a fag, determined to conceal his nervousness. McFadden, looking relaxed, opened his own packet of cigarettes and offered one to Dick.

Dick shook his head and murmured.

'No ta.'

Shuey lit up casually, looking like someone who had arrived, enjoying his new-found celebrity and status as a contender.

"Fuck me," Dick thought. *"What a pitiful excuse for a pitman. The useless twat isn't even face-trained."*

They could hear the murmur of the call in the room and the count being taken. There was a pause of four or five excruciating minutes for Dick then, after a burst of noise from the room next door, the door eased open. Dick moved to return to receive the verdict.

It was Ted.

'I just wanted to let yer both know – the count of hands was too close to call. They're 'aving to do a paper ballot now.'

'Oh, okay,' Dick said. 'Thanks Ted.'

'Yeah, thanks mate,' Shuey echoed.

"Fuck me!" Dick thought.

After a few deep drags on his cigarette, unable to contain himself, Shuey spoke, his accent thickening ingratiatingly.

'Who'd've thought it eh, Dickie? Too close to call. I didnae expect tae gi' ye a run fer yer money.'

Dick grunted, gave a slight nod and smiled sadly.

"In that case, why the fuck did you stand?" He thought.

'Well, now it's "may the best man win", I guess Dick.'

Dick said nothing. What a load of bollocks this man talked.

The paper ballot took an age; handing out the papers, one at a time, people writing the initials of their chosen candidate, thronging forward to drop them into the box on the Chairman's table, checking everyone had voted and then doing the count and a double check.

Shuey devoured two more cigarettes with his chain-smoking.

Dick weakened and had one of his own. He looked out of the window at the youth team on the Welfare's muddy football pitch, stopped for oranges at half-time.

'I wonder who's winning out there, eh Dick?' Shuey said, appearing at his elbow, some of his bombast drifting away as he started to take the possibility of election to the post of Secretary seriously.

'Would you stand for one of the other Committee positions, if I win, like?' Shuey said.

Dick thought about that; the fucking cheek of the question. He wasn't finished yet.

'Let's take things one step at a time, shall we, laddie?'

"Aye, aye," Shuey said, nodding urgently. 'Aye, yer right.'

In the ballroom, Paul was standing with Colin and their mate, Brian Kettle, the third member of a trio that had been friends since their early, primary school years.

Brian had lacked the levels of educational and technical aptitude needed to qualify, like his two mates, for a craft apprenticeship. With a slight cast to his eye and a stammer when he spoke, Brian had relied on Paul for protection from bullies at school, as well as his friendship. As a youngster, Brian had never had the money to get into motorcycling like Colin and Paul and he lacked the coordination and ability to play football to the serious, team level of his two mates but he did play with them in the Welfare's darts team.

Paul muttered to Colin.

'We'd better start drummin' up some support for Flavel or McFadden could swing this.'

'It's a bit late for that, ain't it?' Colin said. 'Anyway, he might be the best man to lead a stand on what's been bothering you – what was it the Coal Board said "high-cost capacity must be eliminated". If they're gonna start tryin' to close pits, McFadden might be your best bet to lead the fight.'

Paul frowned and shook his head.

'Nah, the only thing 'e'll do is cause a load of unnecessary grief at our pit. And if that makes us less efficient it'll make us more likely to be closed.'

'I kn-n-know one thing,' Brian said, "'e's a tired b-bastard. Got n-no work in him. What d-d-d'ye reckon we should er d-do, Paul?'

'We'd better hit a few potential waverers. Just a few votes might make all the difference.' Paul murmured. 'We need a few more to see sense, else we could end up with a Marxist madman as Lodge Secretary.'

Paul gave them a few names each to talk to quickly, sending Brian to speak to miners who would receive him without an undue degree of mickey-taking, before heading for a group sitting at a table at the back of the ballroom. On the way he tapped up Horace.

'Dad, have a quick word with a few o' yer mates. We've got to stop that nutter McFadden swinging this – or it could be the death knell for this pit.'

'Yeah, okay Paul. It is looking a bit too close as it stands.'

The door from the corridor to the ballroom opened enough for Ted to put his head round, looking for them and finding them over by the window.

'Er, could you come back in, gents?'

"This is it." Dick thought.

'Here we go.' Shuey said, rubbing his hands with a hopeful, toothy grin.

Back in the room Dick had no way of knowing. Someone muttered something to Shuey who said a quiet, 'Oh!' as they passed.

Keith looked too bloody serious. This wasn't looking good. When Dick got to him Keith laid his hand on his shoulder and put his mouth next to his ear.

'Congratulations, Dick. You did it again.'

Dick closed his eyes and breathed a slow, silent whistle through pursed lips.

"I'm getting too old for this," he thought.

Arnie announced the result.

"Fuck, that was too close," Dick thought.

He should have been relieved, overjoyed and celebrating but McFadden was the only fucker smiling. But it wasn't the smile of the man having the time of his life, as it had been a few minutes ago. Now he was back to his same, old self, with the same, old, sneering smile.

Only twenty odd votes in it. Hardly the ringing endorsement for which Dick might have hoped; it was almost a vote of no confidence. He would never feel the same about the pit or the Lodge and its members again.

'Maybe next time, eh Dickie?' Dick heard Shuey say, as he accepted the Scotsman's cold, clammy handshake.

Others in the room lent over to shake Dick's hand or came over to pat him gently on the shoulder.

The miserable, fucking bastards, they could all go hang as far as he was concerned.

In the pit's Control Room, Roger Dawkin span round from the console on his chair and looked up as the undermanager walked in.

'All done, Boss?'

'Yep, I've signed all the deputies' reports.' Jim Greaves said. 'No problems.'

'Yeah, I told the Gaffer, when 'e phoned, that everything sounded on track for starting up alright tonight from the reports they phoned in,' Roger said. 'Everything progressin' alright for you on 22's?'

'Yeah, we're on schedule, I just want to try to get ahead 'cause there's no contingency in our plan, so we're up against it 'til we do.'

Jim was in charge of the pit for the weekend anyway, acting for the General Manager. He did this on an alternate basis with Bill Metcalfe, the Services Manager. It had to be a First Class Ticket man in charge of the mine, so Ken Goodall escaped that duty. Colliery General Managers never took a turn and Tim Hope, the rising star, had been relieved from joining the little roster by Mr Everitt, although he did muck in and cover the occasional weekend. Even if he had not been rostered on that weekend, Jim would have come in to expedite progress on setting up the effective system for transferring the chocks on 22's.

During the week, Roger was the regular nightshift Control Room attendant. At weekends, he and the nightshift senior overman, Willie Stokoe, took the Sunday day shift, a five-hour period of overtime for the men and officials underground but a full eight hours for Roger.

'Willie's gone, I take it.'

'Yes, Boss. 'E never 'angs around on a Sunday morning.'

Roger didn't mention that, as he had left, Stokoe had been moaning in his trebly, Geordie accent about the trial of having as an undermanager "some bloody kid who's wet behind the ears and looks like 'e only left school last foockin' week".

'All the Emcor readings alright?' Jim asked, taking a step towards the big metal cabinet housing the gas analysers and the various, moving paper roll charts that recorded the carbon monoxide levels throughout Whitacre Heath's underground districts. These gave early warning of any potential for spontaneous combustion of the coal and other carboniferous rocks, the first indications of a potential 'heating' or fire underground.

'Nothing's moving, all at a normal level, Boss.'

'We need to keep a close eye on the Emcor for 22's. That face has been stopped a few weeks now.'

Jim finished gazing pensively at 22's chart.

'Right,' he said. 'I'm going to call in for a pint at the Horse and Jock' on the way home. You've got the number for it if you need me for anything before I get home.'

'Yeah, you're alright, Boss,' Roger said. 'Have one for me.'

'Yeah, I'll make a point of it. Good health.'

As Jim turned his car out of the main entrance onto the road he noticed the full car park of the Miners' Welfare but its significance was lost on him, he had forgotten that the elections would be in progress in the building on the other side of the road.

In the Miners' Welfare, with the principal officers of the Union appointed they had moved on to select the remaining representatives for the Lodge Committee. Two representatives were standing down due to their approaching retirement dates and two new candidates were selected. Shuey McFadden was retained as a representative. Then they elected the Union's safety representatives and Union's Colliery Consultative Committee members. All of them were re-elections except for the positions held by a couple of miners who were due to retire in the next few months.

Arnie Campion was about to draw the proceedings to a close when Malcolm Kennedy put his hand up.

'Er, Mr Chairman,' he called out. 'I'd like to make another proposal.'

'Can we have another couple of minutes of quiet, gentlemen?' Arnie shouted, calling the meeting back to order. Having read the reaction to his and McFadden's earlier uses of the word Arnie had stopped referring to them as "comrades".

He pointed over to Kennedy and shouted.

'Yes, what was it?'

'Mr Chairman, in view of how close the vote was for Secretary, I'd like tae propose that we make Shuey Deputy Secretary.'

Nothing McFadden could come up with would surprise Dick.

Arnie caught the look of disgust on Dick's face.

'There is no such position in this lodge,' Arnie called back. 'It was a fair election and Dick was elected with a clear majority.'

'Aye, I know that,' Kennedy called back, 'but it was damn close and with all the pressure on the pits and the coalfields from the Board an' all that, we could do wi' some muscle to help the Secretary and President.'

'There's nothing to stop Shuey and the other rep's from providing that additional muscle in their roles as union rep's. Anyway, there is nothing in the NUM Rule Book about having a Deputy Secretary."

'That's true, Mr Chairman,' McFadden called out. 'But that means there's nothin' in the Rule Book that precludes us from hae-in' a Deputy Secretary. What's more, there's local precedent; Cadley Hill appointed one two years ago. Malcolm's got a point, we shouldn't expect an easy ride in the comin' months and years.'

Arnie looked uncertainly at Dick, then shouted back.

'Right, thank you for your comments and suggestions. I suggest Shuey proposes this at the next Lodge Committee meeting and if a majority of the Committee agree with you then it's something that could be introduced in future elections.'

McFadden muttered at speed into Kennedy's ear. Kennedy nodded then shouted.

'Leavin' it 'til then could be too late, given all that's 'appenin'. This is a reasonable proposal. I demand a vote on it.'

The debate continued for ten minutes, with increasingly angry and loud shouts from McFadden's few supporters as they were encouraged by a couple of his activists who were circulating in the crowd.

Arnie, tiring of the argument and keen to wind up the event and get to the bar for a few pints, turned to Dick.

'I'm goin' to see if there's a seconder for the proposal and see what the general mood is in the room.'

'You're what?' Dick growled.

Arnie turned and called, 'Alright, is there a seconder for the proposal?'

'You can't do this,' Paul called out. 'It's a put-up job.'

Arnie shook his head.

'Look, you're entitled to oppose the motion but I'm asking for a seconder.'

A hand shot up and its owner, a mean-faced member of McFadden's clique, yelled out.

'I'll second it!'

The committee member taking notes asked Malcolm Kennedy and the other man to call out their names for the record then Arnie called for the vote.

'The motion is that the Lodge create the position of Deputy Secretary and that Shuey McFadden be appointed to that post, given the high number of votes he registered in the vote for the position of Lodge Secretary. All those in favour, raise your hands.'

Arnie and his two adjudicators counted the hands.

'All those against?' Arnie called.

They counted again but it was clear that there was a majority in favour of creating the proposed position and giving McFadden a chance at the new job.

Arnie announced the result, 'That's a majority in favour of appointing Shuey McFadden to the new position of Deputy Lodge Secretary.'

Men opposed to McFadden's appointment to the role started to shout, curse and boo.

'Fix!' Paul hollered.

Arnie hammered away with his gavel.

Unable to silence the angry pitmen, he bellowed, 'I declare this meeting closed!'

Some of the opponents of the final development remained standing and clamouring.

McFadden and his tiny group of his supporters jostled their way out of the room, with Shuey shaking the hands of a few who had voted for him, on the way. They left the Welfare and headed off to the village pub, the Holly Bush.

Arnie turned to Dick and Keith.

'Sorry about that lads, I couldn't see any other way out of it.'

'Load o' bollocks,' Dick said. 'You were like a fuckin' idiot. You shoulda stuck to your guns.'

'Well I didn't get any support from you. You were conspicuous in your silence.'

'I couldn't speak out against it. I'd have looked partisan.'

'Of course you could, you're the Lodge Secretary, newly re-elected, in case it's escaped yer notice. And it didn't stop McFadden from speaking out. Anyway, you should be celebrating, not moaning. An old hand like you, you should be able to keep a young whipper-snapper like 'im under control.'

'You're fucking joking. 'E's been enough of a nuisance as an ordinary rep'. There'll be some chickens coming home to roost for you as well as us with that bit of bungling, Arnie lad.'

Colin and Brian stood either side of Paul, who remained motionless, scowling in the direction of McFadden and his associates as they processed out of the room.

'I don't believe that,' Colin muttered. 'How did he pull that off?'

Paul said nothing.

'C-can they g-g-get away er with that, Paul.' Brian said.

'Wait 'ere. I'll be back in a minute.'

Paul pushed his way through against the flow of the mass of bodies drifting towards the exit. Arriving at the long, committee table, he found Arnie struggling to placate Dick.

Paul waited for a pause in their exchanges.

'That can't be right, what you just allowed to 'appen,' he said. 'Just 'cause the members agree to introducing a new post of Deputy Secretary don't mean that McFadden gets it automatically. That deputy post is a completely different post to the one 'e stood for earlier. The members should 'ave 'ad the opportunity to vote for other candidates for that position.'

'You make a good point, son,' Arnie said, 'But the proposal was specific in respect of it advocating Shuey as the candidate for the new post and that was seconded and received a majority of the votes. If you want a chance at it, why don't you stand against 'im next time?'

'It's nothing to do with that,' Paul snapped. 'I'm just disputing the lack of due process that's resulted in a stitch-up.'

'Your point is noted, son.'

'So, what are you going to do about it?'

'Like I say, son, I don't agree that there's been a lack of due process in what we've done.'

'What you've done, more like. Did you plan this with McFadden?'

'You want to be careful what you say or you might appear to be making unfounded allegations.'

'I'm not making allegations, I'm askin' you a straight question.'

'Look, the first I knew about that proposal was when I heard it made in here, just now. I would have thought that was pretty evident.'

Paul could tell he was wasting his time. He turned and stormed back to where Colin and Brian stood looking disappointed for him.

'Jesus.' Arnie groaned. 'Spare us from fucking barrack-room lawyers.'

'That's not fair,' Keith said. 'What 'e said was right about the process. And about the stitch-up.'

Arnie shrugged and shook his head as he gathered up his papers and his gavel. He looked up.

'Look Dick, let me buy you and Keith a pint to celebrate your re-election.'

'Celebrate, be fucked,' Dick snarled. 'I've 'ad enough of you and this place for one fucking day.'

He grabbed his notebook and pen from the table and stormed away, with Keith scurrying in his wake, after him.

Establishing the System

B y the time Jim arrived at the pit to start the new week, the Pickrose haulage motors had been delivered, despatched underground and transported all the way to the inbye end of 22's Tailgate. When Paul Wood reported to him in the dirty offices, Jim told him to take the lead in installing them in the positions he had chosen.

The curved track sections had been completed and installed on Friday night, providing the chocks with a direct route off the old face, back along the seventy metres of the tailgate and round and down onto the new face. That morning, Jim would have men deployed to lay sleepers and rail through the whole length of 22B's new face-line and the short distance from the tailgate down 22A's, to the next chocks to be drawn off the old face.

Jim got a message from Mad Jack advising him that he could expect the return wheels for the continuous rope haulage through the new face to be delivered that morning. He arranged for Gerald Chipman to have them transported down the shaft as soon as they arrived and then get a message underground to Les Parker so that he could to travel in with them, using his authority, as Senior Overman, to expedite their transport all the way.

When Jim went to see Mad Jack to give him a pat on the back, he delighted Jim by telling him that the three low-loader trollies they needed would be delivered that morning, on loan from neighbouring Birch Coppice Colliery.

06:40
Jim hurried into the dirty offices in his pit-black and met Joss Kemp, the dayshift district deputy for 22's, that week. They made a quick exit, rushed

across to the pit top and rode the shaft together. Joss immediately struck Jim as a good, steady, old-hand style of official.

When they arrived at 22's, Jim explained to Joss how he wanted the blackboard, labelled "22's Chock Transfer", hung at the district's meeting station and kept up to date to keep the three shifts of men focused on hitting each week's target for the transfer of chocks.

The six days that Jim had devoted to establishing an effective system of salvage, transport and installation had seemed endless. He had been constantly aware of time rushing by without a single chock being moved. The new arrangements would have to be a total success to justify the stand-down and make up for all the lost time. They had two races to win; getting the new face turning coal in time for the seemingly arbitrary but immovable target date and getting it moving away in order to bury the most recent, fallen ground in the waste behind the old face before the Warwickshire Coalfield's extraordinary susceptibility to spontaneous combustion caused 22's to catch fire.

For the first of these imperatives, Jim was sympathetic to the pressures that a man, even as elevated as Rawlinson, was under. The Coal Board's spotlight was on the performance of every area and that meant that an Area Director had to apply that same scrutiny to every one of his pits. As a career mining engineer, Mr Rawlinson would not want to see any pit closed but Jim knew that the Area's top man would have no alternative, that he would be forced by the Board to be ruthless in his use of the axe if Jim and his men failed to hit their deadline.

For the rest of the dayshift, Jim worked with the team of men laying track though the new face-line. He paused, at times, to go to the telephone to check on the progress of the rest of the equipment he was awaiting.

As the shift-end approached, Jim told Joss to give a sufficient number of his men a half-shift of overtime to maintain progress until the afternoon-shift arrived. Then he phoned 22's afternoon-shift deputy, Gary Wallender, updated him on the position and the work in progress and told him how his men needed to be deployed.

Jim stayed with Gary and his team, when they arrived at the face an hour and a half later, to make sure they got into the same working rhythm in laying track that the men on the dayshift had achieved.

The next time Jim looked at his watch he saw it was nearly five o'clock. After a brief discussion with Gary he headed out of the tailgate and made his way out of the mine.

He had started entering 22's District, every day, by walking up the maingate with the airflow of the intake and leaving by walking out through the tailgate. This gave him the best chance of sensing any smell that might indicate the first signs of a heating.

Looking at 22's Emcor reading, every time he was in the Control Room, and getting the attendant to check it for him whenever he was on the phone 'to Control' had become compulsions for him.

Monday: 19:52

Jim had only been at home for forty minutes when he received a call from the pit. Mike Taylor, the Afternoon-shift Senior Overman, described the position at the end of the afternoon-shift and what Gary had kept men working on, during overtime.

As soon as Jim and Gabriella had eaten their evening meal, the phone rang again and Jim answered the phone to Willie Stokoe, the Nightshift Senior Overman. Willie's insubordinate surliness meant that Jim never felt confident that instructions he issued to him for work on nights would be passed on fully to the more junior nightshift officials. Consequently, these phone calls always lasted an inordinate time whilst Jim reiterated his expectations and extracted confirmation from Willie that he had assimilated them all. On this occasion, Jim remained so unassured that he told Willie to get Billy Thompson, 22's nightshift deputy for the week, to call him when he arrived in the dirty offices so that Jim could make sure that nothing was missed.

Tuesday, 4 October, 1983: 06:10

The next morning, Jim walked into the communal room in his pit-black and was greeted by the mixed throng of days and nights officials. He struggled through to where a black-faced Billy Thompson who was just starting to brief his dayshift colleague, Joss Kemp.

The nightshift men had installed the return wheels at both ends of the new face's endless haulage the new face, both haulage motors were powered up and their lock-out systems in place throughout both faces; the lock-out boxes, coupled up by signal cables, being needed to stop and start the haulages' operations, remote from their haulages' motors, along the length of each haulage run.

'That's great,' Jim said. 'So we're all ready for the ropemen. I've got them booked to go in this morning and splice the rope on the endless haulage, so we should have a working transport system through the new face by the end of dayshift.'

In common with most pits, the ropemen who maintained and installed the steel cables of the underground haulages at Whitacre Heath were ex-matelots who had acquired their rope-splicing skills during earlier careers in the Royal Navy.

Two chock-transporting low-loaders were delivered to the inbye end of the tailgate on the dayshift, so, by midday, when Jim phoned to brief Mike Taylor on the surface he was able to tell him, 'By the time your shift gets here we'll be ready to start transporting chocks.'

'That's great news, Boss,' Mike said. 'So you'll want a team loading on the old face and pulling chocks off with the direct haulage and another team transporting chocks down the face and unloading?'

'No, for most of the next three shifts we'll just have men getting the chocks that have already been salvaged moving. There's eleven of 'em in the tailgate, blocking up the transfer route between the old and new face-lines. We need to create a clear run.'

'Righto Boss, that makes sense.'

Two hours later, the afternoon-shift men arrived to relieve the dayshift's overtimers. Gary Wallender finished deploying all of his team at the district meeting station then followed them in to find his Undermanager getting the men working, determined to get a flying start, now they were able to move chocks at last.

First, there was the awkward job of shifting the chocks cluttering up the tailgate out of the way. Each chock had to be dragged off the rails into the track-side, then a low-loader brought alongside it. Jim got them to use a pair of levered, block and tackle pull-lifts, anchored to the road's arch supports, to ease each chock up, progressively, whilst building up a platform of wooden chock blocks under it to raise its base to the same level as its low-loader, before drawing it across onto the vehicle. As they lifted the first chock, its great weight caused the arch supports they were using to anchor the pull-lifts to buckle and twist down.

It was nearly six o'clock before they had it in position on its low-loader.

'Bloody hell, Boss,' one of the men said to Jim. 'That took some doin'.'

'Yeah, you're right,' Jim said. 'We've just got to push on and get these chocks shifted out of the way. We'll load the next two at the same time as each other.'

The first, chock–laden low-loader was clipped, front and back, to the steel rope of the tailgate's endless haulage that had supplied the old face

81

throughout its life. When it arrived at the curve onto the new face-line, its front clip was removed and then it was eased slowly round until it was in position for the front clip to be attached to the rope of the new endless haulage through 22B's new face-line then the rear clip was removed from the tailgate rope. The low-loaders movement was controlled by operating the lock-out box on the road which stopped the motor's operation, remotely. This would be how all of the remaining one hundred and sixty-four chocks would be manoeuvred onto the new face-line.

The low-loader bearing the first powered roof support rumbled steadily through the face like a juggernaut. By the end of the afternoon-shift's overtime, it was in a position for the chock to be unloaded then turned ninety degrees into position on the face and they were well on the way to having the next pair of chocks in the tailgate loaded onto low-loaders.

It raised the men's spirits to see the immediate improvement from having a proper transport system. As Jim left the district he was relieved to be moving chocks but impatient to accelerate the rate of progress.

The clock was ticking.

Bonus

Thursday, 6 October, 1983: 14:25

Dick Favel and Keith Deeming were at the desk in the NUM office, ready to make themselves available for men with union business coming off days. Dick was keen to wrap it up quickly.

'We need to make sharp if we can,' he said. 'I want to get off 'ome,'

Shuey McFadden bustled in, freshly bathed.

'Hey Dickie, about this special bonus payment that Greaves 'as introduced on 22's,' he said.

'What about it?' Dick growled.

'These lads from down the South End of the pit 'ave 'eard about it and they ain't 'appy,' Shuey said, jerking his head towards the group of men standing in the corridor outside the union office, still wearing their pit-black.

"So you've been stirring them up," Dick thought.

'What's it got to do wi' them? It's a special incentive to get the equipment transferred from the old face to the new. The quicker we get that new face going the sooner the productivity bonus'll go up for everyone when 22B's retreat face kicks in.' *[Note 13]*

'It's alright for the men workin' on that salvage job but it's only fair that it should be paid across the board,' Shuey said. 'It should be added onto the general productivity bonus payments.'

Before he could reply, a couple of men started chipping in, through the open door into the office, insisting that Dick do something about it. He couldn't be bothered to argue with them; if he did it would only make it look like he was a soft touch, not prepared to push as hard for their interests as McFadden. He got up from his seat.

'I tell yer what I'll do,' he called out to them as he stepped over to the door. 'I'll 'ave a word with the Undermanager and let yer know.'

'Which undermanager?' McFadden snarled. 'There's no point in talking to that young fucker.'

'I'll talk to the Statutory Undermanager, to Goodall.'

McFadden looked as though he might say something else to Dick then stopped himself and turned to address the gang he had mustered.

'Right lads, I say we leave it to Dick, here, to see what he can sort out with Goodall. Come back and see me, same time tomorrow, and I'll let you know the outcome.'

McFadden turned to Dick and said, 'And don't let the bastard soft-soap you.'

'Fuck off,' Dick muttered as he turned away to sit back down at the desk.

Friday, 7 October, 1983: 07:15

The next morning, the dayshift officials had gone underground and Jim Greaves was in his own dirty office when he heard Dick Flavel arrive in Ken Goodall's office. He tried to ignore their conversation and finish reading and signing his deputies' shift reports.

He couldn't hear much of what was being said because Dick had given Ken's office door a shove and sat down with his back to it but he still heard snatches of Ken's responses. When he heard the words "special payment" from Ken it confirmed his suspicion about the purpose of Dick's visit.

Dressed in his pit-black, with his knee-pads on, Jim was still able to move silently on the rubber soles of his pit boots as he walked across the red-tiled floor of the communal office to the door into Ken's office to make sure his colleague did not give any ground on the matter.

He eased the door open and leant against its frame. When Ken glanced up at him, Dick swung round and registered his arrival.

'Good morning Dick,' Jim said. 'Are you alright?'

'Yeah,' Dick grunted. 'Fine.'

'Dick was just talkin' about the special incentive payment on 22's, weren't yer Dick?' Ken said. 'P'rhaps you'd like to tell Mr Greaves what you were sayin'.'

Dick gave a long sigh.

'I came to ask Mr Goodall to get it included in the productivity bonus. It's causin' trouble with the men elsewhere in the pit. They take the view that 22's men were already getting' the pit productivity bonus and that what they've been offered should be included in that and shared equally. It's a matter of what's fair. And I was just sayin', it should 'ave been negotiated with the Union, not just introduced.'

Jim was determined to be diplomatic in his response.

'It's not normal practice to discuss the incentive bonuses for roadway drivages with the union and this is like that, given that it's development work not production work,' he said. '22's men won't be getting the productivity bonus as well. Their payments will be completely dependent on how well they perform on getting the chocks onto the new face. 22's incentive is relatively generous but that's because getting the new face installed on time is critical for the pit and everyone who works in it.'

'That's as may be but it won't do the pit much good if the rest of the men walk out will it?' Dick said.

'You not going to start that "stopping the wheels" bollocks again, are you?' Jim snapped.

Dick screwed up his eyes and sighed again.

'It's not meant as a threat.' He opened his eyes and looked at Jim. 'I don't want it in this instance any more than you do.'

'Come on, Mr Greaves,' Ken coaxed. 'We can talk this through calmly together.'

'I'm perfectly calm, Mr Goodall,' Jim said, before looking back at Dick. 'But you need to act in the men's interests; you need to make sure they realise that if we don't get 22's turning coal on the first of December, this pit and their jobs will be history. The best thing the men elsewhere in the pit can do is get stuck in and earn their bonuses by turning plenty of coal to keep us in profit and keep the pit open 'til we get 22's producing.'

Jim looked down at Dick, who sat glowering, mutely, up at him. He couldn't imagine what the Union Secretary was thinking.

In fact, Dick knew he had no way of doing anything more about the matter. He didn't even disagree with what Jim had said. But he knew that McFadden would use it as more evidence that he was too weak in his dealings with the colliery's management. He was completely boxed in.

'We're honouring all the incentive agreements elsewhere in the pit, this is something that only affects the men on 22's.' Jim said. 'No one's losing out.'

'Right, I'll tell 'em and see what they say,' Dick said.

He pushed himself up, out of his chair and walked out of the office.

Ken left it a few moments to make sure that Dick would be out of earshot then murmured to Jim.

'You have to give 'em a bit of ground, Jim. Let 'em win one occasionally.'

'"Let 'em win one occasionally"? I'll let 'em win one when they're right. When they're wrong they'll get fuck all out of me. And he was wrong.'

Jim stomped back to his own office to finish reading the deputies' reports before getting off down the pit to 22's.

'Stroppy, young pup,' Ken muttered.

When he had finished his mug of tea, Ken put his helmet on, pushed his chair back and stood up and grabbed his Davy lamp and yardstick. He walked to the door of his office and called out.

'Come on Jim, let's get down that pit.'

When McFadden phoned the Union office from underground, in the middle of dayshift, Dick was forced to admit that he had failed to secure any concession on 22's incentive payments. He knew Shuey would have been anticipating this. He whined and moaned on about Dick's failure.

'I can tell yer, this ain't gonna play well with the men who've taken issue with this payment,' he said as he concluded his griping before hanging up.

'No,' Dick muttered to himself. 'Not by the time you've finished stirrin' 'em up again.'

Dick had always likened his role as Lodge Secretary to that of a politician. Now it felt that, although he still held office, McFadden had become the Leader of the Official Opposition whose role consisted of criticising and obstructing him and opposing every, single thing he said or tried to do.

For years he had been able to criticise the colliery's management and take issue with what they did. Now he had someone doing the same to him, with more determination and on a more persistent, daily basis. Dick could see no way out of the situation; he knew there was no way that McFadden was going to desist and his problem was aggravated by the more robust stance being taken by the colliery's management, ever since Greaves' arrival at the pit.

Dick laid the phone receiver back down on its cradle. He was dreading having to try to explain the outcome to the delegation that McFadden would bring to harangue him when they came up the pit, at the end of dayshift.

He thought about making himself scarce and getting off home but he'd only have to face them all the following week.

He was in an impossible position; he just couldn't win anymore.

Progress and Risk

Jim moved quickly down the new face, doubled up under the low roof, to where Paul Wood was kneeling with the three men who had manoeuvred the latest chock into position.

''Mornin' Boss,' Paul said.

''Morning Paul,' Jim said. ''Morning men.'

A cheerful chorus of, "How're yer gannin'", "Get on Boss," and "'Ey up Boss," greeted him.

'The scoreboard's lookin' good,' Paul said.

'Yeah, it is,' Jim said with a frown. 'But it's still not good enough.'

'I thought twenty-seven chocks was a pretty good score for the week,' Paul said with a frown. 'And we're doin' better, now we've got rid o' them chocks clutterin' up the tailgate and got the system workin' properly.'

'Yeah,' Jim said, dropping down onto his kneepads and gazing at the row of chocks on the face. 'We've gone from three a day to just over five. But, if we don't start doing better than last week it'll take us another five, whole weeks to get the chocks transferred. So, we wouldn't be finished until the twentieth of November. That would leave us ten days to strip all the track off the face, get the panzer and the shearers onto the face and built and proved. There's no way we'd be ready and turning coal on the first of December, it would be a good week later than that if we carry on as we are.' Jim said. 'And we both know what that would mean.'

Paul nodded, turning his mouth down in his characteristic, dour look.

'There was one thing I've bin thinkin' we might be able to improve on,' Paul said, brightening a little. 'When we get the chocks off the trollies, it still takes us a good while to slew 'em round and get 'em into position and lined up with their marker on the roof. I keep thinkin', I wish we 'ad some kind of turntable – so we could slide 'em off onto it and just spin 'em round and then pull 'em sideways into position. It'd make it a hell of a lot easier and speed things up.'

'It would,' Jim nodded.

Jim hated having to suppress creative ideas but, as he thought about it, he could only see problems in this one.

'It'd be difficult to make something with a bearing beefy enough to take a four tonne chock that would leave enough clearance between the top of the chock and the roof? It would take a hell of a battering every time you pulled that weight off the side of it.'

'I thought of those problems meself. And we'd have to have two o' them. There'd be no point in having one for unloadin' 'em if we didn't speed up loadin' the chocks onto trollies on the old face by the same amount.' Paul said. 'That's why I 'adn't bothered saying anythin', I was trying to think of a way of makin' it work.'

He gazed at the steel platform and ramp that Jim had had made to enable the chocks to be pulled off at the same level the low-loader vehicles, deep in thought.

'I tell you what, though, we could try spinnin' 'em round on that before pullin' 'em off,' he said, pointing at the ramp.

'If we kept the surface of the platform plastered with grease they'd turn a lot easier and we could line the ramp up so they slid straight into position.'

'Brilliant,' Jim said. 'How much grease have you got in the fitters' box?'

'There's enough to have a go at it this shift.'

'Let's give it a try,' Jim said. 'If it works, I'll get a boat-load of grease sent in.'

Work Safe

K en Goodall had flown off to Bermuda on holiday, so Jim was acting as Statutory Undermanager for the mine. He was on 22B's, the new face, mucking-in with the men unloading another chock onto the heavily-greased ramp when he glanced back up the face and spotted Joss Kemp approaching at speed, doubled up under the low roof.

Joss sidled up to Jim and laid a hand on his shoulder.

'Boss, Control's bin on,' Joss muttered. 'You need to go and phone 'em. There's been an accident.'

The eyes of the men in their vicinity swung onto them.

'What's happened?' Jim murmured.

'I don't know, Boss, Jerry wouldn't say,' Joss said, quietly. 'But it sounded serious.'

Jim nodded and rushed down the face to use the phone in the maingate to call the Control Room.

Jerry Malin answered the phone.

'Control.'

'Jerry, it's me,' Jim said.

'Hold on Boss, I've got Mr Everitt for you.'

'Mr Greaves, there's been an accident in the Gracebury Shaft. It looks like a fatality. The police are over there and the HMI are on their way. Seeing as how you're acting as Statutory Undermanager you need to make your way out the pit and get over there with me.'

'I'm down the South End of the pit. Wouldn't it be quicker if I went straight to the shaft?'

'No, the shaft's closed. Get yourself to the pit bottom and out that way. Jerry's laid a loco on to bring you out.'

The remaining shaft of the closed Gracebury Colliery had been retained to act as an auxiliary shaft, a secondary intake airway to keep Whitacre Heath Colliery's adequately ventilated as it continued its sprawling expansion southwards. A few months earlier, Mr Everitt had decided that they should upgrade this shaft to provide a faster access for supplies to the distant South End and improve the mine's overall capacity to wind materials.

The shaft had an inset, five hundred and seventy feet above the pit bottom. The inset and the opening opposite had provided access to the disused, higher levels of the old Gracebury workings.

Three shaftsmen had been deployed there that morning, to move certain items and make the inset safe for the re-commissioning of the shaft. Their chargehand had been Jed Davenport, the most experienced of the three.

Jed and Clive Hamilton needed to shift a big, steel plate, about six feet by eight feet from the opening on one side of the shaft to the inset on the other. From the inset, it would be possible to manoeuvre the plate onto the cage in order to remove it from the shaft.

Both of them wore shaftsmen's harnesses, sturdy garments of belts and straps which went over the shoulders and between the legs and round the waist, encasing the wearer's body. A strap secured a hefty, triangular, steel loop to the left shoulder of each harness. Clipped to the loop was a ten-foot length of chain, strong enough to arrest and hold the weight of a falling man.

Jed had decided that they would loop a canvas lifting-strap under each end of the plate, to enable them to ease it just clear of the floor and then walk it out onto the slimy, damp, fourteen-inch wide ledge and across to the other side, lowering it to rest on the ledge as often as was needed to get it across. The plate's size made it unwieldy but, being thin, its weight would be just manageable for two strong men.

Before they had started to move the plate, Clive attached the safety chain on his harness to a girder at head height across the opening.

'You need to latch on with yer chain, Jed,' Clive said when Jed got into position without attaching his own harness chain to a similarly secure anchorage.

'Nah, I'll be alright,' Jed said. 'This'll only take a minute.'

'You wanna do as Clive says,' Davy Tonks, the third and oldest member of the team said, standing in the inset, on the other side of the shaft. 'Work safe.'

Jed paid no heed to his mate's advice, he just gave him a characteristically roguish smile then got into position to lead the way across the shaft, lifting his end of the plate. Davy shrugged and shook his head as he looked across the shaft at Clive then got ready to help and receive the plate in the inset.

Jed and Clive edged across the narrow ledge, Jed shuffling backwards with his left shoulder against the slimy, grimed wall of the shaft lining, moving the plate barely a foot each time before lowering it to rest, briefly.

They were halfway across when the wobbling plate became unbalanced and toppled over and out into the shaft. The air rushing down the shaft caught its large surface area and added more force to the toppling sheet. As Jed and Clive dropped the plate's lower edge onto the ledge, Jed's right boot slipped away and off the ledge. He scrambled for a couple of seconds then dropped away under the plate, which had come to rest with its lower edge on the ledge and its upper edge against the nearest two guide ropes, installed to keep the cage travelling straight as it passed through the shaft.

Pressing himself back against the shaft lining, Clive saw the horror on Davy's face.

'Is he alright?' He said.

'No!' Davy groaned. "'E's gone.'

'What do you mean?'

Davy looked up at Clive, staring, wide-eyed.

'He's gone down the shaft.'

Jim rushed out of the pit and found Mr Everitt in the main office car park, waiting by the colliery's NCB mini-bus which was ready to take them over to Gracebury's pit-top.

By the time they arrived there, a police inspector and three other officers had arrived in two squad cars. Jed's body had been recovered from the pit bottom and the men down there were ready to have him wound out of the mine on the lower cage.

While the shaftsmen had been working at the inset, the cage had been lowered out of their way, about a hundred and fifty feet below them.

Jim and Mr Everitt waited, with dread, as the cage was raised back to the surface.

The cage slowed as it approached the surface. Jim, Mr Everitt, the banksman and the four policemen saw what looked like a boot and a trouser leg lying on the top of the cage.

The banksman stopped the cage without comment with a single ring to the winding-engineman. It was Jed's lower leg. Removed neatly by Jed's impact with the cage, still clad in his boot and the leg of his overalls. The banksman operated the compressed air ram to slide the shaft gate open and Jim stepped over to the edge of the shaft.

The police officers watched, wide-eyed, as Jim leant over the great drop then let himself fall forward to grab hold of one of the cage's heavy, supporting chains, took hold of the leg then threw himself back upright as he lifted the severed limb back across.

Having realised what would be involved in treating the shaft as a scene of a crime, the police inspector suddenly decided that it would be best to leave the investigation to Her Majesty's Mines Inspectorate, whose inspectors were on the way to site.

Later, back at Whitacre Heath, Jim was interrogated, in the dirty offices, by two HMI inspectors and a senior, regional inspector from the Inspectorate. He worried that, being only a stand in for Ken Goodall, in his absence, he knew too little about what had been planned. Then Clive Hamilton and Davy Tonks were both interviewed, although they were both suffering from severe shock.

'I just can't understand why he didn't secure hisself to the girder across the mouth of the openin', like me.' Clive said. 'I told 'im to do it. If 'e'd 'a' done it, he'd still be alive.'

He kept repeating, like a mantra, 'When Jed disappeared I thought 'e must 'ave 'ad 'old o' the steel plate, that 'e'd be 'angin' underneath it, that we'd be able to get 'im back up. It was only when Davy, standin' on the other side, said "he's gone" as I realised 'e'd fallen down the shaft, that 'e was gone.'

Davy was more distraught. He described the horror of seeing Jed's wide-eyed face as he fell backwards, arms thrusting out in desperation, hands grasping at thin air as he tried, in vain, to grab hold of anything to stop himself as he plunged away down the gaping shaft at the start of the long fall to his death.

The senior inspector suspended their questioning, indicating that they would leave it until Davy had recovered sufficiently to be able to make a coherent statement.

The inspectors reconvened their intense interrogation of Jim. As soon as they had finished he rushed across to the undermanagers' bathroom. He showered and changed with little time for reflection but sure that he would lose his First Class Certificate of Competency for having failed to prevent a fatality on his watch.

With damp hair and the usual, black eye-liner of stubborn coal-dust around his eyelids he joined the colliery's Personnel Manager, Ray Bentinck, back in the mini-bus.

Typical of a man in his role, Ray was a complete contrast to the more serious production men in the management team. Normally, he was as frivolous and jovial as a television game-show host. The seriousness of the day's tragedy sat uncomfortably on him.

Jim knew that it was good practice to take a relative to provide support to the wife of any miner who needed to be informed that her husband had been seriously injured or killed. The driver drove them to the address that Ray had obtained for the sister of Jed's wife and parked the mini-bus outside.

Ray went in alone.

Jim sat in the mini-bus, gazing, eyes unseeing, at the dashboard.

The mini-bus driver glanced across at him.

'A rum do, Boss.'

'Yeah,' Jim murmured.

Ray came out ten minutes later with Jed's sister-in-law, who was struggling to compose herself. Jim climbed out of the mini-bus and Ray introduced him to the young woman but it was obvious that the introduction had not registered with her.

The Whitacre Heath estate had been built by the Coal Board in the 1930's but the majority of its properties had been purchased at bargain prices by their occupants when they had been given the opportunity to buy them, a few years earlier.

As the mini-bus swung into a big, square close of terraced and semi-detached houses, Jim saw a couple of mothers chatting as their children played on the trim, central green with its well-established trees.

Jim thought how cruel it was that the Coal Board, having created this pleasant scene of solid domesticity and provided the job that had supported Jed and his family, was now sending him to inform Jed's wife that she and her children had been robbed of her husband and their dad; that Jed would never be coming home again.

As they approached the house that had been Jed's home, Jim looked at the little tricycle and battered, toy pram on the front lawn. He couldn't imagine what he was expected to say.

On the way to collect Jed's sister in law, Ray had said the only thing he could think of to help.

'It's never easy. There's not much you can say, 'cept tell them they've lost their husband and you're sorry to have to tell her and that he didn't suffer. We'll make sure she's got family and friends round her who can support her, fetch the kids from school and take them somewhere else for tea and things like that.'

"That was stretching it," Jim thought, *""He didn't suffer"."* Not much, just that terrible, terrifying, interminable plunge to oblivion, with his leg ripped off, halfway down.

Jed's name amongst his colleagues had been derived from the initials of his full name, John Errol Davenport. Jim had thought about this, he would have to remember to refer to Jed as "John" and not his colliery nickname when breaking the terrible news to Jed's wife.

A plump, unprepossessing woman slowly opened the front door of the neat, terraced house, horror written across her face. She had seen Jim drop down from the front seat of the canary yellow and royal blue mini-bus with its big, NCB logo and open the side door to let Ray and her sister out of the back.

Jim had expected to find that a fine looking, live-wire like Jed would have been married to an attractive wife. He despised himself for it but couldn't stop the thought; she'd let herself go. He couldn't help doubting that she'd ever find herself as good a catch as the man she'd just lost that day.

The poor woman saw her sister, the colliery's Personnel Manager in his suit and the young man in sports jacket and tie, the unofficial uniform of the colliery undermanager. She knew, in that instant, what it meant, what Jim had come to tell her.

Stepping down from the front step, her knees buckled and she collapsed onto her front doorstep, slumping back against the door post. Her sister shot past Jim as he started his reluctant rush up the path towards her.

Jed's widow opened her mouth to howl. But no sound came out.

A Special Conference

Friday, 21 October, 1983

A rnie Campion had arrived at the hotel in Sheffield the night before.

Normally, he looked forward to an event like this. He'd worked hard to get to his position as regional delegate and had never begrudged the rep's and delegates their perks as he'd been working his way up, so he had never had any qualms about enjoying the lavish hospitality, laid on at the expense of the union members. Despite this, by the time they broke out of the last meeting, he had already decided not to indulge himself in another posh meal in the hotel's restaurant followed by a session its bar and a second night in his luxurious bedroom.

Usually Arnie swanned around "Conference", enjoying the sense of his own importance, revelling in the knowledge that their union was indefatigable. But this Special Delegate Conference had been different, this time there was a serious threat; this time it felt like there was a real will to take them on. And what he had seen and heard at Sheffield and in the days leading up to the conference had left great dents and holes in his confidence.

The NUM's leadership had changed in the last year with the election of Arthur Scargill as National President.

Everyone knew Arthur; he had already been a national name, outside the industry, having benefitted from the massive power base he'd obtained, ten years earlier, on becoming the NUM's President for Yorkshire, a region which covered four, major National Coal Board areas.

Arnie had always viewed Scargill as a shameless self-promoter. Too often he'd seen him sit quietly in meetings of the Union Executive Committee, contributing little, a little over-awed, if anything. But when the business was concluded, you could count on Arthur to be the first man out of the building, making a statement to the press. In Arnie's view, this had inflated the power and influence he appeared to wield and enabled him to claim credit for others' initiatives.

It seemed like only yesterday that Arnie had seen a poster at Coventry's Miners' Welfare advertising, "A chance to listen to Arthur Scargill". Some colliery wag had scrawled on it, "Why, what does he play – organ?"

But, a year earlier, Arnie had started to take the Yorkshireman more seriously.

At a press conference, Arthur Scargill had revealed that he had acquired a leaked document. Prepared by the Monopolies and Mergers Commission, this secret Coal Board paper was dynamite; it earmarked a third of the nation's pits for closure over the next ten years.

At around that time, the Coal Board had moved from the pursuit of "coal at any price" to setting a target cost of production for each mine. In the past year, the growing scrutiny on efficiencies had increased the fear of closures throughout the coalfields. For many pits, even those like Whitacre Heath, the possibility of being shut down hung over them like an ever-present axe, ready to fall at any time.

During his time as the Chairman of British Steel, Ian MacGregor's programme of plant closures and redundancies had reduced the labour force of British Steel by over sixty per cent and helped establish his uncompromising reputation in Britain.

But it was not only in this country where the Scottish-American businessman had shown his willingness to take a strong, confrontational approach in dealing with unions. In the harsher, industrial relations environment of the United States' mining industry, he had faced down a strike in which striking miners had been killed on the picket lines by shots fired by police. The Government's recent appointment of Mr MacGregor as Chairman of the National Coal Board had been seen by the NUM not just as an indication of its serious intent but as an act of deliberate provocation.

The National Coal Board's rejection of the union's recent pay claim and its refusal to negotiate unless the Union indicated a willingness to accept pit closures had raised the stakes.

There had been a lot of discussion at the conference about the Government's increased coal stocks. They could see them for themselves when they drove past any coal-fired power station; the stockpiles were mountainous. One of the delegates had found out that the CEGB was renting farm land, next to power stations, on which to store even more.

The conference had agreed, unanimously, to reject the Coal Board's response to their wage claim as unacceptable; reaffirm the Union's opposition to pit closures except on grounds of exhaustion; fight any further reduction in manpower levels; and resist the Coal Board's and Government's plans to close another seventy pits over the coming five years.

To counter the Government's continuing attempts to strengthen its position in anticipation of any serious dispute, the Conference voted to bring in a full overtime ban, just over a week later.

The delegates had tried to convey an air of confidence in the amount of power they held that Arnie did not share. As he walked out of the hotel, Arnie couldn't help thinking that Mrs Thatcher and her Tories had the initiative; it was they who were controlling events. It was clear to him that their union was only capable of being reactive.

Riding Out

Sunday, 23 October, 1983

Paul and Colin rarely went out on their bikes together, these days. Colin still rode out on his own but, with his mate being able to keep his bike at his own home, Paul put that down to it being easier for him to go out on impulse. Actually, it was due more to nostalgia; Colin's regret at not being able to spend as much time as he had, in the past, with Paul.

But they were off out together that morning, even though it was only to be the nine-mile run, over to the National Exhibition Centre, to visit the Motorcycle Show.

'You 'eard about the overtime ban, then?' Paul said, wheeling his Norton out of his dad's garage.

Horace nodded

'I could do wi'out it,' he murmured. 'I was 'oping to top up our savings in the time I've got left before I finish.'

'They ain't left us with any choice though, 'ave they?' Paul said. 'They've bin stockpiling coal like mad, even started importin' it, ready to take us on. The Area rep's reckon the Tories 'ave been plannin' this since before they came to power. They've brought in that MacGregor butcher to take us on and impose more closures on us. We can't just think about ourselves, we could be fighting for the industry, let alone Whitacre.'

'Martin deserves the same opportunity we had; to be a pitman, like his dad, grandad and great-grandfather before 'im.'

'Hmmm,' Horace said.

Actually, at the time that Paul had been coming up to school leaving-age, Horace had been hoping that his son would escape the pit, that his promise at school would enable him to find a career outside the mining industry.

It had amazed Horace when Paul had made it clear that he was determined to go down the pit. It had been an even greater surprise to find that Paul thought so much of Horace that it was inspired, in no small part, by a desire to follow in his dad's footsteps. That, and Paul's eventual agreement to go into the more sophisticated, fitting side at the pit, had made it easier for Horace and Pat to accept their son's chosen career path. But he wasn't keen on the prospect of his grandson ending up down there as well.

'What'll it do to the transfer job on 22's?' Horace said.

'It was going to be tight without the ban but we started averagin' two chocks a shift last week. When the ban kicks in the rate'll drop right back down again.' Paul said. 'We'd got seventy-one chocks transferred by the end o' dayshift, yesterday. That's taken three weeks and we've got over ninety left to shift. I can't see us hittin' the deadline. The only thing we can 'ope for is a stay of execution.'

If Paul had said that to Jim Greaves or Harry Everitt, they would have assured him that any hope of leniency from the South Midlands Area's ruthless, formidable Area Director would prove to be hopelessly optimistic.

Fretting

Monday, 24 October, 1983: 13:55

The summons Jim had received, underground, via Jerry in the Control Room, had been no surprise. He'd known the Gaffer would want to know how he was going to hit the deadline without being able to use overtime. He was still wondering that himself.

Rawlinson wouldn't relax the target for 22B's. There was probably no way he could even if he'd wanted to. And he'd have been expecting them to do it without the amount of overtime they'd been using anyway.

Over the weekend, Jim had been looking at the implications of the overtime ban, in terms of the number of working hours lost each week. It was terrible, a massive hit.

The deadline was challenging enough, even with the overtime. With that gone, it looked impossible.

Fresh from the undermanagers' bathroom, Jim had had three quarters of an hour in Ken Goodall's clean office to revisit his workings. He gathered up his papers and dragged himself down to the Gaffer's office in time for his appointment.

Angela gave him a bright smile as he arrived.

'Good afternoon, Mr Greaves!' She called out.

Her cheerfulness, so inappropriate in the circumstances, jarred with Jim.

'It must be nice to be so carefree,' he thought as she ushered him into the General Manager's office. It was as though she didn't know or care that her job hung, precariously, in the balance, along with everyone else's at Whitacre Heath Colliery, totally dependent on the success or failure of Jim and his team, down on 22's.

'Mr Greaves for you, Mr Everitt.'

Harry Everitt was in his chair at the far end of the wood-panelled room, sitting, as usual, at his desk at the head of his office's great meeting table.

Mr Hope was sitting close to him on the left of the table. The absence, again, of Bill Metcalfe, the Services Manager nominally responsible for the subject to be discussed, was conspicuous.

'Good afternoon, Gaffer, afternoon Boss,' Jim said.

Mr Everitt fixed his penetrating gaze on Jim as he stepped over and took a seat next to Mr Hope.

'Want a cup of tea?' He said.

'Er, yes please, Gaffer.'

Mr Everitt stabbed a button on the office intercom on his desk.

Angela responded immediately.

'Yes, Mr Everitt.'

'Could you do us three teas please, Angie.' Mr Everitt flicked the switch up. 'Right, what have you got fer me?'

'I've been looking at the implications of the overtime ban,' Jim said. 'It'll have a big impact on the rate of transfer and installation.'

'You can say that again,' Mr Everitt said. 'You've bin spendin' it like it was going out of fashion.'

Jim flinched then ploughed on.

'Losing the two days at the weekends and the extra third of a shift after every weekday shift has stripped out over a third of the time we had each week – about forty percent.'

'We improved the rate of transfer of chocks with another couple of innovations and we shifted thirty-six chocks last week. We should keep that up this week, as the last week with overtime and still being able to work overtime at the weekend. But I'm forecasting that when the overtime ban kicks in, on Monday, we'll struggle to do twenty-four chocks a week. And we'll lose at least a couple of shifts that week because I've got to get the machine discs to site, so we won't be able to transfer any chocks while we're doing that.'

'I hope you're not about to start preparin' me for failure, Mr Greaves.'

'No, Gaffer,' Jim said. 'But I've got to find a hell of a lot of time from somewhere.'

'We all agreed that the time I'd allowed was realistic for stripping the haulage track off the face, getting the panzer built and the shearers onto the face, built and re-commissioned. I've just got to make sure that we increase the rate that we do the big, time-consuming job of transferring the chocks, to shave more time off.'

There was a gentle knock then the great oak door to the office opened.

Angela entered with a tea tray poised on one arm, she stepped round to a point opposite Jim and Tim Hope, placed the tray on the table and poured a cup of tea for each of them. She handed them out then strode out with her tray with respectful words of thanks from all three men, who followed her departure with their eyes. When she reached the door, she turned back and returned their gaze.

'You can carry on now,' she said, before leaving and closing the door behind her.

Mr Everitt grunted then carried on poring over the time estimates that Jim had plotted on the sheets of paper he had handed out.

Eventually, Mr Hope broke the silence.

'So you're forecasting that you'll be about five days late, without overtime.'

'Yes, Boss.'

'So you must have been expecting to come in with some time to spare, if you'd been able to keep using overtime,' Mr Everitt said.

'Yes, perhaps two, even three days,' Jim said.

'I think Jim and his team have done a good job accelerating the transfer and his estimate gives us a clear picture of what we're looking at,' Mr Hope said. 'If you can't find any more improvements, Jim, you're just going to have to press as hard as you can on every activity, every shift and I'll make sure you get as much manpower and supervision as you need.'

'Alright, go away and keep the pressure on.' Mr Everitt said. 'But don't forget what's at stake 'ere. If you don't get 22B's turning coal by the first of December, this pit's 'ad it.'

Shaft Exam

The banksman on duty for the first weekend of the overtime ban was a member of staff from an Area, specialist department. He had come out to the pit from the Area Headquarters at Coleorton, on Thursday, and been trained up for the job he was to cover in three quarters of an hour, with four other candidates who had never been trained for underground work.

It was amazing how they were getting away with it. Normally, getting certificated for the jobs they were covering took days, if not weeks.

Graham Dent, the colliery's Deputy Electrical Engineer, was driving the winding-engine. He'd qualified for that role with a crash course of a couple of hours with one of the regular enginemen, the day before.

With impressive self-confidence, the newly-qualified banksman operated the keps that held the cage in position at the top of the shaft, releasing the cage to be lowered. He rang the signal, the bell sounded at the pit top, mimicking the one that Graham would hear in the winding-engine house; five rings: This indicated that the banksman wanted the cage at the pit top to be lowered slowly. Graham would have waited for a corresponding ring of four, from the staff member acting as onsetter in the pit bottom, clearing the cage at the bottom of the shaft to be raised slowly.

The top cage started to lower at a controlled rate. The lower deck dropped below the lip of the shaft then the banksman watched as the top deck lowered away. When the flat roof of the cage was level with the floor of the pit top he gave a single ring on the bell to stop the cage.

Reassuringly, Graham appeared to be alert and in control on the winding-engine and the cage stopped instantly.

Jim had stood watching the cage's move, feeling rough; his head aching from a heavy session in the pub with some mates the night before, his stomach churning from the beer and in response to his circumstances.

Having been brought up to believe that a man should be cool and courageous, he would never admit to fear. But, to be honest, he would have to confess to being a little apprehensive. This feeling arose from three things; the nature of the task he was about to undertake, the horrifying thought of Jed Davenport's recent, awful death and the fact that standing right next to him was the tall, rangy frame of that famous lunatic, Mad Jack.

Jim couldn't help recalling something he had seen Jack do in the coal-winding shaft at Canley, a year earlier. Jack's recklessness had been astonishing, even knowing his reputation for craziness.

It had been just before the end of one nightshift. Jack been on the face where Jim was in charge, as overman. Being keen to get out of the pit before the supply shaft started winding men, Jack had persuaded Jim to join him in riding out from the face on the coal conveyors. In the bottom of Canley's Number 1 Shaft, he had tried but failed to persuade Jim to jump onto the top of the rapid, coal-winding skip with him. Jim had been sure that Jack would be buried by the tons of coal pouring into the skip or fall to his death from the massive container as it rocketed up the half-mile, vertical shaft.

Even without that evidence of how casual Jack was likely to be in the shaft, no one who knew him would have volunteered, willingly, to work alongside him on this job. No one wanted someone with a dodgy sense of humour, right next to them, when they were moving around, inches away from an unprotected, half-mile drop.

But it was Jed's terrible death, two weeks earlier, that dominated Jim's thoughts as Mad Jack spoke up.

'Righto, Jimmy. Lead the way, then.'

Jim hadn't expected that. He'd thought his more experienced and foolhardy colleague would be going first.

The access to the cages was on two levels, the platform to the upper level was reached by a set of steel steps on either side of the pair of tracks used to load supply vehicles onto the cages. The shaft was protected and men were prevented from falling down it by the blue-painted, heavy girder frame, lined with steel mesh, which surrounded it.

With the protective gate slid back, the shaft gaped open, ready to swallow them down, into its terrible depths.

Jim took hold of his silver tally, hanging from the dog-clip on his belt, and unclipped it, to hand it to the temporary banksman.

"E don't need that, Jim,' Jack growled. 'We'll be up and down and in and out o' the shaft too many times to bugger about with us tallies.'

With fumbling fingers, Jim clipped it back onto his belt. He was taking too long, starting to look too reluctant. Three short steps took him to the edge of the shaft where the cage waited, still swinging slightly from being lowered into position.

Normally, when they boarded the cage, a short ramp dropped down to hold the cage steady and provide a bridge across to it but it wasn't possible to use this for shaftsmen as they mounted the top of the cage. Jim glanced down at the black nothingness between the side of the circular shaft and the straight edge of the roof of the cage. He just needed to step cleanly across a gap of about eighteen inches. Anyone could do it.

"I must be insane," Jim thought. *"Standing on the edge of a half mile drop with Mad Jack right behind me. It would be just like him to give me a playful shove when I'm stepping over the drop."*

He readied himself, closing his mind to what he was about to do; it would only take two relatively short steps.

The roof of the cage was orange-brown with rust and covered with a hazardous-looking film of dust and condensation. Jim lifted his right foot high over the chasm and stamped it firmly onto the pitted, steel plate. Mercifully, his boot didn't slip, the surface wasn't quite as treacherous as it appeared. He lunged forward and followed through with his left foot.

He was on the cage.

He heart raced a little, until he had managed to pull the heavy chain of his shaftsman's harness from round his neck, thread it through a link of one of the four great chains that suspended the cage from the winding rope and hook the safety chain onto itself with its clip.

He was secured and safe. If he slipped and fell now it would be to dangle down the side of the cage, on a long chain, over a half mile drop, instead of plummeting away to suffer the type of grim and certain death that had befallen Jed Davenport.

As Jack stepped over to join him on the cage, Jim forced himself to lean over and peer, nonchalantly, down the shaft. He turned the switch on the cap-lamp on his helmet. Even with its strong, focused beam, he couldn't see much of the brick-lined, circular hole before all was dark. Surprisingly, there was no sign of the lights in the pit bottom, they were too far away.

'Here y'are Jimmy, you take the ringer,' Jack said.

He handed Jim a bent hook, attached to an ancient piece of battered steel plate about a foot square, with a simple, steel hammer attached to it by a length of rusty, light chain. He would have to hold the plate with one hand and strike it with the hammer in the other to signal to the banksmen the moves they required for the cage in the shaft while they were in the upper half of the shaft and to the onsetter, in the pit bottom, when they were in its lower half. This meant Jim would be unable to use either hand to hold on and steady himself during their long ride through the shaft

'Righto Jim,' Jack said, as he secured himself with practiced confidence to the cage's chains. 'Ring 'im off.'

Jim struck the plate, four times, with the hammer, the banksman relayed the signal to the winding-engineman, the onsetter responded from the pit bottom and the cage started its steady descent.

Jim's first shift as a shaftsman was underway.

Called Out

Tuesday, 15 November, 1983: 02:54

'W' hy do they always say the same thing?' Gabriella said, lying back in bed, her eyes screwed up against the light spilling from Jim's bedside lamp.

'Why, what do they say?' he murmured, buttoning his shirt.

'They always say, "It's the pit, me duck".'

'Well, it is the pit. That's what we call it.'

'I know that,' she snapped. 'But they don't need to say it. Who else would be ringing us at a quarter to three in the morning? When that phone goes in the middle of the night, I always know it's going to be the pit.'

'Hmmm.'

As soon as he had put the phone down, he had told her why he had been called out then gone for a quick shower. It never felt incongruous to do this just he was about to leave for the pit to pull on his filthy pit-black. Whatever the time, he found that a quick shower, before going to the pit, always refreshed him and cleared his head.

Now, invigorated by the hot water, he was wide awake and focused on what Roger Dawkin, the Control Room attendant, had told him.

'You ought to have the phone on your side of the bed,' Gabriella said. 'Then you could answer them, instead of me.'

'That makes sense,' he said as he made his way round to her side of the bed.

He bent over to kiss her.

'But I think they like to talk to you.'

'Huh!' She snorted. She put her arm round his neck.

'You love getting these calls, don't you,' she said, scowling up at him.

It was what irritated her most when he was called out to an incident; he was always so keen and eager to go and face whatever danger had arisen, to rush out of their life together, fired up by the challenge to come, while she was left behind, with no idea of what he would be doing, what risks he would be taking.

'I'd rather not get them but, when I do, I have to be positive.'

'Huh! You take care.'

'Okay. But you don't have to worry, I'll have my guardian angel with me,' Jim said, patting the Saint Barbara pendant on the gold chain round his neck.

'You always say that. Just remember what you're doing is dangerous. Don't go doing anything silly.'

'And call me when you get out of the pit,' she said as he rushed out of their bedroom.

Jim parked his car in the main office car park. It was bounded on three sides by the colliery's old buildings with their Coal Board blue-painted doors and windows, redbrick walls and black slate roofs.

The fitting and electrical workshops were on the west side, with the winding-engine houses and headgear of the two shafts towering up behind them. The south side was bounded by a sprawling block consisting of the dirty offices, lamp room, pithead baths, deployment centre and canteen. On the east side was the old, red-brick main office building, the light from the Control Room's big windows a beacon in the middle of the long row of dark offices.

With his hair still damp from his shower, the cold, night air chilled Jim's head as he hurried over to the back entrance at the south end of the long office block.

Through the big, Control Room window, he could see Roger, seated at the big communications and conveyor control console. The Nightshift Senior Overman, Willie Stokoe, the official in charge of the pit on nights, stood with his fat back to the window, holding court, leaning on the other side of the console.

As Jim walked into the Control Room, Willie's face bore the contemptuous smirk that Jim had known he would see. With, myopic, little eyes set in a pink, fleshy face behind round, wire-framed glasses, Willie Stokoe was known, universally, to the colliery's men and officials as the Pig.

'So, what d'ye reckon, Meestair Greaves?' Willie needled in his high-pitched, Geordie accent.

"What answer does he expect to that?" Jim thought as he walked over to the Emcor panels.

If the way Willie always extended the vowel sounds of the word "mister" into a sneer when addressing Jim was intended to irritate, it was working.

Still considering the insubordinate intent of Stokoe's question, Jim peered at the Emcor graph for 22's. When Roger had phoned, he had reported it as showing the carbon monoxide in the 22's Tailgate as having risen to twelve parts million, from a normal background level of around eight parts. The tiny nib on the recorder had left a trace on the scroll of slow-moving graph paper, housed behind the small, glass pane on the front of the big, green, metal-clad cabinet. The jagged, little line of black ink had been edging steadily to the left for nearly four hours. In the time that it had taken Jim to shower, dress and travel to the pit, the reported level of carbon monoxide appeared to have risen another two points, to 14 parts per million.

This couldn't be explained by any change in the barometric pressure or some short-term, local variation in the ventilation; it looked like there was a heating brewing. And it appeared to be taking hold fast.

Jim needed to get underground and take a look.

'I tell you what I reckon,' he said, straightening up and turning to face Willie. 'I reckon, as Nightshift Senior Overman, you should be down the fucking pit at this time of night. And with a heating brewing, you should have got your fucking self over to that district and taken charge of the initial response.'

'Oooh no, Meestair Greaves.' Willie whined. 'I needed to be on the bank in case we needed to set up Emergency Control.'

'Load of bollocks,' Jim said. 'It's a senior management decision to instigate the Emergency Control Procedure, not yours. Have you had the deputy take a Drager reading?'

'I was just about to when you arrived,' Stokoe said.

'Fucking useless,' Jim said. 'Roger, could you phone Billy Thompson and get him to take some Drager readings in the tailgate, starting from the inbye end next to the old face and working outbye. Tell him to make sure he's near a phone in ten minutes' time so I can talk to him when I'm in my pit-black.'

'Righto Boss,' Roger said. 'Do you want me to make you a coffee before you go down?'

'No thanks, I'll get straight down there.'

When he returned in his overalls, Jim took a Drager hand-pump and an extra box of Drager tubes from the draw of the desk in Ken Goodall's clean office.

The Drager pump was like a small, rubber concertina between two, curved pieces of grey plastic which fitted comfortably in the palm of a man's hand. The user took a Drager tube, a long, thin phial of crystals, broke the ends off and shoved it into the rubber housing on the hand-pump. This had to be fully depressed so that it would draw a full, calibrated sample through the tube of crystals. Drager was a manufacturer of breathalysers and this test worked in a similar way, measuring the level of carbon monoxide in the air instead of testing for alcohol on a driver's breath,

Jim darted back across the corridor into the Control Room to call Billy Thompson. Stokoe had left, to keep out of the undermanager's way.

'The Pig hadn't wanted me to call you, Boss,' Roger said

'Well you did the right thing,' Jim said as he picked up the phone. 'Is Billy at the tailgate end?'

'He should be, Boss. I'll get him for you now.'

'You know Boss, I've bin waiting for someone to put that fat, old sod in his place,' Roger said as he dialled the number. 'It's not just you, yer know. He's awkward with every manager and he's bloody 'orrible to the men and officials. He treats the men like dogs. I can't stand the bastard.'

'As you were givin' him that bollocking my chest was puffing out 'cause I was imaginin' as it were me as were gi'in' it to 'im.'

Many would think that an undermanager should reprimand a subordinate for speaking disrespectfully about a senior official but Jim believed that if you behaved like Stokoe you deserved the consequences.

'Well, I don't suppose it'll make any difference to him,' Jim muttered.

'Maybe not,' Roger said. 'But it made my night.'

Jim switched the phone onto loudspeaker so that Roger could listen in to his conversation and record what Billy reported.

'Hello Billy, what have you got?' Jim said.

'Mornin' Boss, I've taken two Drager readings in the tailgate.'

The Emcor machine analysed a continuous gas sample that took nearly three hours to travel along narrow-gauge, plastic tubes all the way from 22's Tailgate to the surface and across to the Control Room, drawn by the Emcor's pump. Billy's Drager readings had the advantage of being on-site and immediate.

'Both readings showed we've got seventeen parts." Billy said.

Roger shot a look at Jim. Far from abating, the concentration of carbon monoxide in the 22's Tailgate was rising fast. 22's district was in serious danger of catching fire.

As soon as they had completed the call, Jim gave Roger a list of materials that needed to be loaded up and sent down the pit. Then he phoned Ken Goodall. Jim heard the resigned tone in his colleague's voice as he answered the phone by his bed with a croak.

"Ello.'

'Hello, Ken, it's Jim Greaves,' Jim said. 'Sorry to trouble you but I'm at the pit. I've been called out because 22's Emcor readings are rising. The Deputy's getting Drager readings of seventeen parts per million in the tailgate – and they're on the way up. I'm going underground to build a pressure chamber in 22's Maingate.'

'We need to implement the Emergency Plan. I think you'd better come to the pit to manage the surface Emergency Control.'

'Okay, Jim lad,' Ken said. 'You get on down there and I'll get to the pit as soon as I can. Have you let the Gaffer know?'

'No, I'm ringing him next. But I wanted to make sure we'd got the wheels in motion, before I spoke to him.'

Jim phoned the General Manager, apologised for disturbing him then described the indications of a developing fire and informed him of his plans and the arrangements he had made.

'Alright, get Mr Goodall to let me know if anythin' changes radically,' Mr Everitt said.

It took three hours for the materials that Jim had ordered to arrive at the inbye end of 22's Tailgate. In that time, he had travelled to site and started to lead the men he had deployed in creating the pressure chamber in the maingate, the district's intake airway.

He had one team of six men, in the maingate, starting to build two light but effective barriers to the airflow, about four metres apart, just inbye of the new face. A second team of six were building another, similar pair of barriers further inbye, right up by the maingate end of the old face-line.

Each barrier was a rough, wooden frame made of nine-foot split bars, the six-inch diameter, round spars cut through their centres from end to end that were used to timber up falls on a working face. They used the timber to line and criss-cross the road. When it arrived with the rest of the materials, Jim would get them to use brattice cloth, white, plastic-sealed canvas, to create a partial seal across the road on the side from which the air was flowing. Passage would be maintained through these barriers by leaving a small portal in each, with a flap of brattice cloth hanging over it to provide a seal.

The pressure chamber would reduce the airflow in the section of the maingate between the two faces but, more importantly, it would reduce the air pressure in that section. It was the pressure drop from the maingate across to the tailgate which caused the airflow needed through the old face-line to provide air for the salvage men and to remove explosive methane gas which was released by the coal seam. But this pressure drop also caused some of the air to short-circuit through the waste, providing the oxygen that would be feeding the heating.

Running collapsible plastic air-ducting, a yard in diameter, called air troughing, from the outbye side of the outer barrier, all the way through the pressure chamber to the inbye, face side of the inner barrier would enable the airflow to the face to be maintained whilst reducing the pressure drop between the two gate roads. Jim was hoping that this would starve the heating in the waste of oxygen and bring it back under control.

The materials were all delivered up the tailgate supply route, so when the brattice cloth and air troughing arrived at the new face, Jim needed a gang of men to drag it through the face to the maingate, where it was to be used.

Jim had been forced to suspend all of the salvage and installation work involved in the face-to-face transfer of equipment and redeploy all of the men onto fighting the fire. This was going to cost him a precious shift of progress but he would have to worry about that later; his priority had to be arresting the heating's build up, otherwise the fire's growth was only going to accelerate.

The overtime ban prevented the men from stopping-over. His only assistance when the men departed would be from Billy Thompson. Being a colliery official, Billy was a member of NACODS, not NUM, so he would be able to stay with his undermanager to continue the work on the outbye barrier of the pressure chamber.

The travelling time involved in getting out of the pit meant that the men started to leave the district an hour and a quarter before the end of the nightshift, so Jim and Billy would be left without a workforce from six in the morning until well after eight o'clock, when the dayshift men started to arrive.

During the shift, whilst Jim managed the construction of the maingate pressure chamber, Billy had been at the other ends of the new and old face-lines, using his Drager pump to take periodic measurements of the level of carbon monoxide in the tailgate.

As the nightshift team were leaving, Billy took another reading then came back through the face, stinking of smoke, to report quietly to Jim.

'It's up to twenty parts per million, Boss.'

'Fuck,' Jim muttered.

When the dayshift started arriving, Jim went back through the new face-line to the tailgate with Billy to find that the distinctive, unpleasant, heavy smoke stink of the deep-seated fire was much stronger. Jim used his Drager pump to find that the level of carbon monoxide had shot up to twenty-five parts per million. When Jim handed him the Drager tube, Billy looked at the reading and shook his head, gravely.

As well as being an indicator of spontaneous combustion underground, when it moved above a normal, background level in sufficient concentrations, carbon monoxide was a lethal poison. It was for this reason that every man down the NCB's mines carried a shiny, aluminium bodied container, called a self-rescuer, on the narrow, buckled strap attached to his belt. [Note 14]

At this rate, if they didn't arrest the rate of increase, by the end of dayshift the carbon monoxide in the air was likely to exceed fifty parts per million and they would have to consider their options. In the absence of any other way to arrest and reverse the fire, the management's decision was certain to be to abandon the district and seal off both of its gate roads at their outbye ends.

In this instance, not only would this cause the loss of millions of pounds' worth of valuable, Coal Board face equipment, it would result in the failure to bring 22B's retreat face into production and the consequent closure of Whitacre Heath Colliery, itself.

Two hours into the dayshift, the pressure chamber was complete and functioning, with the majority of the airflow to the old face-line rushing through the air-troughing through the pressure chamber, bypassing the length of maingate between the new and the old face-lines.

When Jim was satisfied with the effectiveness of the chamber, he scurried back through the new face-line, followed by Joss Kemp, the dayshift's face deputy for that week. The tailgate air's hot, tarry stink caught in the back of their throats.

Jim took out a fresh Drager tube and broke off both ends. He compressed the hand-pump fully to empty it completely, shoved the glass tube into the pump's housing then let the spring in the pump ease it back open, drawing a full, measured sample of air through the tube of crystals. Jim removed the tube and looked at the reading.

He handed it to Joss.

'Look at that.'

'Seventeen parts,' Joss said. 'Amazing!'

Jim smiled and nodded. 'I'll give the Emergency Control a call and let them know we've pulled it down.'

By the time Jim came back outbye from the phone, Joss had taken another measurement on his own Drager pump.

'What have you got?' Jim called out, as he got closer to Joss.

'It's down to fifteen now. Well done, Boss.'

They appeared to have bought the pit some more time.

Jim left Joss to re-deploy all of his team back onto the face-to-face transfer work.

'And keep the canary up in the tailgate, until we're sure we've killed the heating.' Jim told him.

On his way out of the mine, Jim made a couple of phone calls to the Control Room to monitor the position and by the time he reached the surface, Joss was reporting carbon monoxide concentrations every quarter of an hour of fourteen parts per million.

It was clear that the fire was still a smouldering presence. But they were holding it; at least for the moment.

Emergency Overtime

Tuesday, 15 November, 1983: 11:27

'How's it going? Bill Metcalfe said.

Jim was still in his pit-black, having gone straight from the shaft to the Control Room to see the trend on 22's Emcor for himself. The Services Manager had taken charge of Emergency Control on the surface when Ken Goodall had finished his unscheduled nightshift.

'It's like drawing teeth without the overtime,' Jim said. 'Where we were doing six chocks a day before the overtime ban, we're struggling to do four chocks a day now. And we've just lost at least another shift tonight, fighting the fire.'

Jim looked at the trace on the Emcor's paper. It had started recording the drop down towards the seventeen parts per million they had measured after establishing the pressure chamber in the maingate but, since then, looked like it might have started to drift up again from that lower base.

'I think I'd better stay on nights to manage the fire-fighting.'

'That's good,' Bill said. "Cause the Gaffer's already decided you were anyway.'

17:42

Bill Metcalfe phoned Jim at home, after the day's management meeting.

'The Gaffer called Dick Flavel into his office this morning and laid it on the line to him.' Bill said. 'He told him if we lose 22's, we lose the pit, so Flavel's agreed to the use of overtime for fire-fighting.'

Jim thought about it.

'That's great,' he said, amazed at the concession. 'But to save the pit we need overtime on finishing off the transfer of equipment onto the new face. Without that, even if we stop the fire from running away, we won't be ready on the first of December and they'll close us anyway.'

'I don't think there's any chance of getting that,' Bill said. 'There's a limit to how far Flavel'll dare to go.'

'Well, he needs to realise that he'll be out of a job, as well as the men, if the pit does get closed,' Jim said.

Bill was right about the limitations on the Lodge Secretary.

Shuey McFadden had burst into the Union office earlier that afternoon to find Dick and Keith Deeming sitting at Dick's desk, discussing how the overtime concession would work. Some of the men had been enjoying themselves, all the time Shuey had been showering and dressing in the pithead baths, by winding him up about Dick's agreement to the use of overtime on 22's.

Tuesday, 15 November, 1983: 14:35

Shuey loomed over Dick and Keith, his eyes wild. Everyone in the corridor outside heard him screech.

'What the fuck d'ye think yer playing at. You've let yourself be conned by Everitt on this one. You're underminin' the national effort to fight pit closures.'

Dick drew a patient breath.

'There's no point fighting to stop this pit from closin' if we let that fire finish it off for us. Go in the Control Room, you can see it for yersen, the CO readings are creepin' up. We've got a fight on us hands 'ere and we've got to step up to the mark and play our part.' Dick said, parroting the words that the Colliery's General Manager had used earlier, when persuading him to cooperate.

'Everitt's bluffin', I bet no one's threatened us with closure.' McFadden spluttered. 'We'd still be profitable even without 22's.'

'But we can't be sure, Shuey,' Keith said, backing up his colleague. 'And anyway, they're saying the Coal Board's taking a critical look at whether each pit'll be profitable in the future, let alone the present, so we don't want to sow any seeds of doubt, now do we? 85's face is expected to be worked out some time this year. We need the replacement capacity or we're finished.'

'That's management's problem to sort out. What's the point in 'avin' an overtime ban if the first time it starts to bite they ask you to lift it and you just capitulate? Why the fuck did you just agree to it off your own bat? You shoulda got me out the pit to discuss it, as Deputy Secretary.'

'I don't need to discuss decisions with you that I've got the authority to make,' Dick shouted back. 'Our constitution makes no mention o' the Deputy Secretary role, so there's nothing that says you need to be consulted on anythin'.'

'Lads, lads,' Keith said patting downwards with both hands. 'Keep your voices down. We don't want the men waitin' outside to hear that sort of talk – or any officials or management that might be passing.'

'I don't gi'e a fuck,' McFadden snarled through gritted teeth. 'Our members should be aware of what I'm saying.'

'The reason they wanted me as Deputy Secretary was 'cause they know 'e's weak as piss,' he said, gesturing at Dick with his thumb. 'They wanted me keepin' tabs on 'im to stop 'im bein' so fucking weak with the Management.'

McFadden span round, yanked the door open and stormed out. Neither Dick nor Keith mentioned the shocked expressions they had glimpsed on the faces of the men waiting in the corridor in the couple of seconds before the door had slammed behind Shuey.

'You'd better let Arnie Campion know what you've agreed to – and why – before Shuey gets to 'im and puts 'is angle on it,' Keith said. 'You don't want the Regional office getting' the wrong impression.'

Dick said nothing.

The door flew open again and Shuey stormed back in, shutting the door more quietly before throwing himself down on one of the chairs round the sides the room.

'On second thoughts I'd better sit in tae make sure you don't make any mare fuck ups,' he said.

Dick's breathing hissed through his nostrils, his face turned crimson.

'Who the fuck do you think you're talking to?' He growled.

'Dick!' Keith said, laying a hand on his forearm. 'You can't do this. We're keeping the men waiting. We've got to start seeing them. Or tell 'em to come back tomorrow.'

'Alright,' Dick muttered, struggling to control his breathing.

'Alright,' he said, still glowering at Shuey. 'Go and fetch the first of 'em in.'

A New Threat to Whitacre Heath

Thursday, 17 November, 1983: 07:12

He had to ask the question, what they were doing wouldn't make sense to anyone.

When Jim had phoned Control at the end of the nightshift, Roger had told him that the Emcor showing the carbon monoxide level in 22's Tailgate had eased up since he had looked at it before going down the pit. Their fire-fighting efforts were failing to arrest the mounting threat from the heating.

And that wasn't the only battle they appeared to be losing. Even with the extra steps he'd taken to accelerate the process, the lack of overtime had resulted in them moving only four chocks onto the new face in the past twenty-four hours.

In a fortnight's time, if fire hadn't already claimed 22's District, Mr Collier, the Production Manager in charge of the Warwickshire Coalfield, would visit 22's Face, on behalf of the Area Director, and find that Jim had fallen hopelessly short of the target of having 22B's in production. With everything that remained to be done, there was simply not enough shifts left.

Sitting alone, on the narrow, old, metal bench that ran along the front of the row of twelve lockers in the undermanager's bathroom Jim steeled his resolve.

08:18

Freshly showered and changed, Jim sat at the table against the wall behind Jerry Malin, the dayshift Control Room attendant, drinking a mug of tea.

Harry Everitt arrived at the pit and called into the control room as usual. When he opened the control room door it hid Jim from his view.

'Good morning, Jerry. 'Ow's everythin'?'

'Morning Mr Everitt,' Jerry said, spinning round from the big, grey Control Room console on his rotating office chair to face him, before glancing in Jim's direction. Jim was on his feet by the time Mr Everitt peered round the door.

'Ah, Mr Greaves. 'Ow's it looking?

'Good morning Gaffer. The CO's still easing up, I'm afraid,' Jim said. 'Actually, I was hoping I could have a quick word with you about it.'

'Oh. Right.' Mr Everitt said. He walked over to the Emcor cabinet, scanned them all but peered most intently at the recorder for 22's.'

The General Manager turned and looked at Jim.

'Gi' me a minute to take me coat off then come on down to my office.'

Jim sat back down to finish his tea, continuing to rehearse what he had to say.

The heating had not been on in the General Manager's office and it was still chilled from the cold of the night as Jim sat down, near to the Gaffer, at his end of the long table.

'What's on your mind then, Mr Greaves? D'ye think we're losing it?'

'We obviously haven't stopped the heating yet, Gaffer, so that's a serious concern but, as of this morning, we've got eleven chocks left to transfer. At the current rate of progress, it's likely to be another three days before we have them all in position. That'll leave us just seven working days, about twenty basic-length shifts, to strip all the haulage track out and get the panzer on, installed and working and the two shearers built and commissioned.'

'And you don't reckon you can do it.'

'No Gaffer. Sorry, it's just not do-able.' Jim said, thinking that his Gaffer knew that as well as he did.

Jim slid the sheet of paper with his revised forecast of the completion times for the remaining major tasks across the table.

'If anything, those durations are on the optimistic side, Gaffer.'

Harry Everitt could see that for himself. Just as he could see that, even with that cautious optimism, they would miss the deadline by almost a week.

Jim prided himself on never raising a problem with his Gaffer without offering a solution.

'The only thing I could think of doing was seeing if you could ask the union to lift the ban on overtime completely, just on 22's, so we can hit the deadline as well as fighting the fire. I know it might be a long shot but I was hoping they would recognise that missing the deadline was as big a threat as the fire to the future of the pit.'

'I don't need you to find things to occupy my time,' Mr Everitt growled.

'I know that, Gaffer. I don't mind suggesting it to them. But I thought it might carry more weight if it came from you.'

'I'm sure it would,' he said, giving Jim a pointed look. 'It's one of the advantages of not being so hard-line with 'em all the time.'

'Yes Gaffer,' Jim said, unconvinced.

'If Deeming and 'Flapper' Flavel agreed to it, it'd still depend on whether the men'd work the overtime offered.'

'If the union'll let us do it, I'll get the men we need, Gaffer.'

'And you reckon we can do it, if we get all the overtimers you need?'

'Yes, Gaffer,' Jim said, with considerably more confidence than he actually possessed.

'Right, I'll see what I can do. Get yer sen off home. I'll get word to you later, on how I fare with our union colleagues.'

'Thanks, Gaffer.'

As Jim got up from the table, Harry Everitt said, 'If you get that face turning by the first of December we'll both get medals as big as a dustbin lids.'

'Yes, Gaffer,' Jim said with a sad smile.

'And if you fail,' Harry said, as Jim reached the door. 'I'll fuckin' sack yer.'

Thursday, 13:20

'Are you going to talk to Shuey first?' Keith said as he pulled up a chair to sit on the other side of the desk from Dick.

That had been the question Dick had asked himself as he considered Mr Everitt's request, knowing that he couldn't refuse it. He had been worrying it over in his mind ever since. He could hold to his previously stated principle, that there was no definition of the Deputy Union Secretary's role so he had no obligation to consult McFadden. That would avoid having to work out how to broach the subject with him. But the prospect of the trouble that would ensue if he left him to find out with the rest of the workforce was unthinkable.

When Mr Everitt had summoned Dick that morning and asked him to agree to lift the overtime ban within 22's, Dick had said he would need to ballot the members on such a momentous step. Mr Everitt had shown him Mr Greaves' forecast. Dick could see for himself that the young undermanager's figures made sense and that taking a couple of days to call and hold a Union meeting would be enough to kill the pit.

The best Dick felt he could offer was to make the case to the Union Committee that afternoon and ask them to decide the matter with a vote. Even that would be difficult to achieve. It could be argued, reasonably, that they should obtain agreement at regional or even national level for deviating from the union's official action. Dick didn't want to involve either, being certain of getting the wrong answer.

At the Nick

A hundred miles away from Whitacre Heath, three young police officers had met up for a pre-shift coffee.

'So when are you moving in?' Lynne said.

'Next week. Julie's taking a couple of days off on my rest days. I've hired a van and we're doing it then. We haven't got too much to move but it'll probably take a couple of trips, after we've given the new place a good, spring clean.'

As often happened at Highbury Police Station, a couple of the younger WPC's were sharing a Formica-topped table in the canteen with their colleague, Neil Bradford.

Tall and slim with dark brown hair and the earnest look of a polytechnic lecturer, the young constable had a nervous, bustling urgency when he moved but a relaxed gentleness when sitting and chatting with friends, like Lynne and Bev. Having grown up with three sisters, back in Yorkshire, Neil was comfortable in the company of girls and in having them as friends.

The young women assured themselves they didn't gravitate towards the young policeman because of his good looks and easy, sociable personality. He was accounted for; three years older than Julie, he'd been with her since she was sixteen and was totally committed to her. That actually helped, because it meant that he would never try it on, unlike so many of his male colleagues at the station. Although, both young women would have to admit that they did harbour the hope that things might change between themselves and Neil, if he and Julie were ever to split up.

It was easy to respect his motivations. He and Julie had similar values, sharing a desire to make a positive contribution to society.

When he left school, he had allowed himself to be persuaded by his mum and dad to go for a safe job in the bank. It was his determination to make a difference that had resulted in him applying to join the Metropolitan Police and leave the bank and his home to follow Julie to London, moving into a little flat with her, when she started training to become a nurse at Hammersmith Hospital.

Having qualified, Julie had taken a job at the Hammersmith for her first two years then obtained a position at Saint Thomas' Hospital, three months earlier.

Moving out to Watford would mean they could work to save up for a deposit and buy their first flat where property would be cheaper. Julie would commute by train and Tube and Neil would have a drive of little more than half an hour between their new home and the police station, with his shift-start and finish times. And he would keep making that journey; London was where he wanted to be, where he could make the biggest contribution.

'Oh no, here comes your partner,' Bev muttered to Neil, looking away, pretending she hadn't seen Carl Bessemer enter the canteen, gaze round the room then start swaggering over, towards their table.

'Come on Brad, I hate to break up your Tupperware party,' Carl said with a leer. 'But it's time we got out on the beat.'

'Just ignore him, I always do,' Neil said with a sad smile as he stood up, stuffed his cap under his arm then downed the last of his coffee. 'See ya later.'

'It's no wonder he's divorced,' Lynne said, when the two men were out of earshot.

'Yeah, twice!' Bev said. 'The only amazing thing is he found two women daft enough to marry him in the first place.'

Carl was older and nominally more senior than Neil, due to his length of service. Out in the corridor, he strode ahead, as Neil followed, fastening his uniform jacket. He turned to call back to partner.

'I don't know what you're waiting for, you could get your leg across either of those two anytime you like. And their mate, Pam. I don't know what you're fucking about at, I'd be up there with any of 'em, like a rat up a drainpipe.'

Neil frowned.

'We're not all like you. They're friends of mine.'

'"Friends of mine", what a load of bollocks. If they're friends of yours they wouldn't mind sharing a few things with you and having some fun. You ought to have a go at that Bev, I bet she's a dirty little minx in the sack.

See you later, Sarge." Carl called out to the desk sergeant as they headed towards the side door from the station out to the cars to start Tuesday's late shift.

"See ya, lads.'

Outside, Neil followed Carl towards their parked car.

'Look, just leave it alone will you?' He said. 'I don't want your lurid fantasies all shift.'

'Alright. I'll keep 'em to myself. But you're right, they are lurid as far as your mate Bev and her tits are concerned.'

"At least the girls didn't have to put up with his nonsense anymore," Neil thought. Having got utterly fed up with Carl's innuendos, obscenities and pestering the young WPC's were content to ignore him, pointedly.

Neil no longer tried to persuade Julie to join him and colleagues for a drink after work if Carl was going to be there. She had made it clear that she shared the views of the young police women in finding Carl tedious and unpleasant. It suited Neil to keep her out of Carl's sight. Julie's most noticeable attributes were her large breasts. He'd found that, if he could keep Julie out of Carl's mind, it reduced the chance of his partner making lewd and explicit remarks about them.

Neil respected his colleague for the occasions when he got things right; weighing up suspicious situations, helping people when it mattered and getting stuck in when there was a bit of aggro.

But Carl Bessemer was still, undeniably, a prat.

Consultation

Dick had been expecting his deputy to explode when he pre-positioned him before the emergency meeting of the Lodge Committee. Instead, his adversary's restraint wrong-footed him.

'You were wrong to lift it for that bit of a fire but the overtime ban is supposed to hit production and now you've let them persuade you not to do that on 22's,' Shuey said. 'Yer now actually underminin' the official and national, industrial action of the National Union of Mineworkers. Big stuff, Mr Flavel. I'm going to pick up that phone now and call Campion and tell him you're bustin' the overtime ban.'

Shuey dialled Arnie Campion's number with a confidence born from familiarity. Arnie answered and Dick and Keith were forced to sit and listen to Shuey reminding the Regional Delegate of Dick's agreement to allow overtime to be used in fighting the fire on 22's and then giving his account of how Dick was intending to take unilateral action and lift the overtime ban to allow non-emergency work to be undertaken.

Shuey listened to Arnie's response.

'Aye. Aye. I know. That's what I told him. D'ye want to speak to Flavel yoursel'?' Shuey grunted and held the phone out to Dick.

'Here y'are. Arnie wants a word wi' you, pal.'

It was no surprise to Dick that Arnie's response was consistent with the one he had received from McFadden. Arnie instructed Dick to inform the management that the overtime ban was official and binding on the Whitacre Heath Lodge and they would not be lifting it under any circumstances.

'In all conscience, Arnie, I can't do that. Everitt don't worry easily but Rawlinson's got 'im worried this time. He means it when 'e says that this pit'll shut if we don't have that face in production on the first of next month. You can tell the way the Coal Board's talking, they're just looking for any excuse to shut pits and I'm not going to give 'em one to shut ourn.'

Thursday, 15:27

In the smoke-filled union office that afternoon, Dick had made the proposal to the eleven other members of the Committee. Thereafter, Shuey had argued passionately against breaking ranks with the rest of NUM.

'Of course it's damagin',' Shuey said, wrapping up his impromptu address. 'The Special Conference of the Delegates of the NUM decided democratically to take this official action to strike back at what the Government is planning and what the Board is doing.'

'This fight is only just beginning. If we and other pits start capitulatin' before we've even struck a blow it will gi'e the Tories and the Board *carte blanche* to dae whatever they want. This is bigger than us, this is the age-old class struggle being kicked off again by Thatcher and the rest of the fucking, capitalist Tories.'

Dick sighed.

'There's no point getting into that class war stuff,' he said. 'I don't want to break ranks with the National Union but I have to urge you to do what's right for our members and keep this pit open.'

'Warwickshire's a special case; there's more fires in Warwickshire pits than in the rest of the country put together, by a factor of twenty. This kind of emergency is virtually unknown elsewhere in the country but we live with the possibility of faces catching fire all the time. We 'ave to get that face working and bury the fire.'

'All I'm askin' is that you make our pit the priority for the next fortnight. After that, there'll be no one more solid in stickin' to the ban than Dick Flavel.' Dick said. 'This is a simple vote: Save Whitacre Heath or let the spon' comb' and the deadline beat us and let it shut.'

Shuey might have succeeded in getting himself into the new, influential position of Deputy Secretary but, apart from the two recently-elected replacements for retiring members, all the rest of the Committee had worked with Dick for years. They thought more like him and had always followed his lead. And neither of the two newcomers were a place-man of McFadden's so they voted with the rest.

Dick secured the mandate he was looking for, with McFadden the only person opposing his proposal.

'Well, I hope you'se lot can live with yersel's, that's all I can say,'

McFadden said. 'You'll all have that on your conscience. And I'm gonna be urgin' our members to show solidarity with the rest of the union and act on their conscience and not work your fuckin' overtime.'

After the meeting Keith phoned the General Manager. Angela connected his call.

'Mr Everitt, the Lodge Committee has agreed to recommend to the men that they work overtime on 22's to achieve the deadline.'

'Well done, Dick. You've done the right thing. They've made the right choice.' Mr Everitt said. 'How do we get it going? Are you gonna talk to the men?'

'Yeah, I'll see the afternoon men at the end o' their shift and come back again before the start of the night and day shifts to explain the position to the charge-hands and as many men as I can get round. I'll try to get the rep's to join me in spreadin' the word.'

Harry Everitt noticed how drained the Union Secretary sounded.

'Thank you for securing that concession for Whitacre Heath, Dick,' Harry said.

Privately, Dick despaired at the disunity that McFadden was determined to promote and dreaded the prospect of having his so-called Deputy alongside him, undermining him as he spoke to the members before the start of each shift, over the next twenty-four hours.

It was six o'clock in the evening before Ken Goodall phoned Jim Greaves to let him know that they would be able to reinstate the use of overtime on 22's transfer work.

The Gaffer had fulfilled his part of the deal, now Jim had exactly two weeks in which to deliver a working face for him or find out how serious his threat of the sack had been.

Pest

L ate in their shift, Carl swung the panda car onto the forecourt of the filling station.

'I'm just gonna pick up a Mars Bar and a packet of ciggies.'

He pulled up next to the forecourt shop and threw the driver's door open.

'Do you want anythin'?'

'No, I'm alright, thanks,' Neil said.

A dark crimson Mini Metro pulled up at the set of pumps nearest the road. Police cars always caught the eye, by design, and the driver, a young, blonde woman in her early twenties, glanced over at Neil and the panda car. She got out of her car and closed the driver's door. Neil could see she was short and neat but curvy in tight, light blue jeans with a light-patterned, fluffy jumper on top and matching, woollen leg-warmers sagging round her calves.

She walked to the back of her car and removed the filler cap. She put a couple of gallons of petrol in the tank of her car and then ducked down and leant across to the passenger seat to grab her purse before tripping across the forecourt and into the building.

When she re-emerged, Carl drifted out behind her, watching her hurry over to her car rather than looking where he was going. He stopped and stood next to the driver's side of the panda car, gazing over at her. When she was settled back in her car, seat belt fastened, she started her car.

Carl yanked his door open and clambered in, starting the engine as soon as he'd closed it. The young woman's Mini Metro pulled away and stopped to give way at the exit onto the wide thoroughfare, indicating left and pointing towards the City.

Carl drew up right behind her, not bothering with his seat belt.

When the Metro pulled out, Carl made sure they were right on its tail. Neil looked across at Carl. He could see, from the intensity on his partner's face, that he was on a mission.

They trailed the car south, towards Islington. Twice Carl pulled up right behind the car when it stopped at red lights.

'Come on Bessie, don't do it,' Neil said.

'Oh, stop moanin'. I only want to check her out and have a chat.'

This happened too often when these types of circumstances arose; nothing much happening to break up the shift, a car with an attractive, young, woman driver on her own and P.C. Carl Bessemer fancying his chances.

Aware of the panda car following close behind her, Rosie Palmer was intent on doing it all by the book.

She had stuck exactly to the thirty mile an hour limit; not too fast, not too slow. She had used her mirror and indicated left to signal her intention before changing lane to stay out of a right filter lane, in order to go straight on. The police car had stayed with her since following her out of the garage. It was right on her bumper while she waited for the lights to change.

When the amber light illuminated, Rosie checked both sides carefully before pulling away, steadily but positively. She was still accelerating in a controlled way when the blue light, on top of the panda car, started flashing, as she had known it would.

She slowed, found a sensible length of road in which to stop and pulled up at the kerbside.

In her door mirror she could see the panda car, its blue light still spinning. She saw the door swing open and the driver get out and put on his uniform cap as he strode towards her side of her car.

Rosie wound down her window and waited, determined to be patient and respectful and keep what she said to a minimum.

The policeman was tall and lean. Wiry, reddish-blonde hair showed around the edge of his cap. He lacked the reassuring air of a policeman you could trust.

Carl put his hands on the young woman's driver's door and bent down to get a close look at her face.

'Good evening, miss.'

'Hello, officer.'

'Is this your car?' He said, eyebrows raised.

'Yes.'

'Could you get out, please?'

Rosie didn't answer. She sighed quietly to herself, opened the door and stepped out of the little car.

'Could you follow me to the back of the car please?'

Rosie did as she had been asked.

'Don't look so worried,' he said.

'I'm not worried,' Rosie murmured.

"Bored and irritated, maybe" she thought. *"But not worried."*

'Now, you appear to have an intermittent fault on your offside sidelight. It was off when I stopped you but it's working again at the moment.'

Rosie said nothing as she looked up at policeman, refusing to respond. His partner had emerged from the passenger side of the panda car. As he looked at her she saw the apologetic look in his eyes. Her lips pursed. The second policeman had the grace to look at the ground in embarrassment before dashing a frustrated glance at his colleague.

'Could you show me your licence please, Miss?' The first policeman said.

Rosie walked round to the passenger-side door, opened it and took out her handbag, conscious of his eyes on her all the time. When she had found her purse and pulled out her licence she returned and handed it to him.

He opened her licence and studied it.

'Rosemarie Palmer. Oh, so you live in Northampton.'

'Yes.'

'I've got friends up near there. Perhaps we could meet up some time – go out for a drink.'

'I'm afraid not, Officer, I have a boyfriend. I'm on my way to see him, now. And I don't see other men.'

The other policeman took the last three steps, to close with them. He lifted Rosie's driving licence from his partner's hand, folded it up and handed it back to her.

'Thank you, Miss,' Neil said. 'We're sorry to have bothered you. Have a safe journey and a nice evening.'

Rosie took a deep, slow breath.

'Thank you, officer. Goodnight.'

She walked back to her driver's door, got into the car and closed the door. She slotted her licence back in her purse and muttered to herself.

'What a prat.'

Neil strode round and got into the driver's side of the panda car.

Carl went to the passenger side, opened the door.

'What's going on?'

'I'm driving,' Neil snapped. 'I've had enough of your nonsense for one day.'

Carl slumped into the passenger seat, swung his legs in and closed the door.

'What's up with you?' Carl whined.

'I keep telling you. You can't do stuff like that. It's not fair on an innocent member of the public.'

'She didn't look that innocent to me. I bet she was gaggin' for it.'

'Don't talk such nonsense!' Neil snapped. 'The only reason she didn't tell you what she thought of you was 'cause you're in uniform. You let the Met' down when you behave like that and you could get us both into trouble for harassment or something if she complained.'

'Fuck me, Bradders, lighten up. It was only a bit of fun. She knew that.'

'No she didn't. She just knew you're a prat.'

Using the Overtime

Being on nights gave Jim the opportunity to move round 22's District, explaining to all of the men on the last nightshift of week the management's agreement with the union and the need for men to work overtime in order to hit the deadline and save the pit.

With the expense of family Christmases approaching, the majority were pleased to suspend their participation in the ban for a fortnight and earn some extra money.

Jim had to refuse a number of potentially incendiary requests from men on behalf of their mates in other parts of the mine who were keen to be allowed to benefit from the temporary, localised suspension of the ban by coming over to work overtime on 22's.

Earlier, on the surface, Dick Flavel's biggest fear had proved groundless. Dick knew that Shuey had been determined to attend, in order to undermine his efforts, by coaxing, cajoling and bullying men not to accept overtime. By eight fifteen, when Shuey needed to have been at the mine if he was going to catch the afternoon men coming off shift, he had already been in the Holly Bush for over an hour and a half.

As he'd looked out through the rain-spattered window pane at the cold, dark night outside, it hadn't looked at all inviting. Instead of heading for the pit, he'd bought his fifth pint, telling himself he could fight a rear-guard action on this matter when it suited him.

In the meantime, he consoled himself with the thought that he had made sure that the Region was aware of what Flavel had done and that they knew that the recently-elected Deputy Secretary was toeing the union line and opposing Flavel's maverick actions.

Saturday, 19 November, 1983: 05:30

With the use of overtime reinstated on 22's, the last powered roof support had been manoeuvred into position by snap-time on Friday night, halfway through the weekend's first overtime shift.

Jim stopped over until the end of the nightshift to make sure the men were re-deployed swiftly and that they adapted to the new, working process involved in stripping out the rail and sleepers of the haulage track as they worked back through the face, laying out the big steel pans to enable the assembly of the armoured face conveyor to get underway.

They soon established a rhythm to the work with four men getting around each pan and inserting the specially-fabricated, hooked hand-tools, which Paul Wood had had made, into the housings on the pan for bolting on the pan-sides and the face-side toe-plates, which also needed to be got into position to be installed as the panzer was re-built. They used the hand-tools to lift the pans off the trolleys which had transported them through the face and then manhandle them into position. *[Note 1.2]*

There were one hundred and thirty of the heavy objects to get into place and link up and the need for the men to bend over beneath the low roof as they heaved each pan into place made the work arduous and more likely to result in back injuries.

This was the type of hard, physical work, in which progress was apparent and at which facemen excelled. They got stuck into the task with enthusiasm, as Jim exhorted them to get more than half a dozen into place in what was left of their shift.

Jim left the new face, just before the end of Saturday's dayshift, with his natural optimism restored for the first time in many days.

Winter Training

The big, old training room was not inviting, in itself. It was badly in need of refurbishment, with its faded walls, paint flaking from its rusting, steel window frames and its furnishings of work-weary desks and chairs. But at least it was warm and, for sixty policemen, that made it preferable to going back out into the bitter wind that was gusting drizzle across the wide-open, flat expanse, outside.

When the briefing ended, a few tried to cheer themselves and their colleagues with humorous exchanges but most retained reluctant, resigned expressions as they heaved themselves from their chairs.

They flocked around the long, trestle tables at the side of the room to recover their coats and helmets from where they had been told to pile them on entry.

It had been a relief to get in there, to throw off their greatcoats and get their hands wrapped around plastic cups of hot coffee. Now it was time to get back out into the cold and put into practice what they had just learnt about the Metropolitan Police Force's latest methods for riot control and the quelling of civil disobedience.

The training room was built into the side of an old, aircraft hangar. Its door opened into a scruffy, little passage that discharged them into the grime and damp chill of its cavernous space.

As instructed, they made their ways to the three police vans, parked in the shelter of the hanger, that they had seen as they had piled off the two ancient, double-decker buses that had brought them down from Inner London to the former RAF base.

The sergeants at the vans issued each of them with a long, heavy baton, a round, black helmet with a hinged, plastic visor and a full-length, transparent plastic, riot shield.

Carl Bessemer homed in on Neil Bradford as they formed a body, in three ranks, on the sergeant nominated as right-marker. Carl repeated what he had said the last time they had undergone riot-response training, six months earlier.

'It's like playin' at being bleedin' Roman soldiers.'

On the orders, they did a right turn and marched out of the hanger and over to the big area of wet tarmac where they would practice the drills.

They had seen the two dozen other coppers, there to act as the mob, dressed in a mixture of jeans and slacks with civvie jackets, pile out of another room in the hanger and head out onto the old runway. The small group waited for them, looking like nothing less than a bus-load of off-duty coppers on a day out. Neil knew from their previous training, in the spring, that they would look more like rioters when they started to misbehave; there were bound to be some idiots among them who would be looking forward to providing a convincing simulation of aggro.

The inspector shouted, 'Right wheel!'

When the marching body had snaked round he halted them and gave the order, 'Left turn.'

The three ranks now stretched across the former airfield's old taxiing route that, for the purpose of their exercise, would be Electric Avenue, Brixton.

The sergeant instructor stood in front of them and bellowed out a few reminders of the purpose of the simulation and some key points on how to perform in their ranks.

In the brief lull that followed, Carl muttered to Neil next to him.

'Why the fuck do they have to bring us out here at this time of year? It's brass monkeys out here.'

Neil said nothing but he did think it strange that they were doing this training in the cold of November. It did seem more sensible to undertake this kind of training in the spring; everyone knew that the silly season for disturbances and unrest, like the recent troubles in Toxteth and Brixton, was the summer when the conditions were right for street-fighting mobs.

Neil knew that their play-fighting couldn't prepare them for the ferocity of what would be involved in any real-life clash. In the event that he was called on to apply the skills they were refreshing, he would be determined to do his duty. But he couldn't help hoping that Law and, particularly, order would improve after each report of civil unrest and that there would be no need for such confrontations in the future.

The controlling mind of the opposition had called his rabble together to receive their own last few instructions. They suddenly broke out of their huddle, jogged towards the uniformed unit, took up aggressive stances and started their catcalling.

Carl stopped musing about the strange timing of the day's exercise when the tennis balls the rioters were throwing, with impressive accuracy, started bouncing off the shields and helmets they were aimed at.

The play-fight was underway.

A Fight to the End

Wednesday Nightshift, 24 November, 1983: 00:51

The dust on the smooth, steel plates of the conveyor pans made the surface treacherous. Jim's boots slipped sideways with every other step, as he rushed, bent double, up through the face to the tailgate.

With the panzer and full complement of chocks in place throughout its face-line, 22B's was starting to look and feel like a mechanised coalface, fuelling Jim's impatience to see it all done and ready for the scrutiny of Warwickshire's Production Manager in a week's time.

In the tailgate, Jim got stuck-in with the six men Joss Kemp had deployed to drag the panzer's bulky tailgate drive section with a pair of rams to marry it up with the rest of the conveyor. This unit and its mirror image, already installed at the other end of the face, would provide the motive power to the flights and chains that would haul away the coal cut by the face's shearers. *[Note 15]*

Friday 25, 03:02

By the middle of the following nightshift, Paul Wood and his small team of fitters and facemen had completed the re-assembly of the tailgate gearhead, with its gearbox and motor installed and ready to run. It was time to start installing the panzer chain. This had been stripped off the old face in bulky, six-metre lengths which were difficult and hazardous to handle. Jim had arranged for monorail to be slung from the arch supports in the road at the tailgate end, to enable the men to unload the vehicles carrying the awkward bundles of steel flights and chain, rapidly and safely, straight onto the panzer.

Six men, working in pairs, lashed a light chain onto each length when it was delivered then rushed it down the face, ducking down low, under the front of the chocks' heavy, steel canopies, as they dragged it over the empty pans. *[Note 16]*

At the same time, pairs of men were working with fitters, on their hands and knees right under the chocks, to install the panzer's pan-sides which stopped the panzer's coal load from rolling off the face conveyor and into the crawling route through the chocks.

At the end of the shift, Jim spoke to Joss.

'Have you managed to get the men you need to stop-over.'

'Yeah, all manned up, Boss.'

'Good. We need to keep encouraging them. I hear McFadden's still trying to stop them working it.'

'Yeah, some of 'em mentioned it but I reckon the opportunity to earn some more money in time for Christmas was always going to prove more persuasive than that idiot.'

'Yeah, you could be right,' Jim said.

Although he did hope that it was also due to a degree of genuine commitment and a determination to prevent their pit from closing.

McFadden's continuing agitation was causing Dick Flavel to live in dread of the times around the start and end of each dayshift. Without fail, his tormentor would saunter into the union office to sneer at him and castigate him for what he spoke of as Dick's betrayal of the union and its class struggle to save the industry. Knowing how lame his defences sounded in comparison to McFadden's bold, revolutionary statements, Dick had ceased trying to justify himself. It exasperated him to know that this would be appearing to confirm McFadden's accusations in the minds of his deputy's audiences.

Saturday, 26 November, 1983: 11:20

Jim had stayed on the face at the end of Friday's nightshift to continue to force the pace on the installation of the panzer chain.

As the end of Saturday's dayshift approached, he needed an official to stop over to provide the statutory safety supervision for the men who would be stopping to work additional overtime at the shift's end. Joss Kemp would normally have been in charge on 22's, on a Sunday morning, but he was away that day, attending the baptism of a relative's baby, so the supervision was being provided by two officials who normally worked elsewhere underground.

Jim crawled off the face into the tailgate and found the first of these two men; Tom Gibbons, an unimpressive, Grade II deputy from 56's, the other East Side district.

Jim approached him.

'Tom, will you stop over for me to cover the men stopping over at the end of the shift?'

'No, it's Saturday afternoon,' the deputy said, turning away. 'I'm goin' 'om'.'

'Come on, it's only for an extra hour and a half,' Jim said. 'We need men to be working every minute we can find on 22's if we're going to keep the pit open.

'Bollocks to it,' Gibbons said. 'I'm goin' 'ome, my family's more important to me than this fucking pit.'

'This pit pays the money that keeps your family,' Jim snarled at him. 'You can't think that much of them if you're not prepared to fight for their livelihoods. You might as well admit it, it's not your family you want to rush off to, it's the fucking Holly Bush.'

Joss Kemp's substitute that morning was Vic Worrell. 'Warbler' Worrell was a deputy who had been acting, surprisingly, in the role of overman on his own, regular district. His nickname arose from his keen membership of the Nuneaton Choral Society and his predilection for inflicting vocal solos on reluctant, captive audiences at the Campville Working Men's Club. Jim had heard another pitman describe these performances as being redolent of the bellowing of a sick calf.

Short and stout, Warbler's bulging eyes, framed by heavy, black spectacles, and the pout of his fleshy lips gave him a permanent, petulant look which was consistent with his attitude. This was in particular evidence when Jim stopped him as he strutted out from the old face to leave the district.

'Vic, we need to make sure we keep the panzer chain going onto the face, will you stop and cover the overtimers?'

'No!' Warbler huffed. 'Time's time!'

Jim glowered at him, his fatigue from having spent more than twelve hours grafting underground, overnight, had shortened his fuse. He made no attempt to prevent the facemen around them from hearing him.

'That is either a statement of the fucking obvious, because time is obviously time or it's the sort of bollocks you'd expect from some Bolshie union rep'. Either way, it's a fucking stupid comment and not what I expect to hear from an acting overman when the pit's up against it.'

Jim turned to include Tom Gibbons in his comment.

'As far as I'm concerned, you can both fuck off. I'll stay and act as deputy and supervise the men.'

Worrell opened his mouth to speak.

'And you needn't bother saying it,' Jim said. 'It's perfectly legal 'cause I still have a valid deputy's certificate. And if you want to complain about a manager doing officials' work, go and see Les Parker and find out what he thinks about your attitudes.'

Warbler and his colleague said no more, being as confident as Jim that their Union Secretary, Jim's senior overman, would support his undermanager's action.

As the two officials skulked away from the face Jim wondered who could have appointed men like them and Willy Stokoe to their positions. With little or no loyalty to the pit, derided by the men to their faces as well as behind their backs, they were incapable of pushing on the progress of work and displayed no leadership aptitude, whatsoever.

It was late that Saturday afternoon, after Jim travelled out of the pit with the overtimers, when the fitters coupled up the panzer chain and were able to give it a successful test run.

Sunday, 27 November, 1983: 00:20

On the Saturday nightshift, Jim was back down the pit, working with the men using a Sylvester device and rams to haul the massive bedframe for the maingate shearer onto the face. The length and weight of this huge chassis made this a frustrating and time-consuming job, requiring much craft to negotiate the corner and manoeuvre the great lump of steel round and into the confines of the face. *[Note 17]*

Once it was on the panzer it would need to be drawn down the face until it was alongside the maingate shearer's big, coal-cutting discs. The discs for both shearers were in their stables on the face, one for the tailgate shearer and two waiting to be built onto the double-ended, maingate shearer, in-cut on the face. *[Note 18]*

Paul Wood crawled out of the last chock on the face, got to his feet and drifted over to stand next to Jim who was on the edge of the team moving the bedframe. The great, lengthy lump of steel had become stuck against the steel arch leg, on the far side of the road, and the base of the last chock on the face from which Paul had just crawled out. Looking at it, Paul felt his concerns snow-balling. Reluctantly he raised them with his Undermanager.

'Boss, have you had a look at where we are against that project plan we drew up?'

Jim was just wishing that he had modelled the moving of the bedframe onto the face to check that it would actually be possible. He was about to dismiss Paul's question as a distraction but stopped himself, stood quietly, looking at the jammed bedframe and thought about its implications.

'To tell you the truth, I haven't looked at it for days,' Jim murmured. 'A bit of a balls-up really, I suppose.'

'Not really, Boss. You don't want to beat yourself up about it, you've been livin' down 'ere. You can't expect to do everything.'

'I suppose I've just been thinking that it didn't matter where we were on the plan, we just needed to stay focused on driving the progress and trying to get everything across the finishing line, in time,' Jim said.

He called out to the men working on the bedframe.

'Hold up, lads. Let's try and pull it back out. We're going to have to create a bit more space somehow.'

As the men moved quickly to set themselves up to free the bedframe and haul it back, Jim looked at Paul.

'Let's have a think about it. We have to be ready for production in three days, that's nine shifts and nine half-shifts of overtime.'

'Boss,' Paul said quietly, turning away from the rest of the men, 'I can't see how we can get both machines finished and all of the pan-sides on by Thursday dayshift.'

They discussed the times involved in finishing getting the bedframes and then the machine sections into position on the face and then building, testing and commissioning both machines.

'Under normal circumstances, I wouldn't expect us to be able to do all that before well into the weekend,' Paul said.

'We can't create any more time and throwing more men at it won't help, there's only so many of us can get round the jobs at any time,' Jim said. 'The only thing we can do is keep pressing on as fast as possible and hope that everything goes smoothly from now on.'

Paul looked at Jim.

'We can't afford to lose this one, can we Boss?'

'No, we can't,' Jim said, shaking his head. He frowned. 'Having a fortnight where we couldn't use overtime has cost us too much time. It looks like that might be going to kill us.'

'Are you stayin' on nights this week, Boss?' Paul said.

'No, I'll be in again tomorrow night but I was going to stop through and work some or all of the dayshift and then get back onto days for Collier's visit on Thursday. The Gaffer'll want me here to take whatever flack's coming on this.'

Over the next two days, they pressed on with installing the pan-sides and getting the machine sections into position and the maingate shearer built up.

The two shearers' bedframes had to be hauled into position, one behind the other on the face, then the maingate shearer's machine sections were pulled up to and onto the tailgate bedframe first, in the right order, then hauled along onto the maingate shearer's own bedframe; the first gearhead to drive one disc, the power-pack, the haulage section then, finally, the second gearhead for the second disc.

Three groups of men worked on removing the upright girders of the square-work that had supported the face-line while it was headed-out and during the installation work, opening up the face ready to be cut. They ran the panzer in reverse to transport the girders back up to the tailgate to remove them from the face.

Whenever possible, Jim took the opportunity to have the panzer run continuously, to allow the chain and flights to clean out the conveyor's races in order to minimise the resistance on the motors if and when the face started turning coal.

Wednesday, 30 November, 1983: 12.57
Towards the end of Wednesday's dayshift, Jim reviewed the position with Paul. It was looking desperate.

The maingate machine had taken nearly seven shifts to build and now, with boring predictability, disaster had struck; there was a defect on the haulage section that was supposed to move the shearer through the face. Paul was hopeful of effecting a repair, even if it only lasted for the duration of the visit, but there were still forty pan-sides left to fit and the re-building of the tailgate machine was only half complete.

'If we're lucky, we'll get the maingate machine going long enough for the big boss's visit,' Paul said. 'But there's no way we'll have the tailgate machine working.'

'Will you stop on overtime again to keep the pressure on?' Jim said.

'Yeah, I'll stop over, Boss.'

'I'd better get out the pit and let the Gaffer know the position.'

On the surface, after a quick bath and change and having darted into the Control Room to snatch a look at the Emcor to check 22's CO trend, Jim went straight down to Angela's office.

'Hello Angie,' Jim said. 'Can I have a word with the Gaffer?'

Angela gave the young manager a long look.

'Yes, he's in there now. Go straight in.'

Jim knocked and entered and shut the door carefully behind himself.

He stood by the door, twenty feet away from the General Manager who was at the other end of the long table.

Jim gazed down at the huge, colliery plan spread over the table under its layer of thick, transparent plastic. Despite his best efforts, all that, a century of coal production at Whitacre Heath, the pit itself, was all about to become history, all due to him.

Mr Everitt looked up from the papers on his desk and gazed at him.

'I wanted to let you know how we're fixed on 22's, Gaffer.'

The General Manager's head dipped as he considered him over the top of his spectacles.

'And?'

'It's not looking good, Gaffer.'

'Whad'ya mean, "not looking good"?'

"As you know, we've got all the chocks on and the panzer's running. We've got the maingate shearer built now but it's got a defect in its haulage section.'

'Yeah. I 'eard.'

'I'm still hoping we can get it running long enough for it to be able to limp through the face for the duration of Mr Collier's visit. The tailgate shearer's sections will be in position on its bedframe but it's not going to be commissioned in time. And then there's the tailgate A-frame.' Jim said, referring to the powered roof support that pushed the conveyor over at that end of the face and provided the platform on which the rippers who operated that end of the face would work. 'That's not going to be in position and re-built in time.' [Note 19]

Jim was braced for the serious bollocking that had to come.

'Come 'ere,' the Gaffer growled.

Jim walked down the side of the table.

'Sit down.'

Jim pulled the leather-backed chair out, dropped down onto it and slumped back, utterly drained, now that he'd admitted defeat.

'What are you worried about?'

143

Jim frowned in confusion.

'Well – I haven't done it. We're not ready. Mr Collier's visit tomorrow…'
Jim said, not bothering to mention the inevitable consequence.

'You'll have the maingate shearer running, yer say.'

'Hopefully, Gaffer. But you know what it's like with a haulage section
fault, you can never be sure. If it needs a complete, new section it could
take us two, more likely three more days, getting it there and fitting it.'

'Right, well just make sure it's fixed. Then tomorrow I'll bring Lol
Collier up the maingate. I'll make sure we've got Mr Hope with us and
I'll get him to let you know when we're enterin' the district. You wait at
the inbye end and, when you see our lights approaching, let the belt away
and start the face up. Get 'em cutting down the face slowly and then keep
it goin' until it reaches the maingate. Can you make sure that happens for
me?'

'Yes, Gaffer. But what about Mr Collier, he'll see…'

'You leave Lol to me. Of course he'll see the tailgate shearer's not ready.
And he'll see the tailgate 'A'-frame ain't there neither. But you just make
sure that all them pan-sides are on the face, even if they ain't properly
secured, up past the shearers. Then, whilst Lol's on the face, keep that
maingate machine making its way down the face while we're crawlin' up to
it and 'til we're well past it, then keep that fucking panzer runnin' 'til we're
off the face and away.'

'Lol's no fool, he knows more about pit-work than us pair o' fuckers
put together. But he ain't going to be here lookin' to shut one of his own
pits, he'll be rootin' for us. We just have to show him what he needs to see,
so he can tell Ralphie we made it.'

Lol's Visit

Wednesday, 30 November, 1983: 15:20

Jim walked out of the main office building and across the car park and pit yard to the fitting shop. The Colliery General Manager had just transformed the situation from hopeless to possible. When your Gaffer did something like that, Jim thought, it reminded you of why you'd die for him.

It remained too much of a longshot but he still had time to make preparations and put in place contingencies to give the Gaffer's plan the best possible chance of success. First, he went and found Mad Jack and persuaded him to double-back on nights, with him and Paul Wood, to try to get the maingate shearer going. After that, he went home for a couple of hours' sleep.

Gabriella prepared their meal quietly and woke him for dinner at nine o'clock.

As he left the house Gabriella came to the door and kissed him.

'So will you be ready for the big day in the morning?' She said.

Jim stood in their open, front doorway, Gabriella still holding onto his arms as he looked out, trying to find an answer from the dark of the night. He wished he could express confidence in his ability and that of his team but decided not to tempt fate.

'I hope so,' he said, tapping the wood of the doorframe, for luck. 'Anyway, we'll know in the morning.'

'Well no one could have done any more, so make sure you come out of the pit and come home as soon as Mr Collier's gone. And give me a ring at work and let me know you've done it.

Wednesday, 22:45
That night, Mad Jack went down the pit and travelled in to 22B's face with Paul and six other fitters and apprentices; an astonishing number of

145

mechanical staff to see in one place but consistent with Jack's reputation for never going anywhere underground without a large entourage.

Jim hung around the cramped area on the face where the maingate shearer occupied virtually all of the space on the face-side. The big, thick, steel plate that covered the machine's haulage section had been removed to expose its inner workings. The fitters squeezed in to work in the tight clearance between the top of the machine and the roof beams of the powered roof support whilst Mad Jack lolled in the crawling route, occasionally making suggestions in the relaxed, gravelly tones of his deep, North Warwickshire accent.

The work of Jack and the fitters was a black art to a mining engineer like Jim, so he could make no useful contribution, only offer encouragement. The long hours of work and lack of sleep, over the last few weeks, made it a battle for him to keep his eyes open.

Despite their best efforts, as the end of the nightshift approached they had still not fixed the shearer's haulage section; its discs would spin but it would still not move in either direction.

The two machine drivers were joshing Jack.

'These lads have got the right idea, keeping Mad Jack out the way while they're workin'?' The lead driver said.

'Arr, yer right there,' his mate replied.

"D'y'ear? 'E nearly did fer a mate of 'is, Larry Chaplin, a couple o' months back. He wuz walkin' up the road an' saw Larry a-workin' on the engine of 'is Land Rover, under its bonnet. 'E snuck up on him and stuck 'is 'and in the driver's winder and pressed the 'orn. 'E deafened 'is mate, Larry. Made 'im jump art 'is skin. 'Is 'ead flew up 'n' knocked the bonnet off its rod. The fucking, great, heavy lump crashed dahn on the back of 'is y'ead.'

'Wuz 'e alright?'

'Nah! It broke his fuckin' neck!'

Jack gave a single snort of amusement as the facemen and fitters all guffawed.

'Wot y'on about, 'e's out of 'ospital now.' Jack called out when they quietened, setting them all off again.

Jim wasn't laughing. He could hold back no longer, he crawled over the fitters' legs to get to Jack.

'Jack, it's getting late. I was hoping we could let the Gaffer know we were set up and ready for turning when he gets to the pit.'

'Yeah, I know that, Jimmy. I'm keepin' an eye on the time. Leave it wi' me and the lads. I just want to check summut out.' Jack said, giving Jim a wink.

As he had done on several occasions in the past, Jim forced himself to keep faith in his eccentric colleague and moved down the face to pretend to check on other preparations.

Forty minutes into the nightshift's overtime, Jim went back to Jack to press for an answer. Before Jim had time to speak, Jack crawled away from the fitters, draped over the machine, towards him.

'Right,' Jack said. 'We might have a way of getting it going in the short term.'

'That's all we need.' Jim said, looking hopeful, hoping, at the same time, that he wasn't about to be on the receiving end of another of Mad Jack's daft jokes.

'We shoulda found it before, Jimmy, but it looks like we've got a stickin' piston on the pilot valve in the haulage section.'

Jim didn't have a clue what Jack was talking about but didn't bother to admit to his ignorance, he just gave Jack an emphatic, encouraging nod.

'So, there must be something you can do to find a work-around for that,' he said, putting his faith in Jack's masterly reputation for botching up temporary, mechanical expedients.

'Yeah, there is. We need to clean it up and do a complete oil change, to get all the debris out as'll be in the hydraulic circuit. We can get the machine runnin' but it won't last long and the piston's worn so it won't be too efficient and that debris in the oil is tellin' us the section won't last long.'

'That doesn't matter, as long as we turn coal for an hour this morning, that's all we need. You can have it for a week after that, as far as I'm concerned.' Jim said. 'Have you got enough oil and everything you need to do the oil change?'

'Yeah, we've got that but we need some emery cloth to clean up the piston with,' Jack said.

'What? You mean you've got all these fitters with you and none of you have brought any emery cloth in with you.'

'Yeah, it's a bit of a bollocks up, really,' Jack chuckled. 'But we 'ad no way of knowing as that'd be what we'd need.'

'Bloody hell, Jack! So what are you going to do?'

'One of the lads has bin and phoned the bank and we've got a young apprentice on nights, young Brett Redfern's comin' down the pit straight away to bring us some,' Jack said. 'In the meantime, we'll get that oil change done.'

'Fuck me, Jack, you don't half like cutting things tight. '

'We'll be fine,' Jack said, airily. 'Have I ever let you down before?'

'No,' Jim conceded. 'But you've come fucking close on a good few occasions. And this'd be the wrong time for a first.'

'There's only one big downside though. We're likely to completely shag the haulage section.'

'We're always shagging machine sections with this BJD rubbish. Let's give it a try.'

Mad Jack and Paul ignored the undermanager as they discussed how to complete the repair, then Paul sent two other fitters to go and rob some metal tubing out of the Tailgate shearer's haulage section.

Two hours into the dayshift, Jack and his team had exceeded their most recent estimate by an hour and the machine was still not working.

'I don't want to pester you and piss you off, Jack,' Jim said. 'But we do have to have this fucking machine moving in the next hour or we've had it.'

'Okay Jimmy,' Jack growled with a lazy smile. 'Leave it wi' me.'

After another couple of gentle reminders from Jim over the next hour, the cover of the haulage section was still off with the emergency spare-part surgery still in progress.

Joss Kemp, the face deputy on days that week, called over the face tannoy system.

'Come in, Mr Greaves.'

Jim crawled to the nearest tannoy, pressed the tiny, rubber covered "Speak" button and shouted.

'Hello, Joss.'

'Boss, I've just had a call from the Control Room. They say that Mr Hope's phoned up the pit to say that he and the visitors have entered the maingate and to let you know.'

'Okay thanks,' Jim shouted back over the tannoy.

He crawled back up to the maingate machine, cursing quietly to himself.

'Jack, did you hear that? That's it.' Jim said. 'They'll be on the face in thirty minutes. I've got to go down and meet them. You don't want to have that machine in bits and still fucked when the Gaffer gets here.'

'Typical ain't it lads?' Jack said to his fitters, 'As soon as there's some senior bods around the undermanager fucks off and leaves you to go and do some serious arse-licking.'

'Bollocks,' Jim called back, crawling away, leaving the two machine drivers to help the fitting staff. Jim knew that, if they completed the repair, it would still take them a quarter of an hour to get the big steel plate cover, which they had removed to open up the haulage section, back into position and bolted down. Like so often in the past, he just had to trust in the wizardry of that legendary nutter, Mad Jack.

The dayshift chockers and snaker were on the tailgate side of the machine, lying idle in the chocks' crawling route, ready to snake the panzer over and advance the roof supports if the machine ever started, so Jim encountered no one as he crawled miserably down the face to the maingate.

'Here he is,' one of the maingate rippers said to his mates as Jim emerged into the road on his hands and knees. 'On his way to the labour exchange.'

'That's it then Boss, it'll be the sack for you when the Gaffer catches up wi' ye,' another said, cheerily.

'Yeah,' Jim said, forcing a wry smile. 'It won't be the first time.'

It was a ridiculous position to be in; hoping that the shearer would be repaired and that, thereafter, with no test run, all of the face's newly installed equipment would operate and turn coal with the future of the pit depending on it.

A thought struck him, he'd missed something out in all the intensity of his drive to complete on time; he'd forgotten to seek the support of his guardian angel. As usual, he felt reproached for his neglect.

He moved away from the rippers and past the panzer driver. In his desperation, he took hold of the pendant round his neck, closed his eyes and called on Saint Barbara for a miracle. When he opened his eyes the approaching cap-lights of the visitors were in sight, as they made their up the walking route alongside the maingate's belt conveyor.

'Take the lockout off the conveyor, Peter, and ring it off.' Jim said to the panzer driver.

The button lad operating the conveyor belt had been coached to be ready, so when the young panzer driver turned the switch and pressed the signal button on the steel lock-out box, the belt surged away. For a moment, Jim thought of the belt as a magic carpet; he wished he could just jump onto it and fly away from the looming doom and embarrassment facing him, down there on 22's.

Mr Hope was leading the way as the visiting party arrived. He gave

Jim his usual cheerful and charming smile and greeted him with a relaxed confidence that Jim couldn't share. Mr Collier was behind Mr Everitt and Bill Metcalfe was at the tail-end, looking out of place, underground.

'Good morning Mr Greaves,' Mr Everitt said. 'You know Mr Greaves, don't you, Boss?'

'Yes of course I do,' Mr Collier said, shaking Jim's hand. 'Good morning, Jim.'

Jim could never imagine how Lol Collier, with his gentlemanly civility, had progressed through the tough ranks of mine management, been a colliery general manager and ascended to his current position as head of the coalfield.

Mr Collier asked Jim about the state of the spontaneous combustion and the efforts to control it. Then he asked about how the face-line had been headed out. Pleased to have the opportunity to kill more time, Jim explained how a Dosco Dint-header had been used for this. [Note 20]

The new face-line had started from a square-work junction that had been formed in the maingate and it had been driven through to connect with the other square-work junction in the tailgate. Pitmen used the word "thirling" as the term for when two roads connected in this way, underground.

Having exhausted that topic, Mr Everitt told Jim to lead the way onto the face. Jim winked at the panzer driver as he passed him and he started the stageloader, the armoured conveyor which received the coal from the panzer and carried it over to load it onto the maingate conveyor belt. Over the tannoy system, the electronic two-note warble of the stageloader's pre-start alarm kicked in, followed by a similar, lower-pitched warning on the face for the start of the panzer. The electric motors on the two, steel conveyors started up then their empty chains started their clattering racket.

As soon as they had both stepped delicately over the potentially lethal panzer, ducked down and crawled onto the face, in the crawling route in the chocks, Mr Everitt spoke to Jim as quietly as he could over noise of the maingate panzer drive's big, electric motor and the rattle of the empty chain conveyor.

'Is that machine gonna go?'

'They were still finishing fixing it when I left them, but hopefully it'll start up.' Jim said. 'They knew we were ready for them as soon as the panzer struck up.'

He turned to carry on crawling up the face, grimacing as his embarrassment became acute.

As they made their way, slowly, up the crawling route in the chocks, the panzer continued running, conspicuously empty.

On and on they crawled. Jim gave himself over to despair. He had to accept it, they'd had it; so near and yet so far.

Then the noise of the panzer chain, further up the face, changed from its sharp clatter to a softer hiss, lower in volume. It was the distinctive sound of the panzer when its metallic noises were muffled by a load of coal.

Just in time, they were away.

Mr Everitt made no comment.

They crawled on through another ten chocks, with the black stream of coal flowing down the face, and encountered the face deputy, Joss Kemp, crawling down the face, now that the machine was away.

'Fucking well done,' Jim said into Joss's his ear as he passed.

Joss smiled and winked at him.

Mr Everitt paused to introduce Joss to Mr Collier.

Another forty chocks further up the face, they met the shearer coming down the face. The Number One machine driver, controlling the shearer's leading and higher disc, was taking care to prevent it from striking any of the horizontal steel girders of the square-work that had supported the roof during the installation of the face and which now stuck out, from over the tops of the chocks' canopies.

Jim crawled round the first driver and waited, in the shaking ground, watching the sparks flying from the disc's picks as they struck a band of fool's gold in the seam, while his gaffer bellowed, in conversation with the passing machine man. Jim was unable to hear what either shouted to each other over the roar of the shearer's discs as their great bits tore the coal off the face and threw it into the continuous pile being carried away on the panzer. Mr Everitt finished whatever he was saying to the machine man then carried on crawling up the face.

'Keep movin',' he shouted at Jim. 'We don't want to 'ang about on 'ere.'

Jim scrambled past the second machine driver. Paul and another of the fitters were trailing the suspect machine.

'Where's Jack?' Jim shouted.

'He made sharp, as soon as the machine was going.' Paul called back. 'He left us to put the cover back on the haulage section.'

Jim nodded. That was no surprise; the crafty, old beggar had managed to evade the senior visitors.

Fifteen metres on they encountered the snaker who was using the snaking rams, one built into the base of each chock, to ram over the panzer into the area of fresh floor opened up by the shearer. Beyond him, Jim could see the two chockers dropping individual chocks and drawing them in. The partially completed face was coming alive and had started on its journey back along its two gate roads.

Jim's spirits started to rise. He clutched gratefully at the gold pendant dangling from his neck. However hard you worked and however competent you were, you still needed to be blessed with good fortune or, even better, a watchful guardian angel. And men; good men.

Mr Everitt kept them moving quickly through the last forty yards of the face, in which many of the pan-sides were not secured to the pans. Each chock's snaking ram was supposed to be coupled to the panzer by a clevis on the clevis-rail built into the pan-sides. Where the pan-sides were not secure the chocks were a couple of feet back from the panzer and not yet attached, making the incompleteness of the installation apparent to even the most casual investigation.

Mr Collier followed Mr Everitt's lead and crawled with determination to the tailgate, turning a blind eye to the state of the pan-sides in this last length of the face.

When they reached the tailgate end, Jim crawled through the last chock and out from under the lip of the roof of the face, praying that Mr Collier wouldn't mention the state of the last pan-sides.

Jim was expecting the usual, senior managers' summaries of their levels of satisfaction, or otherwise, with what they had seen but Mr Everitt was determined to keep Warwickshire's Production Manager moving.

'Are you coming out with us, Jim?' Mr Collier said

'No, if it's alright with you, I'll carry on here, Boss.'

'Right, let's get goin',' Mr Everitt said, turning and striding off down the tailgate.

Lol Collier paused and gave Jim another benevolent smile.

'Thank you Jim. You've done a good job bringing it into production on time.' He said, with a distinct twinkle in his eye. 'Keep up the fight, now, and get that fire put to bed.'

Tim Hope winked at Jim.

Bill Metcalfe breezed past.

'See you later, Jim,' he said.

Jim went back, got onto his knees and crawled back onto the face. When the shearer reached the maingate he told them to turn round and cut back up the face to leave it as close as possible to the tailgate supply road, ready for the haulage section change it needed. As soon as that second cut was completed, he stopped the production and got the face team to revert to installation work and the fitters to finding a permanent fix for the struggling maingate shearer. With the team redeployed, he headed off on the ninety-minute journey to the pit bottom, to get a ride up the pit.

On the surface he phoned Gabriella and told her the good news. Despite his tiredness, he told her that he was determined that they would get out to Lichfield that evening, for a few beers and a curry in the Bengal Restaurant, to celebrate.

Before he left the pit, he called in at the General Manager's Office, having obtained Angela's assurance that Mr Collier had left the mine. He told Mr Everitt that it would be well into the following week before the face would start turning properly.

'That don't matter,' the General Manager said. 'The important thing is the Production Manager for Warwickshire can go back to 'is boss and tell 'im 'e's seen that face in production. But you don't want to hang about, you want to get that face away properly. You've still got to beat that fire. And the best way to do that is to bury it.'

With the face-to-face transfer complete, it was apparent that the General Manager wanted Jim to stay in charge of the new face and get it producing properly before he was going to be moving him anywhere else.

Dicky Heart

Saturday, 10 December, 1983: 18:10

Dick flopped back in his armchair, sick with fatigue. He hadn't slept properly for weeks.

When he thought about it, he hadn't slept well since that debacle of an election, two and half months earlier. He was drinking more than usual but even that wasn't helping him relax or get to sleep at night. And he'd started chain-smoking.

He knew the problem; it was stress and McFadden was the cause.

Dick and Shirley had been over to Coventry for a few hours of Christmas shopping. All afternoon, he'd been unable to concentrate on present-buying. All he'd been able to think of were the threats that McFadden posed to Whitacre Heath but particularly to him and his post as Secretary. Every day at the pit, McFadden undermined his authority and demeaned his standing in the eyes of the membership.

When they'd arrived back home, Shirley had seen how grey Dick looked, in his face, and told him to go and sit down while she made him a cup of tea.

The house was warm, the coal-fired boiler for the central heating was still lit when they got back. Dick had left Shirley to turn up the thermostat and operate the rake on the boiler to liven it up.

He was still struggling to get his breath after lifting in all the bags of presents from the car. The sickness and discomfort felt like a bit of indigestion.

Then it changed; the chest pain was a sudden clamp which tightened its grip. He groaned, death had its claws into him. The burning pain flooded from his chest into his arms and back then shot into his jaw.

His teeth clenched fast, he groaned. He was dying.

McFadden had beaten him.

Even in his agony, as the blackness swallowed him, it grieved Dick to picture that bastard laughing at the news and stepping into his position.

Sunday, 08:45
Bill Metcalfe was in charge of the pit that weekend. When he arrived on Sunday morning, he became the first member of the management team to hear of Dick's heart attack. Roger Dawkin, the Control Room attendant on the dayshift, had heard the news from Keith Deeming.

"Im and Shirley had been out Christmas shopping,' Roger said.

Shirley was the pit's Canteen Manageress. Dick had left his wife to move in with her, a couple of years earlier.

"When they got back, Dick had a bit of turn, so Shirley told him to sit down. She was makin' 'im a cuppa tea when she 'eard 'im groan. She ran in and found 'im unconscious. She called the ambulance and they managed to get 'im to intensive care and that's still the position, as far as Keith knew.'

'Poor old bugger,' Bill said.

'Yeah, you never know when it might 'appen to you, do you?'

Bill nodded, looking thoughtful.

'Makes you wonder whether McFadden will try to seize control if Dick pegs it,' he said. 'If he does, we'll be in for some fun.'

Monday, 06:35
The first time that Keith Deeming spoke about Dick's heart attack with Shuey McFadden was on the Monday morning, before the start of dayshift.

'Yeah, it's a shame. Mind you, he does carry a lot o' weight.' Shuey said. 'Quite appropriate really; Dickie Flavel – an' his dicky heart.'

Keith frowned at McFadden and raised his upper lip.

Buried

The men on 22B's knew their face had moved out from its face start-line when the chocks no longer supported its roof girders and their lining of corrugated tins had all dropped into the waste behind. The roof was now composed, completely, of a fifteen-inch band of friable shale. Above that was the broken ground of the caved waste, left behind by 22A's, the worked-out advancing face. The face start-line collapsed behind the chocks.

Despite this, when Jim arrived at the pit that morning, he was greeted by the news that the Emcor was showing that the level of carbon monoxide on 22's had started to climb again.

Jim suspected that the collapse of the roof behind the retreat face must have opened up air routes to feed the heating that was still alive in the old waste. He decided there was no need to change their strategy; as always with a deep-seated heating, it was important to try to bury the fire, progressively, by keeping on turning the face over to keep the waste caving and compacting behind it.

The face had retreated about fifteen yards from its start-line. The old face-line and the two gate roads inbye of the retreating face remained open, providing an air course which might draw air through the old waste. When they had gone a little further they would be able to build a stopping wall in the tailgate, to reduce the air leakage. In the meantime, they needed a quick, interim measure to buy some more time. *[Note 21]*

When Jim went down the pit, he travelled in to 22B's Face to look at the junction of the tailgate road with what had been the new face-line. He knew that if they could cause that higher and wider opening to collapse it would increase the resistance to the airflow considerably and make a real difference.

The junction's square-work consisted of two large RSJ girders, each with a web depth of ten inches, which spanned the road in support of smaller girders, perpendicular to them, at metre intervals, with webs of eight inches, the latter having been lined with corrugated tins to support the roof. The two great RSJs were supported at each end by a wooden chock. *[Note 22]*

Jim thought that all they would need to do to cause the junction to collapse was to make the chocks unstable. He decided to try to achieve this by wrapping a chain around a wooden, chock block in the middle of each chock and pulling them out with rams.

That was the theory.

When they tried it, they pulled the blocks they had tackled out of three of the chocks. However, in the inbye chock on the left hand side of the road, the chock block just swivelled out from under one side whilst remaining in place, still supporting the other block.

All four wooden chocks bowed slightly and the outer of the two big cross beam girders slumped towards its right-hand end. Several corrugated sheets dropped out, clattering and banging as they fell to the dirt floor and stone dropped from between the smaller girders.

But the junction remained standing.

The group of eight men, their face deputy and Undermanager looked at it in silence.

'I reckon we can leave Warwickshire to do the rest,' Joss Kemp said to Jim. 'She'll come down when she's ready.'

Jim thought about it. It was not the decisive result he had wanted; they had made no difference to the size of the opening that still yawned behind the face, helping to keep the fire alive.

'I think we need to have another go at it before we've retreated too far away from it,' he said. 'We really need to do something to make a difference to that bloody heating and we won't get the stoppings on behind us for another week or two.'

'So what do you wanna do?' Joss said.

'I'll go and reset the chains on the chock blocks on the other side from the gaps we've opened up in the chocks.'

Joss looked aghast at his undermanager.

'I wouldn't, Boss. You can hear Warwickshire, she's on the move. We've upset her now.'

'He's right, B-boss,' Brian Kettle, one of the tailgate rippers, said. 'Er, don't d-do it.'

Jim thought again. If it did come, that great structure of steel would crash down on him under the weight of the many thousands of tonnes of rock on it, with little or no warning. Going out under it was not an attractive prospect but he wanted to see it down and collapsed behind them, as he had pictured it, not still standing or only partially collapsed.

'Push the rams out again, lads,' Jim called out. 'I'll re-set the chains.'

Jim set off across the junction, treading softly, as if hoping this would prevent his footsteps from unsettling anything. He grabbed the bulky, right-hand chain, heaved some slack and wrapped it around the back block of the creaking, wooden chock then secured the chain to itself with the hook on its end. He moved quickly and lightly across to the left side and did the same to the chock from which they had only partially removed a block. All the time, the junction's steelwork groaned and worked around him. He wrapped the chain in and out of three wooden blocks in this chock.

He walked back across the junction, trying not to look nervous as the men looked on. In theory, he was on the safer side of the junction now, but if it came down without warning he would still be flattened.

He got the third chain and wrapped it around the next block of the outer, left hand chock. He heard the sound of splintering timber behind him and knew that the fourth chock was failing. He darted to his left, back to the group. The right-hand end of the nearer of the big girders fell on its right side as the chock collapsed.

'Pull away.' Jim shouted and the three men operating the controls on the rams that were attached to chock blocks drew their rams in so the chains became taught. First the wooden chock on the left end of the inbye girder then the chock on its right end lost blocks as the rams pulled in. The inbye cross girder plunged down. The remaining outbye chock on the left now had all of the junction's weight thrown onto it. It bellied out and collapsed. The junction caved in and was lost to sight in a roaring, billowing fog of dust.

The gang of men cheered.

'That fucked it,' one of them shouted.

As the dust started to clear, Joss chuckled and turned to Jim.

'That was very daintily done, Boss, if I might say so, the way you danced around that junction.'

'I bet you'd've been bloody dancing if you'd been with me,' Jim panted, still fizzing with adrenalin.

Breaking Up

Friday, 23 December, 1983

It was to be a token appearance for Jim, that morning. He'd gone down the pit, with the men, at the start of dayshift for a quick trip round 22's and out; he didn't want to hang about on this occasion.

Jack Charlton, the General Manager at Jim's previous pit, had never let the men up early on the last working day before Christmas. As a consequence, Canley Colliery's men had always been bitter in their knowledge that other pits would be "giving their men an hour". In contrast, Harry Everitt had announced that the three shifts at Whitacre would be allowed out of the pit an hour before their normal shift-ends.

It was unusual for Jim to disagree with his former gaffer on anything but he approved of Mr Everitt's concession, knowing that it generated goodwill amongst the workforce and showed them that their hard work, throughout the year, had been appreciated.

The facemen and the button lads operating the conveyors would be expected to work through their twenty-minute snap time but other, outbye workers, like the supply lads, would be having a quick, Christmas party. With the benefit of his earlier experience as an official, Jim knew that it was important not to end up having to choose between policing these celebrations or turning a blind eye. It was better to just get out of the way and let them get on with it.

He walked up the maingate, watching the continuous, lumpy, black stream flowing by on the belt; the last Whitacre Heath coal of 1983.

When he got to the inbye end, he greeted the panzer driver then the maingate rippers then clambered over the panzer and ducked down under the first powered roof support. He crawled through quickly, inspecting the face and wishing the men a Happy Christmas and all the best for the forthcoming New Year as he passed.

He crawled off the face at the tailgate, had a quick, few words with Billy Thompson, the face deputy, then wrapped it up.

'Righto Billy, I'm going to leave your district. I'm heading out of the pit.'

A good thirty-five minutes later, after a brisk march, Jim arrived at the outbye end of the tailgate to find 22's supply lads, a couple of dinters, a fitter and an electrician preparing for their party. An old, wooden, ventilation door, laid on chock blocks to raise it above the thick, stone dust on the floor, was being laid as the table for their subterranean feast.

The oldest of the supply lads, a late-starter underground, in his early thirties, had brought in an old, floral dress, slipped it over his pit-black, put a tatty, curly, raven-coloured wig on, under his pit helmet, and plastered his mouth with red lipstick. He was obviously intent on providing his mates with a 'feminine' presence at their office party, despite the dark stubble on his chin.

The Law prohibited alcohol from being brought onto colliery premises, except when expressed permission had been granted by the Colliery General Manager, so Jim thought it best to accelerate his withdrawal when one of the other supply lads offered him a swig from an unlabelled, glass bottle. Its contents were either cold, milk-less tea or, more likely, neat whiskey. He became even keener to be on his way when he saw the female impersonator whip a sprig of mistletoe out of his snap bag.

Jim exchanged quick compliments of the season with the men and departed at speed.

Walking out along the main return airway, he reached the air crossing and cut through to the intake airway. The air pressure difference slammed the fourth and last pair of heavy, wooden air-doors closed behind him with a deep thud, as he walked the last few yards to the crossing's junction with the main intake. He jumped straight on top of the coal that was still streaming on the main belt from both 22's and 56's faces. He rode out to just before the East Loader, where the coal poured into mine cars on the Loco Road, where he vaulted off the speeding belt, landing alongside the belt structure with practised agility.

He went and spent a few minutes with the loader man, in his cabin above the Loco Road, chatting about how the coal-turning had been going that morning, looking forward to the forthcoming festivities and listening to the man's pit gossip until a loco arrived to pick up a full train of coal and draw it off to the pit bottom, having delivered a train of empties to the back of the loader. Jim strode down the Loco Road to catch a ride and got the driver's permission to ride, illegally, in the loco's rear cab, just in front of the leading mine car full of coal.

In the pit bottom, Jim met up with a small group of engineers, overmen, senior overmen and his colleague, Ken Goodall.

After five minutes, Les Parker, exercised his authority as Senior Overman and instructed the onsetter to tell the banksman to stop winding supplies, given that it was now noon. The cages were prepared for manriding and the waiting group filed onto the two decks of the cage for their own, early exit from the pit.

When all the dayshift officials were out of the mine, bathed and dressed with their reporting completed, they went over to join the General Manager in his office for his annual party.

This year, Harry Everitt had invited Warwickshire's Production Manager to join them. Jim suspected that the two senior men had agreed that this would be a good occasion on which to celebrate the recent salvation of the pit, achieved by Jim's first strip of coal off 22B's, during Mr Collier's most recent visit.

The General Manager's Office: 17:43
It was dark outside, all of the sausage rolls had been devoured, the white bread of the last few, Red Leicester cheese and onion sandwiches were starting to curl up as the managers, overmen and deputies continued to make inroads into the crates of Bass Blue Label light ale, stacked at the end of the room.

Harry Everitt was drinking whiskey and water. It had not been difficult for him to persuade Joey Bondar to join him in a glass when the overman had arrived from the pithead baths. Since then Joey had been quaffing bottles of light ale with whiskey chasers. Jim could only assume that the Gaffer was deliberately trying to get Joey plastered as quickly as possible. Having accepted Joey's invitations to join him for a pint after a couple of weekend dayshifts, Jim knew that it didn't take much to lower his overman's inhibitions.

Extra chairs had been brought in and placed round the office's walls. The room was packed with men and buzzing with banter when Joey bellowed out to call them all to order, the timbre of his voice even coarser than usual from all the whiskey he had downed.

Having secured their total attention by silencing anyone who had tried

161

to continue conversing with threats and coarse insults, he announced that he had decided to entertain them with a joke. In slurred words, he proceeded to relate the tale of the proud mother who was travelling on a train with her new baby.

In his story, a man was sitting opposite the young woman. The man was reading his newspaper but every so often he would lower it, appear to peer over at the mother and baby, then raise it again. Each time the paper hid the man's face, the young mother was sure that she heard him stifling a chortle before tittering to himself.

Eventually, the young mother couldn't help taking exception to this repeated rudeness. The next time the man lowered his newspaper to look over at them, she spoke out.

"Excuse me, but would you mind telling me what you find so amusing?"

At this point in Joey's story, Harry Everitt, who was sitting behind him on a chair by the wall, listed towards Lol Collier, seated next to him, to confide something in his ear. Lol, nodded and then leant sideways towards Harry to murmur his reply.

Determined not to be interrupted, Joey swung round and glared down at Mr Collier.

'Oi! Shut up Shit-Face!' he hollered, as he loomed over the Production Manager. 'I'm tellin' a joke.'

Jim was both stunned and appalled by his overman's impertinence towards the top man of the Warwickshire Coalfield. He was even more surprised by the response.

'Oh,' Mr Collier said, with a kindly smile, his face a picture of contrition. 'Sorry, Joey.'

Joey nodded at him, evidently satisfied. Swaying noticeably, he turned back to address the rest of his audience and continue his yarn.

'So the young mother says to the man,' Joey switched his voice to falsetto: '"Excuse me, but would you mind telling me what yer keep sniggerin' at?"'

'The man starts chuckling and says, "I'm terribly sorry, me duck, but I can't help it. It's just that – that's the ugliest baby I've ever seen in me life!"'

'The young woman jumps up, bursts into tears and says, "You 'eartless beast! I've never been so insulted in all my life!"'

'"Oldin' the little baby in one arm she 'eaves back the door o' the compartment and starts runnin' down the corridor, 'eadin' for the guard's van to throw 'erself and 'er baby off the back of the speedin' train and end it all.'

'A couple o' carriages down, she bumps into the train guard, comin' the other way, checkin' tickets.'

'Seein' 'ow upset she is, 'e says, "You poor woman. Whatever can be the matter?"'

'She says, "I've just 'ad the worst insult you could imagine. It was so cruel I'm going to throw meself off the back of the train and end it all."'

'The train guard puts his arm round her and says, "Now, nothing can be as bad as that, me dear. Now you come with me, back to the guard's van, and I'll pour you a nice cup of tea from my flask."

'"-And, I tell you what, I'll 'ave a look in me bag – and I might even have a banana for your pet monkey".'

The hilarity that followed the end of the joke did not detract from Jim's sense of wonder at Joey having got away so easily with such gross insubordination to someone as senior as Mr Collier.

After another half an hour, the beers and whiskey caused Joey to turn affectionate. Having been on the receiving end of this type of behaviour when he had had a pint with Joey in the past, Jim could see what was coming. Joey was standing alongside Mr Collier's chair towering over and talking to the Production Manager who was showing a surprising degree of tolerance for his nonsense.

Eventually, Jim heard Joey say to Mr Collier, 'You're bloody lovely, you are.'

Then Joey bent down.

'I luv you,' he mumbled before giving Mr Collier a sloppy kiss on the top of his bald head.

Jim cringed but was amazed to see Mr Collier beam contentedly in response.

A little later, Mr Everitt called Jim over to join him and the Production Manager in conversation.

'Me and Mr Collier were just talking about your infamous Canley colleague,' Mr Everitt said.

'Mad Jack?' Jim said, with a smile.

'Yes,' Mr Collier said. 'I'd heard about him when he was at Canley but always wondered if there was any truth in all the tales, Jim.'

'Oh yes,' Jim said. 'He really is mad, Boss.'

Jim went on to give an account of a particular, recent incident from

163

amongst Mad Jack's legendary exploits, involving his allegiance to Nuneaton Borough Football Club.

One Saturday afternoon, before the start of a Home match, the winner of the club's Annual Grand Draw was to be presented with his prize; a brand new car from a manufacturer in nearby Coventry.

Home and Away players lined up, in their teams, in the centre of the pitch, as the presenters were introduced.

'Ladies and gentlemen, please give a warm round of applause for Mr Brian Coxon, the Chairman of Nuneaton Borough,' the announcer called out over the ground's public address system.

'And please welcome Mr Daryll Thomas, Marketing Director of Peugeot Talbot UK, who is here to make the presentation for Nuneaton Borough's Grand Prize Draw.'

The two dignitaries walked out, the Chairman of the Club waving to acknowledge the applause and the handsome Mr Thomas of Peugeot Talbot, looking like a self-conscious double for the Hollywood heart-throb, Gregory Peck.

A photographer from the Coventry Evening Telegraph stomped along behind them, with his bulky camera.

The prize car motored in from the main gate in the corner of the ground, to park near the patch of mud around the centre spot. The driver got out of the car, in light blue, Peugeot Talbot overalls, and handed over the car keys to Mr Thomas.

'And now, ladies and gentlemen, a warm round of applause, please, for the lucky winner of this fantastic prize,' the announcer called out. 'Mr Jack Lloyd!'

As they waited for the Jack to make his way from the stands to the centre of the pitch, the announcer's voice continued to boom out.

'And on the pitch, in the centre circle, we can see today's prize, a new Talbot Solara. This beautiful, four-door saloon, a truly generous donation by the Peugeot Talbot Motor Company, has a modified floor pan, aft of the rear seats, to accommodate a huge boot and give it a whole half-inch more rear legroom than the hatchback!'

'Sporting the new, sleeker, front end and coming in 1.3 and 1.6-litre versions, this is a very competitively-priced car. But our lucky winner, today, gets his for free!'

'The new Solara is available in some appealing colours but Mr Lloyd's prize is the 1.3 litre version, in the blue of Nuneaton Borough's home colours. A wonderful prize for one of the Borough's longstanding, loyal supporters.'

Some Home supporters were already aware of Jack's win but for most there who knew him, it was a surprise to hear that he had won this highly desirable prize.

It was also a surprise to see that, as he walked out to take ownership of the car, waving to all corners of the ground, accepting the crowd's applause like a celebrity, he was accompanied by a little man in a mackintosh and flat cap.

'Who's Jack got wi' 'im?' One of the Canley pitmen in the stands said to his mate. 'D'ye reckon it's 'is minder?'

'More likely 'is fuckin' psychiatrist,' his mate replied.

Jack and his companion arrived in the centre of the pitch to be greeted by the Club's Chairman and Mr Thomas. The ceremony was only to be a few seconds, in order to get the afternoon's game underway without taxing the patience of the fans.

On behalf of the Peugeot Talbot Motor Company and with a firm handshake, the dashing Mr Thomas handed Jack the car's keys and ownership document, made out in Jack's name.

The Telegraph's photographer got them to hold their poses, indicating with his hand the way they should move, to make sure he got a nice shot of Mad Jack, beaming next to his prize with the Chairman, Mr Thomas and the little man in the raincoat, for Monday's paper, the reluctance of the last member of the group at being included in the tableau being particularly apparent.

As soon as the photographer indicated he had finished, Jack turned to the man in the mac', drew the ownership document out of its brown envelope with a flourish, pulled a pen from the inside pocket of his NCB donkey jacket and leant on the bonnet of the car to write on it. He tore a strip from the document and handed it to the little man, along with the keys to the car. The man handed Jack a thick bundle of cash, shook his hand to end the series of transactions then got into the car and drove off the pitch and out of the ground's main gate.

The Club Chairman and his guest followed Jack, in silence with furrowed brows, as he ambled back towards the stand, head down, counting his sheaf of notes.

The bemused fans of both sides, still trying to come to terms with what they had just seen, gave only subdued applause a couple of minutes later when the referee blew the whistle for the kick-off.

'That's not a true story is it, Jim?' Lol chuckled.

'Well, I wasn't there but I know men who were. And I've seen him do far worse than that!'

The Whitacre Heath management's and officials' session continued for another hour. Jim stayed until the end, when all of the crates of beer and the last whiskey bottle were emptied, then repaired to the Miners' Welfare with Joey, a few other officials and Ken Goodall.

When Gabriella came to join them in the Welfare, she only had a quick half of bitter with them before deciding it was time to extricate Jim. He walked out with her, into the dark of the evening and down the steps from the entrance, with exaggerated steadiness, satisfied that the colliery's leadership had drawn its working year to a close with due ceremony.

Christmas at Whitacre

Christmas Day, 25 December, 1983: 12:20

Jim had volunteered to take charge of the mine on Christmas Day and Boxing Day. He had worked every Christmas morning when he was an official, having found it a good way to generate a thirst for a pre-dinner pint and the appetite for a big, turkey dinner.

In his latest role as an undermanager, it had an additional advantage; it meant that in return for enabling Bill Metcalfe to spend Christmas morning at home with his wife and children, Jim would be rostered off on the first of January, leaving him clear to recover from partying on thirty-first of December into the early hours of the New Year.

The only work to be undertaken was the statutory, twenty-four-hourly inspections. The officials had gone down the pit earlier than usual and none of them had hung around on their inspections, so they were all out of the pit in good time.

When they arrived in the dirty office in their pit-black to dash off their statutory reports, Jim had the kettle boiled ready for those who wanted coffee and the tea-pot brewing for the rest. But, most of them were like Jim, focused on getting to the pub or the Miners' Welfare before returning home for their Christmas dinners. So there were no takers for coffee or tea and few bothered to snatch one of the mince pies Jim had brought in.

The Control Room was manned round the clock, even at Christmas, to maintain the watch on the environmental monitoring. When the officials had all gone to the pithead baths, Jim called in on Roger Dawkin, to let him know there were no problems reported by the officials and that he was leaving the mine.

'All done, Boss?'

'Yeah, all done. Everything quiet on the Emcors?'

'They're all level,' Roger said. 'Don't worry yourself, nothing's gonna catch fire today, it never does at Christmas.'

'Bloody 'ell, Roger, don't say that,' Jim said. 'We don't want to go tempting fate.'

'What time will you get away?'

'Oh, I'll be fine,' Roger said. 'Jerry's doing the afternoon-shift an' 'e said 'e'd come in an hour early so I could do the same as you. I'm gonna grab a quick pint at the Welfare, on the way home for Christmas lunch.'

'Righto, I'll see you tomorrow.' Jim said. 'A happy Christmas to you and the family.'

'And the same to you. Boss,' Roger called after Jim as he stepped out into the corridor.

"And let's hope it's a prosperous New Year for all of us,' Jim called back.

Twenty minutes later, Jim had called in at home to collect Gabriella and they were in the packed, little, smoke-filled bar of the Horse and Jockey at Bentley, roasting by the fierce heat of its open fire, ablaze with prime house-coal from Baddesley pit, less than half a mile down the road.

Four Whitacre men were standing, crammed into the corner, at the end of the bar nearer to the fire. Jim had expected to see them in there, having recognised Horace Wood's Reliant three-wheeler, parked outside the pub. Horace and Paul and Paul's mates, Brian Kettle and Colin Beighton had obviously all crushed into Paul's dad's little fibre-glass car to take the trip from Whitacre Heath to the Horse and Jockey.

Jim introduced Gabriella to the four men.

'So what brings you to the Horse and Jock, Horace? Are you on a Christmas Day pub crawl?'

'No, Boss,' Horace said quietly, with a gentle smile. 'Me and the missus take a little drive over here every so often. It's the finest pint of Bass in the area, as yer know. But me and Paul and the lads always call over for a quick couple of 'alves on Christmas mornin'. We thought we'd see you in 'ere.'

Jim was familiar with the pitmen's occasional practice of drinking rounds of half-pints, having done it, himself, with the men from his mother's side of the family, one Christmas over in South Derbyshire, when he was in his early teens. It appeared to indicate an interest in moderation but, whereas a pint tended to last them for a steadier six or seven pulls, a half-pint would be downed in a couple of healthy swigs, so it only served to accelerate the rate at which the beer was sunk.

As one o'clock approached, outside, the brightness of the sun, shining in the clear, blue sky, was already starting to fade.

Brian turned away, in the crush next to the bar, and bought a last round of halves. Without asking Jim and Gabriella he included them and handed them a half pint each without a word.

'Thank you, Brian,' Gabriella said.

'Thanks Brian,' Jim said, raising his glass. 'Happy Christmas, men!'

Kath, the fierce, old landlady called time, promptly, and started to harass the pitmen, 'Come on, drink up and get yourselves home.'

Jim and Gabriella chuckled at the lost causes of some of the regulars.

'Come on Kath, one last round.'

'No,' she barked. 'It's Christmas Day, you should all be at 'ome with yer families – and your wives cookin' yer dinners.'

As they left the pub, Jim savoured the lingering taste of hops, brought to life on his tongue by the chill of the fresh, winter air.

He and Gabriella called out festive good wishes to Horace and Paul and his mates and all the other pitmen leaving the pub, cheered by the beer, looking forward to their dinner and some time off together over the holiday; as always, anticipating the forthcoming year with optimism.

Bitter

Saturday, 7 January, 1984: 07:22

Only one in six of the National Coal Board's management actually worked at collieries, the other eighty-five percent being employed at the various area headquarters and a couple of research establishments.

Managers at the company's pits were required to work however many hours, at whatever time of the day and week, they were needed. For this, they were paid a Colliery Allowance of roughly two thousand pounds a year. So, for managing the pit at weekends and on holidays, like Christmas and New Year, responding to being called out to emergencies and working long days, at whatever hours were required, during the working week, they received a bonus of just under forty pounds per week.

When the NUM had introduced its overtime ban, two months earlier, the colliery officials' union, NACODS, had continued to work weekend shifts, undertaking the daily safety inspections mandated within the Mines & Quarries Act.

There was no point in having officials willing to undertake underground inspections if there were no banksmen and onsetters to let the men on and off the cages and no enginemen to operate the winders. And there was essential safety work and maintenance to be done to keep the pits open.

The Coal Board addressed these gaps by calling on its managers to undertake weekend overtime and do the essential, NUM jobs. No manager played their part with more diligence in this than Jim Greaves.

Jim had always been happy to graft in return for additional money, and weekend work during the NUM overtime band was being remunerated at a rate of thirty-seven pounds a shift. Seven or eight shifts a month made a big difference, when you were young and had extended yourself as far as possible with your mortgage.

That morning, Jim looked out from the warmth of his dirty office.at the freezing darkness outside, feeling the effects of the previous night's beers. His heart sank as he heard Mad Jack approaching, even before his rangy form bounced into the dirty office where Jim sat at the big table, opposite Les Parker.

"Morning Les,' Jack sang out. 'Come on, Mr Greaves. The sooner we get started, the sooner we're done.'

'I suppose you're right,' Jim muttered, making a show of lifting his mug to finish his tea, putting the job off for as long as possible.

Jim was wearing more than his usual pit-black this morning. On his legs he had put on a pair of old jeans over a pair of Gabriella's tights before pulling on his orange workwear trousers. On his upper body he was wearing a tee-shirt, then an old, long-sleeved shirt under his workwear shirt. He had pulled a heavy wool jumper over that before struggling his heavily padded arms into the sleeves of his orange workwear jacket. He had become the Michelin Man before he stepped into the shaftsman's harness.

His arms were so stiff with cladding that he could barely bend them and had to dip his head towards the mug in his hand. He drained it and placed it on the table.

'Alright, let's do it,' he said with a complete lack of enthusiasm as he stood up, threw the sturdy chain of his harness over his shoulder, put his helmet on and grabbed his leather gloves.

'Enjoy yourselves,' Les said with a cackle.

'Yeah,' Jim murmured. 'Thanks a lot.'

Both shafts were circular, brick-lined and fifteen feet across.

They inspected the warmer, upcast Number 2 Shaft first, from the top of both of its cages, before leaving the shelter of the pithead building to make their way up the six flights of steel stairs to the top of the Number 2 Headgear. As their ascent took them above the height of the big winding-engine house, it exposed them to a brisk, chilly breeze as they climbed.

They stepped off the last flight of stairs onto the top level of the headgear, two hundred feet above the ground.

Pausing to catch their breath, as it formed dense, white clouds in front of their faces, they looked out to the east. The dawn was bringing no warmth but it was lightening the sky, lifting Jim's spirits a little as it started to reveal the surrounding, rural landscape. It had no discernible effect on Jack who remained irrepressibly cheerful, whatever the circumstances.

Jim knew the job now, having partnered Jack on numerous occasions.

He opened the hinged cover to expose the bearing on his side of the first of the great, twenty-foot diameter, pithead wheels whilst, on the other side of the wheel, Jack did the same on the other bearing.

Jack called the banksman on his walkie-talkie to get him to signal to the manager operating the winding-engine to run the cages through the shaft while they watched the bearings' motions. When they had satisfied themselves that the bearings were in good condition, they flipped their covers closed and moved round to do the same on the other wheel.

With this done, Jack radioed the banksman and told him to run the cages through more steadily while he and Jim each inspected one of the two winding ropes to make sure there were no loose strands or other visible damage to them.

Jack radioed the banksman and told him to inform the onsetter at the pit bottom and the winding-engineman that the inspections of Number 2 Shaft and its ropes and headgear were complete and they should all move to take up their roles on Number 1 Shaft and its winding-engine.

Having climbed down, Jim and Jack strolled round to Number 1 Headgear and started the two hundred and sixty-four step climb to the top, with Jack maintaining a chirpy commentary.

By the time they had completed a satisfactory inspection of the winding wheel bearings and ropes, they were able to enjoy the view, from their towering platform, of the surrounding, North Warwickshire countryside, in the dim, grey light of the cold, winter morning.

'You know, when Terry Bond started as Colliery Mechanical Engineer at Canley, the Gaffer was settin' out 'is duties,' Jack said, leaning on the icy, steel rail, round the top of the headgear's high platform.

"E said one thing as 'e expected his Mechanical Engineer to do was personally inspect the bearin's on the 'eadgear. Terry told 'im 'e wouldn't be able to do that 'cause he suffered from vertigo. Old Jackie Charlton told 'im 'e didn't give a fuck, if 'e was gonna be Canley's Colliery Mechanical Engineer, it was Terry's duty and that was that.'

'That afternoon, Terry dropped a requisition note in with Sheila for the Gaffer to sign – for a pair of binoculars.'

Jim looked at Jack and chuckled. He cheered up for a few seconds until he remembered that, now, it was time for the part of the shift that he'd been dreading; inspecting and maintaining Number 1 Shaft. In the downcast shaft, the air would be howling down the confined opening at a gale-force, fifty cubic metres a second.

The force of the air as it poured down the shaft made it even more interesting when they came to step onto and off the cage over the two-foot gap and the half-mile drop beneath it.

When the banksman had rung off the cage and it had dropped below the lip of the shaft they were subjected to the air's full power. This morning, at eight degrees below zero, was the coldest Jim had experienced on this job. All his cladding provided no defence against the extreme chill factor of the blast that played on them.

Their exposure to the severe cold was prolonged by the cage's deliberate progress down the shaft as the winding-engineman lowered them as slowly as possible to enable them to inspect all of the brickwork that formed the shaft lining with due rigour.

Condensation formed on the shaft-lining as the air compressed during its descent of the shaft. Every thirty feet there were garlands, square, metal gutters which collected the water that ran down the shaft wall. The garlands were connected to each other by metal down-pipes.

This shift, luck was against them. One of the down-pipes had become disconnected from the garland it served. Jim hit the steel plate he was holding to signal to the onsetter to give a single ring to the winder to hold them in the shaft.

With the cage as their unsecured, narrow working platform moving gently under them in response to their efforts, Jim and Jack struggled to wrestle the pipe back into position and secure it. Despite the freezing cold, the water in the garland had not frozen at this level and a free-flowing flood cascaded over them as they worked. They were forced to return to the surface to fetch a replacement rubber gasket from the fitting shop and then climb back onto the cage and descend again, to get back under the freezing shower to reseal and secure the down-pipe.

Further down they had to use the sledge-hammers Jack had brought with them to remove the icicles that had formed on the lower garlands. Thick as trees at their tops and eight to ten feet long, they whistled like bombs falling as Jack and Jim knocked them off. Jim hoped the onsetter in the pit bottom was remembering to stand well back to avoid being struck by a lump from the icy missiles as they shattered as they crashed into the pit bottom.

When their cage landed in the pit bottom at the end of its creeping descent, Jim and Jack were wound back to the pit-top to swap cages and inspect the other side of the shaft on the second cage. As the winding-engineman ran the cages through the shaft, Jim shivered violently while they stood waiting, Jack's continuous, happy chatter doing nothing to distract him from his cold and sodden misery.

By the time they had completed the second slow descent of Number 1 Shaft, Jim was so cold that he could barely move. He was relieved when Jack called down to the onsetter.

'Right, rap 'im off for a fast ride.'

The cage accelerated out of the pit bottom with Jim holding on tight to one of the cage's big, supporting chains.

At the pit top, they unhitched their safety chains then stepped carefully off the second cage. The banksman signalled to the winder to land the cage properly and Jim and Jack got into the cage to go back down the shaft.

As soon as they had alighted in the pit bottom, the onsetter signalled for the cage to be raised so they could shovel all of the broken ice from the great, fallen icicles out of the shaft bottom and into a mine car in order to make it possible for the cages to land fully, in the bottom.

Their last job was to descend to inspect the sump under each shaft bottom. This was like an extension, down below the shaft, into which all of the many thousands of gallons of water pumped out of the pit each day were delivered.

Sump inspections were supposed to be done in pairs, for safety, to give any man who slipped and fell a chance of being saved from drowning in the freezing, deep waters of the sump but Jim agreed with Jack that they would save time by working unaccompanied and doing both sumps at the same time.

Jack disappeared off, to walk round, through the air-doors into Number 2 Pit Bottom to inspect that shaft's sump while Jim raised the battered, steel trapdoor to access Number 1 Shaft's sump. Forgetting to lash himself on with the chain of his shaftsman's harness, he disappeared into it, climbing down the slippery rods that formed the steps of the rusty step-ladder.

The intense darkness of the sump-hole and the blackness of the shaft's lining made the depths of the sump waters inky black. The eeriness of the place was heightened by Jim's knowledge that a pit bottom deputy had fallen into the sump and perished in it thirty years earlier. No one would ever know whether that had been an accident or a case of suicide.

Jim was relieved when he tested the sump's submersible pump and proved that it lowered the water level at the appropriate rate. He rushed back up the ladder and out of the opening and dropped the steel trapdoor with a loud clang, glad to be heading for the cage ride out of the mine.

On the way up the shaft then walking across to the dirty offices, Jack remained buoyant. Jim stomped along, silently, with him, his shoulders hunched up and locked with cold.

Back in the dirty office, Les Parker said, 'I've just made a pot of tea, if you want one.'

'I'll have one with yer, thanks Les,' Jack said.

'I won't bother, thanks Les,' Jim said. 'I'm going for a bath, I'm bloody frozen.'

All for thirty-seven quid.

Jim stood under a hot shower in the undermanagers' bathroom for three quarters of an hour, trying to get some heat back into the core of his body. When he was dried and dressed, he phoned Gabriella and told her he was driving straight to the Horse and Jockey at Bentley and suggested she drive up the four miles from their home to meet him there.

In the pub's cramped, old bar with its scrubbed, wooden tables and red-tiled floor, Jim stood as close as he could to the blazing coal-fire but still couldn't get the cold out of his bones. The other customers were mainly Baddesley and Birch Coppice pitmen but, although some of them had been out ferreting for rabbits, the overtime ban meant that none had worked a weekend shift that day.

As Jim picked up his third pint of draught Bass, the morning felt worthwhile at last.

'I was ready for this,' he said to Gabriella, holding up his glass and gazing at it appreciatively.

'I reckon I've earned it this morning.'

Flashpoint

As Britain entered the Twentieth Century, one in ten of its workforce worked down its coalmines. When the coal industry was Nationalised, in 1947, there were still three-quarters of a million miners employed down over eight hundred pits. By 1983, this had been reduced to two hundred and forty thousand miners at one hundred and ninety mines.

These had still been significant numbers and, with the expansionist "Plan for Coal" endorsed by all three political parties, for young men like Paul Wood and Jim Greaves, the future had seemed assured as they had started their careers in mining, respectively, in the 1960's and late '70's.

Pitmen of all ages and levels understood that pit closures were a natural feature of their occupation; mines exhausted their workable reserves and then they closed. But closures in the relatively efficient and productive mines of the South Midlands were less frequent.

When the threat of widespread closures was revealed, Jim and many in their Area's management heard the news and assumed that the axe would be falling elsewhere, probably on mines in areas with thinner seams, lower quality coal and lower productivity.

Jim's and Paul's careers had started in the last years of "coal at any price", a ridiculous, national slogan which served only to justify slack and inefficient production. Since then the bar had been raised progressively and radically.

The first change had been that pits had been targeted to produce an individually-specified annual volume at, or below, a specified price per tonne. Soon after, this had developed into a demand for pits to break-even, something no other European coal industry was even attempting, all of them being subsidised.

By 1984, British collieries were expected to return a profit or face closure.

Mining engineers are not easily daunted and, throughout the industry, they had knuckled down to beat their targets and keep their mines alive.

Jim had always seen the importance of developing an economic coal industry for Britain and been keen to play his part in driving up the production and efficiency of the pits at which he had been employed. Like many managers in his Coal Board area, it seemed to him that, as long as they kept clearing the thresholds they were set, then with determination, skill and innovation they would be able to keep their South Midlands pits open.

Even when the Conservative Government appointed Ian MacGregor as Chairman of the NCB, the man who had laid about with his axe so heavily at British Steel, colliery managers still had faith in their country's need for coal and their own ability to raise performance to keep sufficient pits open for a thriving industry.

The miners had gone on strike over their pay claim in 1972, with some justification and public sympathy because their comparatively tough and dangerous jobs had become so poorly remunerated. On that occasion they had won.

When they went on strike for another significant pay rise in 1974, the rest of the country was more inclined to believe that the miners were simply exploiting the power revealed by their strike, two years earlier. This time, Ted Heath, the lethargic, Conservative Prime Minister, had gone to the country, to seek a more convincing mandate with which to take on the miners, posing this question to the electorate:

"Who runs this country, the miners or the Government?"

Sensing his lack of confidence and resolve, the electorate decided that the answer was that it was clearly not Heath and his Tories and so Harold Wilson had been able to form a Labour Government.

Rather than accepting that their party had been evicted from office due to their leader's weakness in calling an unnecessary General Election, many Tories had blamed the miners for bringing down their Government.

Six years later, in 1980, the Tories were back in power, having defeated striking steelworkers. The next year, Margaret Thatcher's Conservative Government had instructed the Coal Board to start implementing pit closures. The announcement of these, without consultation, had caused uproar in every coalfield, including even the famously moderate Nottinghamshire Area. With Lewis Merthyr in South Wales as one of the pits on the list, South Wales Area's miners had walked out and sent flying pickets to other coalfields.

Within a day, a national miners' strike had been threatened. Recognising that this was a fight they could not win, the Prime Minister and her cabinet had backed down. It had appeared to many that the Conservatives had found their limit and would step back from taking on King Coal.

The Conservatives' introduction of legislation forcing unions to hold strike ballots and outlawing secondary picketing was popular amongst an electorate that had seen over-powerful, militant and destructive trades unions bring the country to its knees in the 1960's and '70's. The Tories went on to introduce penalties for secondary action by trade unions, including the sequestration of union funds. The TUC had resolved to call a general strike if any union was actually threatened with this measure but its resolution had failed with the first test case.

The Stockport Messenger dispute of 1983, between Eddie Shah and the National Graphical Association printers' union, arose from the union's opposition to new, labour-saving technology at Shah's print-works, at Warrington. At the time, a number of major companies had supported Shah, actively, in this dispute, providing funding to enable him to make use of the new, industrial relations Law and take successful court action against the NGA for organising "secondary action". Shah's action had been successful and some believed that his victory had given the Tories the green light it needed to take on the miners.

The Government's planned conflict with the NUM had two advantages, as far as some Tories were concerned; it would be a major step towards breaking the unions' economic stranglehold at the same time as avenging their perceived grievances from 1974 and before.

The Conservatives were re-elected for a second term in 1983, after the country's victory in the Falklands Conflict. By 1984, with two NUM ballots for proposed action against pit closures having been defeated, the Tories believed the balance of power was moving in their favour.

Thursday, 1 March, 1984
The National Coal Board informed miners at Cortonwood Colliery in Yorkshire that their pit was going to close, even though their mine still had workable coal reserves. The workforce immediately went on strike.

The closure announcement did not seem particularly surprising to Jim Greaves. Between the miners' bloodless victory over Thatcher in 1981 and

March 1984, forty pits had been closed. Warwickshire's own Newdigate Pit had been closed in 1981, due to its relatively high extraction costs, even though it still had workable reserves.

For Jim, it was reassuring at the management meeting at Whitacre Heath, that afternoon, that the older heads amongst the management team agreed that it would be safe to assume that any closures arising from the NCB's separate announcement, that day, of an intended reduction of four million tonnes in coal production would be for the same reason.

Friday, 2 March, 1984

The next day, the NCB ended this extraordinary week by announcing the closure of another five pits; two in Yorkshire, one in Durham, one in Kent and one in Scotland.

Everyone in the industry was stunned; six pit closures, in four different areas, announced on two consecutive days without any warning or consultation.

Monday, 5 March, 1984

With a mandate to take industrial action from a ballot, Yorkshire's NUM announced that its members would strike from the ninth of March to fight the sudden bout of closures.

Tuesday, 6 March, 1984

The NCB chairman, Ian McGregor, now played his hand. He declared the 1970's agreement on the management of closures obsolete and, without any consultation or warning, announced the planned closure of another twenty pits with the loss of yet another twenty thousand jobs.

Jim and his colleagues in Whitacre Heath's management team could not believe the lack of subtlety in this approach. Meanwhile, nationally, the NUM saw it as an act of the most deliberate and blatant provocation.

Following the lead of Yorkshire's NUM, the Scottish Area NUM Council called on its members to join the strike on the ninth of March. By then, Scotsmen at Polmaise had already been on strike for three weeks against the Board's sudden announcement of its intention to close their pit.

Monday, 12 March, 1984: 06:51

Shuey McFadden flew into the Union office.

He was furious with himself; if he'd arrived earlier he could have beaten Keith Deeming to the chair behind the big desk.

He threw himself down onto one of the metal-framed chairs round the wall of the office then jumped back up, grabbed the chair, hoisted it across the room and slammed it down in front of the desk with a clatter.

Initially, Mr Everitt had refused to allow Shuey to take on the full-time duties of the Lodge Secretary in Dick Flavel's absence. However, when it became clear that Dick's sick leave was going to be protracted, if not final, he had felt obliged to relent and allow McFadden to start spending all of his time on the bank.

'I can't work out what we're daein',' Shuey said. 'What are we doin' here?'

''Ow d'ye mean?' Keith said, with a bemused frown and slight shake of his head.

'Monday mornin', more than half the country's miners start the week on strike, fightin' back against an oppressive regime that's hell-bent on crushin' the unions and wreckin' our industry and, at this pit, we start another week as if everythin's hunky-fucking-dory.'

'You know as well as I do, it's 'cause of what was agreed at the National Exec' Committee's meeting on Thursday,' Keith said. 'They decided that the areas would make up their own minds – and that's what we're doing. The Midlands, Nottinghamshire and Wales haven't come out yet.'

'It comes to something when we're as weak as those self-seekin' Nott's bastards. They're not expectin' any cuts so they don't strike.'

'Well, there might be something to be said for being like them,' Keith said. 'They're like the rest of us in the Midlands. They've got a reputation for being moderate, less militant. Perhaps that counts in your favour when they're considering closures.'

'Fuck me, you sound as bad as those fuckin' Taffies. They were soon sendin' their flying pickets out when that Merthyr pit was threatened with closure a few years back.'

'Arnie says their Exec' member was saying that they didn't want to be the first to come out, after Cortonwood was announced. And he said they're still moaning about Yorkshire not supportin' 'em and comin' out when Lewis Merthyr was threatened.' Keith said. 'Someone on the South Wales area executive suggested holding mass meetings to try and get a firm mandate from miners not to cross the picket lines, seein' as 'ow their members were respectin' them. Apparently, the Regional President, and their General Secretary said it was too risky.'

'So what?' Shuey said. 'I don't know what you're on about.'

'They were worried about which way the vote would go,' Keith said. 'Don't you see? Even a militant area like South Wales isn't confident that the men'd vote for a strike or even to keep on not crossing picket lines.'

'Heathfield was sayin' recently that he didn't think younger miners would strike over pit closures.' Keith said, referring to the union's National General Secretary. 'Too many of 'em 'ave mortgages, now.'

'I don't know what they're bothered about,' Shuey said. 'Half of the country's out – solid. The other half, we have to picket out – and picket hard, if needs be.'

Keith thought for a while, remembering what Dick had said after their confrontation with Mr Greaves, shortly after the young undermanager's arrival at Whitacre Heath.

'We have to be careful that we're not falling into a trap here. The strategy was to keep the overtime ban on and hold out until next winter when the coal stocks'll be down before calling a strike. If we let them push us into a full-scale strike, now, we'll all be out for all the months of summer weather when the demand for coal's lower.'

'So what do we do? Let 'em walk all over us?' Shuey shouted. 'They've declared war. We can't sit by while they just do what they want. The overtime ban's been in place for six months. Coal production's been down by a third and coal stocks are down, according to the National Exec'.'

'That's as may be,' Keith said. 'But I don't know where they've bin gettin' their figures from. You 'eard yourself, in the Colliery Consultative Committee, two weeks ago, Whitacre's had no reduction in output. The only thing bannin' overtime's done is save the pit a shit-load of money on overtime. Hope was sayin' as Everitt's thinking of never reinstatin' it, except for emergencies, 'cause all bannin' it's done is improve the pit's profitability.'

'Anyway, what I'm saying is we need to be careful. I went round to see Dick yesterday.' Keith said. 'He was saying we're lettin' the wrong messages get out.'

Shuey glowered at the mention of the designated holder of the Lodge Secretary's role.

'That's 'cause the press and the media are part of the right-wing establishment and they're pushing out the Tories' propaganda,' Shuey snarled.

'No it's not! It's 'cause we're us own worst enemies at times.' Keith snapped. 'Four years ago, when Scargill got that leaked copy of the Coal

Board's secret plan targeting ninety-five pits for closure, as plausible as he can sound at times, few outside the industry believed it or had any sympathy. And why? 'Cause 'e's always bangin' on about bringin' down the bloody Government, so everyone thought it was all to do with that, as usual.'

'In the news, the Government and Coal Board are going on about how we're calling a strike without a national ballot. We need to get the message across that Thatcher's decision to close pits on an area basis means that last week's Exec' Meeting was right to authorise strike action on an area by area basis, in line with Rule 41 of the Constitution.' Keith said. 'We need to get the message across that it's all been done in accordance in the Union's democratic principles.'

'Fuck democratic principles,' Shuey screeched. 'The fuckin' Government and Coal Board have chucked them in the bin. This is a fight for our industry and the defence of the working class.'

Keith thought for a while.

'They're ready for it and we're not. It's a fight that the Tories have been preparin' for fer years.' He said. 'We've kept tryin' to negotiate throughout the ban and the Tories 'aven't let the Coal Board meet us. Ten years on, they still harbour a grudge for that idiot Heath gettin' hisself thrown out in '74.'

'For some, it goes back even further; they say when they announced the twenty closures, last week, one o' the Cabinet said, "This'll teach the bastards for 1926". They hate us!'

'They backed off over Lewis Merthyr in 1980 but they've been preparin' 'emselves for war for the last four years.' Keith said. 'Have you heard of the Ridley Plan?'

'No,' Shuey said, with a pout.

'It's the Tories blueprint for taking us on. It was prepared in the late '70's by Nicholas Ridley, one of Thatcher's right-hand men.' Keith said. 'It came up with all the Laws they've brought in to curb union power, increasing police powers, gettin' away from relyin' on coal and stock-pilin' it, ready for a long strike. They've done all that and more.'

'Scargill's mole at the NCB HQ at Hobart House says that the big pay rises they've given to the coppers and power generation workers is all to get 'em on-side, ready for a bust-up. And they've got the supply-lines open, importin' coal, and got the NCB to set up contracts with loads of little haulage contractors, every man with an old tipper truck, so they don't have to rely on rail if the train drivers come out and support us by refusin' to move coal.'

'There's nae surprise in any of that,' Shuey sneered. 'We know we've got a fight on our hands and that those bastards are provokin' us. But that's why we have to picket hard and get a hundred percent of the membership oot. They're just usin' the Law as a weapon against us, so we'll have to disregard it, picket the ports and the steel works and the power stations. We'll have to go for a general strike if that's what it takes. It's either that or let them decimate the whole industry.'

'All I'm sayin' is that there may be no right answer for us but we can't go alienating the whole country,' Keith said. 'If we do, we'll lose the war.'

'We'll lose the fucker if we capitulate without any fight,' Shuey bellowed, his seat flying back as he sprang from it and stormed out of the office.

Monday, 12 March, 1984. 15:55

Whitacre Heath's daily management meeting had been brought forward by an hour. Jim arrived in Mr. Everitt's office a few minutes early to make sure he didn't miss what they were assembling to hear; the BBC broadcast of the verdict of the NUM's National Executive on whether to call a national strike of the NCB's miners.

Angela, had brought in a large, transistor radio and placed it on a small table by the wall, next to an electric socket. After a little difficulty, she managed to tune into Radio 1. "Street Dance" by Break Machine crackled out of its speaker.

'Bloody hell,' Phil Grant, the Electrical Engineer, said, grimacing in response to the whistling, synthesizer sounds and the fast rattle of drum machine. 'It sounds like the panzer's runnin' out of race.'

The rest of the management team chuckled, despite the gravity of the occasion.

As the song ended, Steve Wright's voice came from the radio, *'You're listening to Radio 1 and it's time to go over to the Newsroom for the Four O'clock News.'*

'The National Union of Mineworkers has called for a national strike.'

'That's it, we're at war,' Harry Everitt said, grimly. 'The stupid bastards.'

After the other headlines, the newsreader said, *'Following the six-month overtime ban, Arthur Scargill, the President of the National Union of Mineworkers, has declared an unofficial national strike in response to planned pit closures. The NUM has said it is determined to remain on strike until the plan to close the mines is withdrawn.'*

As soon as the news item was completed, Mr. Everitt murmured sadly. 'Switch it off would you, Mr. Greaves.'

As Jim stood up from the table and went over to the radio, he was conscious of the historic moment of what they had just heard.

'Well, if it goes on for long, it'll give you plenty of time to work on your handicap Gaffer.' Phil Grant said.

'You may be right, Phil,' the Gaffer said, with a mirthless smile. 'But we could still've done wi'out it.'

'Well that's that,' Horace said. 'Now we'll 'ave to see what our Area does.'

Paul had dropped in at his parent's house for a cup of tea so he and his dad could hear the announcement together.

'Well, hopefully, Nottinghamshire and us'll throw us lot in with the rest of the country,' Paul said.

'I'd bin 'oping they'd find a way out,' Horace said.

'There's no way out with the Tories and this MacGregor character,' Paul growled. 'We've got to stand and fight or 'e'll do the same to us as 'e did to our steel industry.'

'I 'ad enough in the last two strikes, ten years ago.' Horace murmured. 'I never thought we'd see them days again in my time. I thought we'd all learnt us lesson and there'd be no more trouble like this.'

'Well, it's not our doin', Dad.'

Pat was standing round the corner from them, hidden in the doorway. She stepped back into the kitchen, looked out on the daffodils in their back-garden, bobbing in the breeze in the dull light of the early spring afternoon and shook her head, lips pursed.

Wednesday, 14 March, 1984

Two days after the NUM's strike call, the NCB admitted that one hundred and thirty-two out of one hundred and seventy-four pits were out. The working pits were in the NCB's Nottinghamshire, North Derbyshire and South Midlands Areas.

Within two days, the NCB would announce that it had been reduced to only eleven collieries working normally.

The Miners' Strike was on.

PART TWO

Strike

Strike Ballot

Sunday, 18 March, 1984: 10:24

The leadership of the Midlands Region of the National Union of Mineworkers followed Nottinghamshire's example, leaving their lodges to decide, pit by pit, whether to work or strike.

In Warwickshire, the Canley and Baddesley Lodges had met and voted the day before, whilst Birch Coppice's Lodge had convened earlier that morning.

As Colin and Brian followed Paul into the ballroom, at the back of the Miners' Welfare, the three friends were already aware of the results from the Saturday ballots and immediately heard other pitmen discussing the outcome of that morning's meeting at Birch Coppice. Keith Deeming had phoned the Birch Coppice Miners' Welfare and obtained the news of the vote from that colliery's Lodge Secretary.

The big ballroom was packed with many more men than had attended the Committee member elections, six months earlier. The miners sat in the rows of chairs and on tables around the edge of the room or leant on the walls and against the locked and shuttered bar at the back of the room.

Keith called the meeting to order.

'Brothers, the Lodge Committee intends to follow the example of the other pits of the Warwickshire Coalfield and allow you, the members, to propose and discuss options and then vote upon them,' Keith shouted.

He started by appraising them of the results of the votes at three of Warwickshire's pits.

Baddesley pit's men had taken the view that it was an unofficial strike; they would work but continue to observe the official overtime ban.

Canley and Birch Coppice had also voted to work but with a proviso, whilst their lodges would not mount picket lines themselves, their members had voted to not cross any line formed by miners from elsewhere.

With those announcements complete, Keith called out.

'Without further ado, I'd like to invite proposals from the floor for the approach that this Lodge should take to the strike.'

Shuey McFadden leapt to his feet to speak from his place at the Committee's table.

The Deputy Lodge Secretary thrust his upper body forward. With his hands placed wide apart, flat on the table, his elbows sticking up, shoulders hunched and eyes bulging, he looked like some great, grotesque spider.

Instead of making a proposal, Shuey launched into reading a speech he had scrawled on several, scruffy scraps of paper.

'Comrades, I think it's important that we look at this matter in context,' he bellowed.

Around the room, men shifted in their seats or changed stance, some folding their arms and looking askance towards the top table.

'All around Britain, miners are takin' action to mount a defence against a Right-wing, capitalist aggressor who's determined to curb the unions' power and bring the working class to heal. The Tories ha'e bin plannin' this for years – it was one of their main aims when they came to power.'

'How many of you know about the Ridley Plan?'

Keith was amazed at his cheek; a few days earlier Shuey hadn't even heard of the document, now he was presenting himself as the expert on it. He parroted Keith's explanation of the Government's plan and its origins and his description of its contents from a few days earlier; the proposal of moving away from coal to reduce the NUM's power, the increased police powers, the new union laws and the stock-piling of coal. He even wrapped in all of the other Government measures that Keith had told him about, like the big pay rises for the police and power generation workers and the arrangements to ship in coal, saying that these were also included in the Plan.

The rest of Shuey's speech replicated everything that had come into his mind as he had ranted at Keith in the Union office. He spoke of the provocation and the Government's use of the Law as a weapon and the need for them to disregard the Law against secondary picketing in order to picket the ports, steel works and power stations. He said that they should take heart from the success of the overtime ban which had forced the Government and "their crony MacGregor" to take action at that time.

After fifteen minutes, Shuey drew his rant to a close.

'Are we gonna stand with our comrades in Yorkshire, Scotland and Wales or are we gonna sit by and let the Tory bastards decimate our industry?'

Shuey sat down, looking around the room, expectantly. If he had been hoping for any applause for his speech, he was to be disappointed.

'Thank you, Shuey,' Keith said, careful to avoid using his title of Deputy Secretary. 'I need to remind you that I called for proposals for the stance we take with respect to the unofficial National Strike. Do you have a proposal for the meeting?'

Shuey jumped back up, his eyes wide and staring.

'Aye, I dae,' he yelled, white flecks of spittle flying from his lips. 'Comrades, I propose that Whitacre Heath Lodge of the National Union of Mineworkers show solidarity with our comrades around the country, that we strike, one hundred per cent, nae crossing of picket lines an' all that.'

'Thank you Shuey,' Keith said. 'Brothers, the proposal is to strike. Do we have a seconder for that proposal?'

One of Shuey's mates, remaining slouched in his chair in the middle of the room, gave a half-wave of his hand.

'Right, thank you,' Keith said, noting the seconder's name.

The second proposal came from Toby Sweet, an aging and respected face-worker.

'I proposed that we follow the example of Birch Coppice and Canley – we don't strike but we don't scab neither by crossing any picket line made up of miners from other pits.'

A number of pitmen raised their hands to second this moderate proposal.

A hand went up in the middle of the room. Jasper Ford, a header involved in driving the main roads to open up new reserves in the pit, always looked angry and aggressive. He always spoke in a remarkably forthright manner, even by the standards of blunt, Warwickshire pitmen.

Keith pointed in his direction, nodded and called out to him.

'Yes, Jasper.'

'That's just a fudge. It's like we don't know us own minds. It's all very well for Yorkshiremen and others to go on strike like McFadden there says but if 'e came from Warwickshire and knew a bit more about pit-work he'd know that within a few weeks o' strikin' we could end up with every face down our pit on fire. We run the risk every pit holiday fortnight, when we don't turn coal off them faces. If we strike and lose all us faces we won't have a pit and we won't have any jobs to go on strike for neither.'

Some men muttered in agreement. Colin and Brian could not hear what Paul was saying but it was obvious from his chuntering that Jasper's stance had infuriated him.

Shuey fumed and muttered to himself but chose not to respond.

'I say, work.' Jasper called out.

'Do we have a seconder for that?' Keith shouted.

'Aye!' A voice rang out from near the exit at the back of the room. Its owner's hand shot up and everyone looked round.

There were no more proposals after these three most obvious options so Keith opened up the meeting to debate the propositions. All three proposals received mixtures of support and criticism.

It was clear that Jasper's point about the particular risk to pits in their area being allowed to stand idle had proved persuasive but there was a real reluctance to have Whitacre viewed as a scab pit by adopting a policy which ignored the call to strike and the plight of those miners whose pits had been targeted for closure.

Like those in the Birch Coppice and Canley Lodges, the general view amongst Whitacre Heath's members was that the strike would be conducted on an area basis and that the more militant areas, like Wales and Yorkshire, would be busy fighting their own battles in their own coalfields, making it unlikely that Whitacre Heath would be troubled by secondary pickets from another area.

Despite his passionate belief that they should be part of this struggle to save the nation's pits and coal industry, Paul held back, confident that his perspective would prevail.

Jim Greaves was sitting at the table in his dirty office, still wearing his pit-black and shaftsman's harness from his weekend shift as he finished reading and signing the deputies' reports for Sunday dayshift. He was keen to get finished, changed, bathed and away. He had to be back at the pit on nights because Mr. Everitt, with the benefit of all his experience, had decided there needed to be someone on site to act for the General Manager, in these early days of the strike.

Jim wanted to call in at the Dog at Dog Lane, have a couple of pints of Bass, get home for Sunday lunch and then be able to relax and have a couple of hours sleep before returning to the pit.

Les Parker, sitting on the other side of the table picked up the handset of the antiquated, black, Bakelite phone and dialed the Control Room number.

'Hello Roger, Les 'ere,' he said. 'Any news from the Miners' Welfare.'

After a pause to listen, Les said, 'Oh.'

Jim stopped writing and looked up trying to read any reaction in Les's hard-bitten features as Roger, on the other end of the phone, explained the ballot's outcome and commented upon it.

'Yes,' Les said. 'Well, we shall see. Thanks Roger.'

Les placed the phone handset back in its cradle and looked thoughtful. Jim looked over at him and raised his eyebrows.

'They've voted to work but not cross picket lines.'

'What the hell does that mean?' Jim snapped. 'They'll come to work, find a picket line and be on strike.'

'Not necessarily, the Lodge has decided that it won't put out a picket line.' Les said. 'In the last strikes, in '72 and '74, the pickets from Yorkshire and Wales never found us. I don't think they even knew we existed, stuck out here, in the countryside. They could find Birch 'cause it's right on the A5 and Baddesley's not too far off the Watling, so they hit them but they never got to us. Might be the same this time.'

'I hope you're right.' Jim said. 'We've got to fight to keep this pit open, not give the Board an excuse to shut it.'

Keith had called for a show of hands from those in favour of each of the proposals. A full commitment to work through the strike had not proved popular, the proposal to strike had been supported by almost forty per cent of those present. As soon as the hands had gone up in response to the call for those in favour of the option of working but not crossing picket lines it had been apparent that this had carried the vote and a relieved murmuring had set up around the room.

Paul shook his head, his disgust evident to Colin and Brian who had been watching for his reaction. He had been disappointed but not surprised when he had looked over at his father and seen Horace raise his hand for the option of working but not crossing pickets.

When Keith confirmed the result of the ballot and thanked them for their attendance, Shuey sprung up and called out to Keith, who was right next to him, so that everyone in the room could hear him.

'Mr. President, can I just say…'

'No, I'm afraid not, Shuey. You had your say earlier. The members have considered the options and the Lodge has made its democratic decision and that's the end of the proceedings.'

The noise in the ballroom swelled as those seated struggled to their feet, started stretching themselves before moving towards the crush of men making their way out of the doors at either side of the ballroom's shuttered bar. Some were ready for a drink in the Welfare's bar, others were heading for the pub or straight home for lunch but no one had any time to stay and listen to any more of Shuey's pontifications on the dispute and their involvement.

By the end of the day, Keith had heard on the news that the Nottinghamshire delegates' conference had voted to return to work. The next day, Arnie told him on the phone that the ballot of their Area's Leicestershire miners had revealed that over ninety per cent of them were against striking.

The National Union of Mineworkers was in disarray.

From the Court of
King Arthur

Monday morning, 19 March, 1984: 01:17

Most months the NCB's newspaper, *Coal News,* consisted of simple articles on mining innovations, tables of production and safety statistics, inter-Area and inter-colliery sports results and human interest stories about employees and their families. However, the March edition always attracted more interest and scrutiny from pitmen because it featured all of the contestants who had progressed through to the final of the national Coal Queen Competition.

Although underground work was the preserve of male employees, many thousands of women worked for the National Coal Board and this contest proved there was no shortage of beauties amongst their number.

Several rows of individual, colour photographs in the centre-pages of this month's paper revealed the shapely winners of all of the area heats, wearing one-piece swimsuits, high-heeled shoes and come-hither smiles.

Jim had heard that this year's South Midlands Area Coal Queen was the assistant to the General Manager's Secretary at his previous pit. He had seized the copy of the latest edition of the paper off the table in the Control Room, sat down and whipped it open at the centre-pages, scanning the photos until he found her.

"Barbie Whittaker, Assistant Secretary, Canley Colliery, 28yrs, 5' 7", 38-24-36."

The familiar, suggestive smile was instantly recognizable. His eyes ranged over her generous curves, her shapely legs, lengthened by the high cut of her yellow swimsuit and towering heels.

He gawked at the image, and thought to himself, *"Phwoar!"*

Mad Jack had called over, dressed in greasy overalls, helmet and boots with the yard-long monkey wrench he had been using on some job dangling from his right-hand. He stood in the Control Room door, talking to Roger Dawkin, having declined the Control Room attendant's offer of a mug of coffee. When he stuck his head round the door, he saw the subject of Jim's attention.

'You wanna leave that stuff alone, Jimmy lad. It'll make yer go blind.'

'Uh-oh,' Roger said, rising from his chair to look over the Control Room console and out of the window at the big, tarmac square, in the dark, outside. 'I don't like the look of this.'

Jack followed his gaze out of the Control Room window.

'Yer right,' he muttered. 'This looks like trouble.'

Jim put his mug down, closed the paper and rose from his chair.

The carpark had been empty, except for Jim's car and the couple of others that were parked in a row by the pithead baths and dirty office block. A red, Commer mini-bus had just turned up, in the middle of the open space, its side emblazoned with the name "Excelsior Van-Hire" and the telephone number "Barnsley 78212".

As its front, passenger door swung open, its side-door slid back. A massive man with a drooping moustache stepped down from the front, allowing a shorter, wiry man, sitting in the middle, to slide across the bench seat and drop down from the vehicle. A gang of men piled out of the side door from the seats in the back of the vehicle and started stretching to ease away the stiffness from their long trip down to Warwickshire. A Barnsley contingent of NUM pickets had just confounded everyone's expectations and found Whitacre Heath.

'That's bloody strange,' Roger said.

'What is?' Mad Jack said.

'Well, they're wearin' their pit-black.'

The same thing had occurred to Jim. It was strange to see striking miners dressed in their workwear. They had obviously had the foresight to bag up their one-piece overalls, leather belts and steel toe-capped pit boots in their laundry nets and take them home from their colliery's pithead baths. It might look strange but it was like combat kit, useful attire if you were expecting a spot of bother or a pitched battle. These men had an air of professional intent; they clearly meant business.

'Well, I'd better go and tell them they're trespassing,' Jim said as he walked past Jack to the door.

Jack shook his head.

'Well, if you're going out there, I'm comin' with yer,' he said, turning to follow him. From his tone, it was clear that he thought it was Jim who was mad, on this occasion.

'Be careful, Boss!' Roger called after them. 'There's no sayin' what they're intendin'.'

They walked, in silence, up the unlit corridor, out of the door and round the end of the building to emerge onto the car park.

The pickets swung round and stared at them. At eleven to two, the odds were hopeless but Jim was still glad to have Jack, as bodyguard, at his shoulder. Although he did worry that it might prove provocative; having Mad Jack still toting his hefty, steel wrench.

The smaller man who had been in the front of the mini-bus strode, confidently, towards them. Jim walked right up to meet him.

Jack stopped, four paces back but ready. The giant picket, who had shambled over behind his leader, stopped, taking up a position mirroring that of Jack's.

The man facing Jim would in his mid-thirties; younger than most of those in his gang. He looked sharp-eyed and intelligent.

'Good morning,' Jim said. 'My name is Greaves. I'm Undermanager at Whitacre Heath Pit, acting for the Colliery General Manager. This is National Coal Board property and it's my duty to tell you that you're trespassing and instruct you to leave.'

'That's alright, Mr. Greaves, you've got yer job to do. You just tell us where yer want us.'

'It's only a couple of hundred yards back up the pit lane to the edge of the pit property but I'll jump in my car and show you.'

'Alright, Boss, whatever yer think's best. Righto lads, do what the undermanager says.' The ringleader called out to his men. 'Everyone back in the wagon.'

'Follow the undermanager's car, Archie,' he called through the van to the man climbing back into the driver's seat.

'Do you want me to come with you?' Jack said quietly.

'Nah,' Jim murmured. 'I should be alright with them.'

Still alert to the possibility of attack, Jack waited and watched as Jim strode over to his car.

Jim unlocked his old, Ford Escort, slid into the driving seat, stuck his key into the ignition and started it up. Switching on the car's lights, he reversed out of his parking space, turned the car round then sped over to the middle of the car park.

When all of the pickets had hauled themselves back into the cramped mini-bus, its lights came on and the driver swung it round to pull up behind Jim's Escort and gave him a flash of headlights.

Jim led them out of the car park for the few seconds' drive to the pit gates. On the right they passed the main entrance to the office building, then the main car park for the pit's workforce. On the left was a terrace of black brick houses that had been occupied by colliery officials until their sale, a couple of decades earlier and, just past the pit gates, was an ash-covered parking bay, fifty yards long and wide enough to take three articulated lorries, side by side.

Jim turned his car round on the road onto which the pit lane opened out. He jumped out of his car and walked round the front of the mini-bus. The ringleader clambered out of the vehicle after his big, right-hand man.

'Right, this is off the pit premises so you're won't be trespassing, for however long you're here before the police arrive,' Jim said.

'Okay Boss.'

All of the men had got out of the mini-bus, except the driver who was ready to drive the vehicle away and park it up somewhere less obvious.

'Now, I'm going back to the pit offices to phone the police to let them know that it looks like we'll have illegal, secondary picketing before the start of dayshift.'

'Okay Boss, we know you 'ave to do that.'

'Right,' Jim said. 'Goodnight lads.'

Jim walked back to his car, thinking that it seemed likely that experience gained at other pits in the last fortnight had given this group of pickets a better understanding than Jim of the terms of engagement and the way this conflict would be conducted.

'Excuse me, Boss,' the ringleader said, walking over to Jim's car.

Jim stopped, turned and waited for him.

'I was just wonderin', is there any chance of a bit o' scrap timber? It's a cold neet and we've come a long way.'

The man's cheek amazed Jim; they were Yorkshiremen, on the other side of the dispute from him, threatening his pit. But then he thought about it. They were still pitmen and, like all pitmen, this man obviously understood the bond that existed between them as well as Jim.

'I tell you what,' Jim said, 'I'll ask the big engineer who was just with me to see if he's got an old oil drum and ask him to drop you some timber with some waste oil to get it started.'

'Thanks, Boss. That'd be champion.'

"But I'm still expecting you to be moved on before you have the chance to get much benefit from it."

The Barnsley-man gave him a nod.

Twenty minutes later, back in the Control Room, having completed his call to the police, Jim heard the roar of an enormous diesel engine. He looked out to see a huge digger speed round from behind the fitting shop, across edge of the car park and off towards the pickets' position. Jack had quickly quartered some old, rotting sleepers using the colliery carpenter's circular saw. He delivered these with several gallons of waste oil and a forty-five-gallon oil drum, with its top cut out, which he had punctured in various places with a hammer and spike to create air holes.

A few minutes later, Jim walked across the corridor into Ken Goodall's clean office and looked out of its window. Flames blazed from the waste oil the pickets had poured over the first of the wood, waving five or six feet above their heads. It was like looking across at the enemy's campfire the night before some historic battle.

Jim had only been back in the Control Room for five minutes when the leader of the pickets appeared at the Control Room window. The big man was with him again but this time he had a hand over one eye.

Roger walked round the big console and slid back the clear, plastic sheet in the bottom of the window-frame, creating an opening through which they could converse.

'Sorry to bother yer, I know this is a bit of a strange request,' the smaller man said, 'but Sammy's got an ember in 'is eye. I know you might not be able to help but we'd be really grateful if you could let us borrow an eye-wash bottle to wash 'is eye out.'

Roger looked round at Jim.

'Boss?'

Jim thought about it. This was alien territory for all of them, the opening hours of a new type of conflict. This dangerous-looking, little Yorkshire militant and his big mate might be engaged in some ruse; if Jim let them into the Control Room and they sabotaged something or ran amok with their comrades he would look worse than a fool.

'Let 'em in Roger and I'll irrigate his eye for him,' Jim said.

If the Yorkshiremen did abuse his trust they would dishonour themselves. He hoped he was right in believing that no true pitman would do that.

Roger told the two pickets to make their way to the top end of the building and enter through the unlocked door, then he went off to the first aid centre in the pithead baths block to fetch the necessary items.

Whilst Roger was out of the room, the pickets' leader looked at Jim.

'So,' he said, "ow many faces d'ye work 'ere?'

Jim frowned, instantly suspicious that he was probing for intelligence to inform their picketing tactics, before deciding this was unlikely.

'We've got three,' he said. 'Two advancing, one retreat.'

'It's all retreat at our pit, the only way to go.'

'And which pit are you from?' Jim said, expecting the man to refuse to divulge such incriminating information.

'We're all from Railsmoor, Barnsley Area.'

'Yeah?' Jim said. 'I've been there. I had a fortnight in your Area when I was on the Engineering Training Scheme. Most of your colliery general managers were a lot younger than the ones down here, in the South Midlands Area. Your General Manager joined us on a couple of nights out on the town in Barnsley, Chris Turnbull. Is he still there?'

'Yeah, 'e's still there. 'E's good old lad, ain't he Sammy.'

'Aye, 'e's alreet,' the big man said, turning his head to look at Jim with his good eye. 'Looks after t'lads, pays good bonus and makes sure yer get the right kit to turn coal.'

Roger returned from the First Aid Centre with a couple of eyewash bottles, a first aid box and some cotton wool and lint. Jim went to the gents' to wash his hands before treating the Yorkshireman's eye. When he returned, he got the big miner to sit down on a chair then told him to pull his eyelids open as far as possible.

He peered into the man's bloodshot, watery eye.

'Yep,' he said. 'You've done a good job there. There's a tiny bit stuck in your eyeball. Do you want me to try to pull it out with tweezers?'

'Yeah,' the big man said. 'If yer could, Boss,'

While Jim worked on him, the leader of the pickets kept talking to Jim and Roger, with enthusiasm, about the faces at Railsmoor pit and the face equipment they used. It made Jim envious when he told them that Railsmoor had Anderson Strathclyde shearers on its faces.

'You're lucky,' Jim said. 'We have to make do with BJD rubbish. "British Jeffrey Diamond", it's a disgrace having our country's name associated with that unreliable junk.'

'Bollocks to it,' Roger said, suddenly. 'If it's alright with you, Boss, I'm gonna make these two lads a cup o' tea.'

He stood poised, looking expectantly at Jim.

Jim stopped and looked up from his ministrations. He looked Roger in the eye then nodded.

'You really need a patch on your eye to prevent infection, until you can get to see a doctor,' Jim said to the big miner.

'Whatever yer think's best, Boss.'

'You'll look even more like a pirate than usual, with an eye patch, Sammy,' his leader said.

It amazed Jim that the pickets' leader was prepared to divulge even first names so casually whilst preparing to engage in illegal, secondary picketing.

As Jim finished dressing the big miner's eye, Roger returned with two mugs of tea for the pickets.

'So how'd you find us?' Roger said. 'We thought you'd have found Birch Coppice before us, it's right on the Watling, the A5.'

'Well, since you asked,' the shorter man said, 'it was your NUM Lodge Secretary. He phoned up Arthur, told him your lads had voted to work but wouldn't be crossin' pickets and asked 'im to send a picket line down. So Arthur collared us and told us to assemble me crew and get down 'ere.'

'It was actually Scargill, himself, who sent you?' Jim exclaimed.

'Yeah, it was. As it 'appened.'

'McFadden!' Roger muttered, shaking his head. 'The devious little bastard.'

'Yeah,' the Barnsley-man said. 'I suppose yer could say that. Actually, he'd promised to meet us down 'ere, when we arrived. You couldn't let us give 'im a ring and fetch 'im round 'ere could yer?'

'Go ahead,' Jim said. 'Get him an outside line, Roger. Get the bastard out of bed. Have you got his phone number?'

'Oh aye, Arthur gave it to me and told me to call 'im if 'e wasn't 'ere when we got 'ere.'

Roger dialed "9" for an outside line then handed the receiver to the leader of the pickets who had taken a scrap of paper bearing Shuey's number from his pocket. The Yorkshireman dialed the number then waited, the phone receiver pressed to his ear.

"E ain't answerin',' he said.

'Let it ring, that bastard'll take some waking, 'e'll have had a belly full o' beer.' Roger said. 'They reckon 'e's drinking the strike funds.'

The Yorkshireman nodded.

'Strike funds?' Jim said.

'It's pourin' in from Birmingham, especially from the Sikh community, for some reason,' Roger said. 'Apparently they've bin collecting in buckets and they've started deliverin' it by the bucket-load. They say that thievin' bastard's started drinkin' it with 'is cronies.'

'Nah, e's not answerin',' the Yorkshire man said, handing the phone back to Roger. 'Yeah, I 'eard that. We've started gettin' funds from the Indians and Paki's in Bradford, up our way.'

'Wait a few minutes but keep trying McFadden's number,' Roger said. 'Get the bastard up.'

On the third attempt, after letting the phone ring for several minutes, they all heard a gruff, sullen, "Ullo.' through the phone's handset.

Roger chortled.

'Ian Flint 'ere. From Barnsley. I've got your picket line at Whitacre Heath Pit. You were supposed to be 'ere to meet us. You need to get yourself round to the pit entrance.'

'Alright, I'll be there in a bit.'

The two Yorkshire men stayed in the Control Room, finishing their mugs of tea.

The one called Ian said, 'I think we'd better be going and see if your man's out there.'

'That bastard'll have gone back to bed, I bet,' Roger said. 'You won't be seein' him.'

After another twenty minutes, the door into the office building at the top of the corridor opened and they heard furtive footsteps. The Control Room door opened a little and Shuey's bloodshot, bulging eyes peered round it. He spotted the two Yorkshiremen then his face grew more malevolent as he locked eyes with Jim.

Ian Flint said, 'McFadden?'

'Aye, aye,' Shuey said with a deep scowl. 'Come on lads, let's go somewhere where we won't be overheard by these 'ere.'

Ian stood up.

'Thank you for sortin' Sam's eye out for him, Boss,' he said. 'And thank you both for the tea. It were very kind.'

He let his big friend lead the way.

Sam turned and said over his leader's head, 'Yeah, thanks for sorting me out, Boss. An' good luck.'

Sam walked out and Ian followed. As soon as the Barnsley men were out in the corridor, McFadden leant back in through the door.

'And you'll be having some trouble before all this lot's over,' he snarled, his shaking hand pointing at Jim.

Before Jim could respond with a "fuck off", Ian Flint had swung round, darted back through the door and grab McFadden's arm.

'Don't you talk to the Undermanager like that,' he barked, his face inches from Shuey's. 'Where we come from, we treat our gaffers wi' respect. An' he's a proper gaffer, so you mind yer mouth.'

Ian nodded at Jim and walked out. Shuey turned away and skulked after him, like a whipped dog.

Jim walked out after them, into the corridor.

'Hey! McFadden!' He called after him. 'Where d'ye think you're going?'

'T' the union office,' Shuey called back.

'Oh, no you're not,' Jim said. 'You're on strike. You've got no right being on colliery premises. And if you don't get off 'em I'll have you arrested for trespass.'

The two Yorkshiremen stopped and looked at McFadden.

'Fuckin'…' McFadden muttered.

'He's got you there, pal,' Ian jeered.

The two Barnsley men cackled at Shuey, as he stormed out of the building.

The First Picket Line

D awn revealed the full extent of McFadden's subterfuge.
Whitacre Heath's Acting NUM Secretary had instructed his
two henchmen, Kennedy and Wykes, to arrive early, encounter
the Barnsley men then comply with the Lodge's decision not to pass pickets
and join him on the line.

Jim observed the early developments, hidden from the crowd on the
pit lane by the gloom of the Undermanager's unlit, clean office.

A few of the early arrivals for dayshift were turned away to head back
home, hoping it would all be over in a day or two so they'd be able to get
back to working normally. But others bolstered the line and, in less than a
quarter of an hour, a mass-picket was mustering.

Roger's voice sang out from across the corridor.

'Boss, are you there?'

'Yes, Roger, I'm here,' Jim shouted back, easing round the desk then
opening the door to slip out of the office and across the corridor to join
Roger in the Control Room.

'I've got them on that police number for you.'

Jim walked through and took the phone receiver.

'Hello.'

'Hello, police,' a man's voice said. 'Identify yourself.'

'My name's Greaves, Undermanager at Whitacre Heath Colliery. Who
am I speaking to?'

'The police,' was the unenlightening response. 'What's your message?'

'I've been in charge overnight at Whitacre Heath. I spoke to someone
on this line nearly five hours ago to report that a group of flying pickets
had arrived here to picket out our dayshift. I asked for a police presence
to move them on, given that secondary picketing's illegal. They've started

picketing, no one's crossed the line and some of our men have joined it. We've got a line of about seventy pickets now. I want to repeat my request that the secondary pickets be moved on in order to make it possible for our dayshift men, who are arriving, to get in to work. My concern is that the picket line is becoming established when we had a chance of getting the men into work today and preventing our pit from going out on strike.'

A silent pause followed.

'Hold the line,' the voice said, after a few seconds.

'Okay,' Jim said but whoever was on the other end of the line was already gone.

Jim's patience was expiring. It was starting to feel like Whitacre was going to be neglected.

After a couple of minutes, the man came back on the line.

'A senior officer will be arriving in a patrol car at your main gate. Can you go out and meet him?'

'I can try. But it depends on whether I can get out through the picket line. When will he be here?'

'I told you. He'll be arriving any time now.'

'Right. I'll go and try to get to him.'

Jim put the phone down and muttered, 'Twat!'

He told Roger what he had been asked to do. Jim had decided, earlier, to keep the main entrance's big, double doors closed and locked, due to its proximity to the picket line, so he had to leave the offices by the door at the top end of the building.

As he walked out, Jim wondered about the mysterious police contact number; who, what and where were they? Would they really be geared up to provide a rapid and significant enough response to deal with a picket line in deepest North Warwickshire, especially when they could expect the worst of the trouble to be in places like the four Yorkshire areas?

He strode round the building, onto the pit lane and took the lonely walk up to the rear of the pickets. Their number had swelled to well over a hundred, with more joining them, by the minute. Men no longer approached with any intention of passing through. The pickets were massed, shoulder to shoulder, four or five deep and blocking the road.

'Excuse me, men!' Jim called out loudly to the rear rank.

Seeing his jacket and tie when they turned, the men parted and he was in amongst them. The horde closed back in behind him.

As he repeated, "Excuse me" the middle and front ranks slowly parted and he was able to squeeze through to step out onto the other side; into the world outside.

Jim walked away, prepared for a blow or a missile to the back of his head. He stopped, thirty yards from the line, and waited in the middle of the pit lane.

The noise was rising as the growing army of pickets talked, laughed and shouted to their mates in their mounting anticipation.

Looking back at them, Jim felt like Gordon of Khartoum.

He saw the big Barnsley man, with the eye-patch he had applied, in the middle of his group of mates, in their orange workwear and pit-boots. The leader of the Yorkshire gang was next to him, running operations, doubtless applying experience gained on picket lines up north, in the previous week. Still briefing those around him, he caught Jim's eye and nodded. Jim nodded back, unsure of what that swift exchange was supposed to have conveyed.

Jim turned back round to see a pristine, white Ford Granada, with a broad, fluorescent orange strip along its length, turn into the lane. It cruised silently to a halt, twenty-five yards short of where he was standing.

As he started striding towards the car, the driver's door flew open and its sole occupant leapt out.

Tall and immaculate in his full-dress uniform, the Chief Superintendent placed and settled his braided cap on his head, smartly, as he marched towards Jim, his strides lengthy and business-like.

'Mr Greaves?' The officer drawled, leaning to one side and craning his neck to gaze over Jim's head at the host of pickets.

'Yes. I was hoping for support from you lot, hours ago. If you'd got here an hour ago we could have got rid of the flying pickets and stopped this line from forming. We need your assistance, now, to stop this lot from preventing our dayshift men from getting in to work.'

'Yes, that's fine,' the policeman said, abstractedly, continuing his assessment of the crowd of pitmen. 'I've got some troops on their way.'

He spun on his heel, marched back to the patrol car, put his hand through the driver's door window and pulled out a radio mic' on a curling lead. He spoke into it and received responses several times, all the time, keeping a calculating eye turned towards the picket line.

Jim couldn't believe it; the Government couldn't have decided to send in the Army like it had at Tonypandy, at the start of the century. That would be the end; British squaddies facing English pitmen. Would they be armed?

Would it be the awful prospect of fixed-bayonets? Or would they just have been issued with pick-axe handles?

The country appeared to have gone completely mad.

Roaring engines approached and three luxury coaches swung off the road and into the pit lane. Jim stepped aside as the leading coach drove, at speed, towards the pickets. The pickets' line broke and they scattered to the far side of the lane, allowing the bus to pass through. The other two coaches slowed as they passed and parked on the side of the pit lane, behind the first. Jim saw from the signs on the back of the coaches that they were from South Wales.

The front doors of the coaches swung open and the tallest coppers Jim had ever seen started pouring out of all three. The badges on their helmets indicated that they were from the Glamorganshire Constabulary. With their helmets on, they all looked to be about eight feet tall.

So they were what the Chief Superintendent had meant when he had spoken of troops.

Without any shouted orders, the policemen walked round to the other side of the coaches, formed a long line and extended their arms to close the gaps between themselves. The coaches moved off and went on to park in the colliery's office car park.

The Chief Superintendent strode over to them then shouted, 'Forward.'

He raised the loud hailer he had brought from the boot of his squad car and bellowed at the pickets.

'Move back off the road.'

The pitmen gave ground and drifted back onto the small, ash-covered lorry park. The pit lane was cleared and lined by the single rank of giants that faced the pickets, towering over them.

With the way cleared, Jim walked back along the pavement, on the other side of the pit lane, and round to the top end entrance of the offices.

A few moments later Ken Goodall arrived and was able to drive in, unimpeded. Having parked up, he entered the main office block and joined Jim at the window in the Undermanager's office to look out at the side-lined mass of pickets.

They could hear the shouts of the miners as they called out entreaties to their work-mates as they came down the lane.

The distinctive figure of a massively over-weight, young pitman appeared and he waddled his way down the lane. Bobby Lowe had enhanced his notoriety and received much opprobrium down the pit for a recent conviction, having been found guilty of stripping off and standing in his front room to expose himself to a trio of teenage girls on their way home from school, after getting drunk at the Miners' Welfare one lunchtime, whilst on a week of nights. Even allowing for his judgment being impaired by drink, Jim still found it incredible that Bobby could have entertained the notion that the unfortunate girls would be interested in having his obese body exposed to them.

Bobby reached the picket line and continued sauntering past on the other side of the lane from the pickets.

'Go on Bobby, keep going,' Ken muttered.

Men in the crowd of pickets were shouting to him:

'Come on Bobby.'

'Come and join us, Booby.'

'Come on, join your mates on the picket line, Bobby.'

As Bobby's elephantine progress continued, he gave the pickets a cheery wave. When he got to the end of the crowd, he waved a casual two-fingers at them, with a carefree grin, as he kept going.

A great chant went up from the furious pickets.

"YOU FAT BASTARD!"

'Good old, Bobby,' Ken said as he and Jim chuckled at the scene.

Ken sat down at his desk and Jim took a chair to discuss the night's events and subsequent developments.

Outside, on the pit lane, three young men from the village were walking up the pit lane. Two were on the Mechanical Apprenticeship Scheme, the third was an apprenticed electrician. Few of the pickets bothered to shout at the apprentices. With the reduced noise, the voice of Alan Walker, a big, heavily-built pitman, rang out.

'If you cross this picket line, m'lad, you needn't bother comin' round our bloody 'ouse ag'in.'

The intended recipient knew the warning was addressed to him.

Brett Redfern stopped and looked at Alan. Staying out on strike would be a breach of the terms of his apprenticeship and would result in his immediate dismissal, ending his chosen career. Brett could only shrug, lamely, in the direction of his girlfriend's father before jogging away, head down, after his mates.

He caught up with Jamie, the apprentice who had embarrassed him with the words he had scrawled in marker pen on his hard hat a few months earlier. Jamie gave Brett a sideways glance.

'Don't worry,' he said. "E don't mean it.'

'Huh, you don't know what 'e's like,' Brett muttered, morosely. "E means it alright. And Tess won't cross 'im when 'e stops 'er from going out wi' me.'

'So that's us finished!'

A Pot of Tea

Friday, 23 March, 1984: 04:45

The three miners sat on the two benches in the central room of the dirty offices, still in their pit-black. They'd come straight there from the shaft, on the last shift of their week of nights.

It was unusual territory for them. Normally, they wouldn't step across the threshold of the officials' offices and certainly wouldn't be invited to sit down in there.

All three men were silent and thoughtful.

Jim had made a fresh pot of tea for them and the officials. He poured four mugs and handed one to each of the miners then sat down, at the other end of the old, worn, iron-framed, wooden bench from Jasper Ford, a nights-regular header.

Jim put his own mug down on the rough, wooden surface of the table under the window behind him, then turned back to look across at the bench opposite where Howard Freer sat, next to Arthur Lambert.

As a button-lad, Howard would normally have been operating one of the belt conveyors underground. Thin and sensitive-looking, he was relatively mature for his junior job, having been a late entrant to the industry. Arthur, an old faceman, was something of a loner.

They were the nightshift element of the tiny nucleus on which the management hoped to build. They had each been allocated to a deputy, each shift, to spend time accompanying them on their inspection of an underground district; token work, intended to keep them engaged and attending.

Howard had been paired up with Stan Townsend, the face deputy. When they had returned, Stan had picked up his district's report book and sidled over to where Jim was pouring hot water from the kettle into the big, dented, aluminium teapot.

'They've bin round to Howard's house twice this week,' Stan had murmured. 'Strikers. Whilst he's been at the pit. First night, Tuesday, they 'ammered on the door and shouted, terrifying his wife and two kiddies, three and five. Last night they 'eaved a brick through 'is window. He's dreadin' what 'e'll find when 'e gets 'ome this morning.'

'Oh,' Jim had nodded. 'Thanks for telling me.'

Jim chatted to the men about what they had seen down the pit that night, where they had to travel to in order to get home and what children they had, amazed at their thoughtful serenity.

There had only been these three men, all week, breaking the strike on nights. On days, it had grown from fifteen men at the start of the week to twenty-four by Friday dayshift. There had been a dozen every day on afternoons. Just four percent of the total work force was working.

Jim respected these men, their conviction and their bravery; they were breaking the strike conscientiously. He respected those miners who were striking conscientiously as well. But it didn't seem to take the same degrees of conviction and bravery to march with the vast majority as it did to stand, virtually alone, on the other side of this fierce, passionate, emergent strife.

Jim had heard that the picketing had been particularly bad, six miles away, at Baddesley pit. He took a particular interest in this old, family pit because his grandfather had worked there. That pit was solid. With the calm, clear leadership of its Lodge Secretary, Joe Windridge, every man working, except for one solitary striker.

The day before, the NUM had thrown huge numbers at Baddesley and the violence of the picketing had been extreme. As one working miner was trying to drive through the pit gate, the pickets had managed to turn his car over, onto its roof, and set it on fire with him still inside it.

The room's sombre atmosphere was in keeping with Jim's mood. That miner should be pitched against miner was an abomination; his father had told him to expect to find himself as a member of a brotherhood linking every man who worked underground in mines wherever they were from in the world. It was heart-breaking to see those bonds being severed between Britain's colliers, impossible to see how they could ever be repaired.

Jim had wanted to allow the men to leave the pit by the back gate but Mr Everitt had been adamant that he should adhere to the Board's policy that working miners were not to be allowed to sneak in or out. They were only to be allowed to arrive and leave by their normal route.

Jim had checked with Roger, in the Control Room. There were only three or four pickets at the moment. Their number wouldn't start to swell until just before six o'clock.

'Right lads,' Jim said. 'You'd better get a bath and get off home. Thank you for coming this week. See you next week.'

The men stood up, slowly, and took their mugs to the sink in the corner of the room. Howard went over to start washing them.

'Don't worry about that Howard,' Jim said. 'I'll do the washing up.'

Having shuffled over to the door, Jasper turned.

'Thanks for the tea, Boss.'

Thanks Boss,' Arthur called back from the corridor.

'Yeah, thanks, Boss,' Howard said. 'See you next week.'

'Yes, I hope so,' Jim said. 'You take care.'

Early Days

During the week, the reinforcements from Barnsley were farmed-out amongst the homes of various Whitacre Heath strikers.

Ian Flint had been sleeping on Paul and Susan Wood's settee the previous week and then gone home for the weekend with the rest of his men. They had all had to get up in the middle of the night in order to travel back down by a tortuous route that avoided the numerous police roadblocks and ambushes between Yorkshire and Whitacre Heath.

With Ian's misgivings about McFadden having only increased since their first meeting, he had made Paul his link into the local Whitacre Heath Lodge.

'You're obviously seriously motivated and committed,' Ian said to Paul, as they stood together on the picket line that morning. 'Why 'aven't you stood for a place on your Committee?'

'I can't be bothered with their petty politics. An' I ain't interested in pushin' meself forward in that. The only things I'm interested in is supporting me family – and protectin' our livelihood and the country's coal industry from the Tories.' Paul said. 'My grandfather was a Geordie. 'E was on the Jarrow March in 1926 when 'e ended up down 'ere. He told me about the way the miners had been treated by the Tories and their lot then. They've made it obvious they're still intent on smashin' the working class, startin' with us.'

'It's certainly as well to know what we're up against,' Ian said.

They stood in silence. Paul feeling the need to raise something but wary, in case the Barnsley man thought him paranoid. He decided to mention it, just in case.

'Brian reckons 'e saw some stranger round the village. E' was wonderin' if 'e could've been a plain-clothes policeman.'

"E could be right,' Ian said. "It's best to be cautious. They might well be after me – as a coordinator. We know they've got Special Branch trying to infiltrate the pit communities. They think they're bein' clever but there's no one in the police force who looks or sounds anything like a pitman.'

'In the first few days our pit was policed by the local coppers and they were relaxed and friendly, 'ad a chat wi' us, like you'd expect from your local bobbies. They broke that up straight away, shipped in a load o' coppers from elsewhere, round the country. They don't want any fraternisation.'

'They say those big Welsh coppers have been billeted at the Army barracks near Nuneaton,' Paul said.

Ian nodded.

'The week before last, hundreds of us travelled across the border into Nottinghamshire in the early hours to try to stop the scabbing at those pits,' he said. 'We all homed in on Harworth and, by the afternoon-shift, we'd picketed it out.'

'The next day, there were even more of us and we travelled down to get the strike spreading across the Nott's coalfield. They'd found a thousand extra coppers, overnight. And within a couple o' days, there was more like ten thousand of 'em, all armed to the teeth and controlled from some central control centre they've set up 'specially somewhere.'

'On the Monday the police had allowed us on't' pit buses takin' the dayshift in, to talk to the lads, to ask 'em to reconsider and come out on strike wi' us,' Ian said. 'Two days later we couldn't get near t'pit gates, let alone get on a bus. The police were one step ahead of us all the time for a few days. 'Til we worked it out.'

'What was it?' Paul said.

'Phone tappin'. You heard the click if they disconnected before you put phone down. We started listenin' fer it.'

Paul said nothing; it sounded unbelievably clandestine.

'When the area coordinator called to let you know where we were to go next day, you'd hear it,' Ian said. 'When you set off, they'd be waiting for yer and start tailin' you somewhere along the way. We've had to start doin' it all by code. So now, we get a piece of paper with letters and numbers on it and you have to decipher it against a code sheet. That's how the lads know where to muster, which pit, what time and date.'

'What're they gonna do about that Court order, you know, that Yorkshire NUM must withdraw its flyin' pickets?' Paul said.

'Ignore it. The Tories are misusing the Law for their own purposes,'

Ian said. 'No one should be in any doubt about what's at stake here. We're up against a Government that cares more about profits than it does about people or the country. We're not just lookin' for a victory for the miners, it's for the whole workin' class. We 'ave to fight on, whatever they throw at us, 'til we've won. The only question is 'ow long it'll take. The strike in '72 lasted seven weeks, '74 was just over a month. This one looks like it'll go on for months – there's more at stake. We're gonna have to dig in, this time.'

By the end of March miners were talking about staying out until Christmas, when the next winter would make the shortages of fuel start to bite.

The Times reported: *"Mrs Margaret Thatcher is willing to spend any amount of money to ensure that the Government is not again defeated by the miners' union"*.

Her Chancellor, Nigel Lawson, shared her determination and said that whatever they would have to spend to beat the miners would be *"a worthwhile investment"*.

On 11 April, there was a shocking development; a majority of the supervisors' union, NACODS, voted to join the strike. The Government was alarmed. But the NACODS leadership failed to call their members out, citing safety grounds and the need to prevent flooding and fires from closing their pits to justify their decision. There was widespread suspicion amongst the NUM's membership that they had been got at by the Government and MacGregor.

The miners saw it as an infamous betrayal by their supervisory colleagues.

Broken Glass

All the demands of the recent months had caused Jim to allow his car to become overdue for a service. Having dropped it off at the garage, the previous evening, it was only as Gabriella swung her Mini into the pit lane that he started to regret having accepted her offer of a lift in to work.

The picket line had been relatively quiet on the previous day but this morning looked completely different.

Gabriella saw the mass of pickets, on the parking bay to the right of the pit lane, and the line of police officers struggling to hold them back.

'What shall I do?' She said. 'Had I better stop?'

'No,' Jim said. 'We'd better keep going.'

He didn't want them to see the car falter and become enflamed by the possibility of a victim but then berated himself; the men wouldn't recognise Gabriella's car. If he had been in the driver's seat, there would have been a better chance of the pickets seeing and recognising him.

He stared straight ahead, attempting to exude the casual confidence he lacked.

In a couple of seconds, they were through, the shouts of the pickets receding behind them as they motored past the office building's main entrance and into the central car park.

Gabriella pulled up. They kissed and Jim got out.

'Are you alright?'

'Of course I am,' she said. 'Why wouldn't I be?'

Jim smiled at her, shut the passenger door and strode off towards the entrance to the lamp room and undermanagers' bathroom in the dirty office block, turning to give her a cheery wave.

He saw the pit's blue and yellow mini-bus with the NCB logo on its side heading for the pit lane in front of her. Inside were the winding-engineman and a deputy and several men who were being transport, overland, to the Gracebury Shaft. Jim convinced himself that Gabriella would be safe, going back through the line, that they'd know who she had dropped off by now and that there would have been enough pitmen in the crowd who would have recognised them both.

Fifteen minutes later, Jim walked into the dirty offices in his pit-black.

'They reckon the picketing's been rough this mornin',' one of the deputies said. 'They threw a load o' cobbles at the Gracebury bus.'

'What?' Jim said.

'Yeah, that's right,' Stan Townsend chipped in. 'One of my men came in to say 'e'd got late, getting in for dayshift, 'cause it'd flared up out there. 'E saw it leavin', covered in dents. The lads inside it had had to get down on the floor to save 'emselves. 'E reckoned they'd peppered it. There wasn't a pane of glass left in it'

Jim's heart went cold.

'Was anyone else involved?' Jim said.

''Ow do you mean, Boss?' Stan said.

'I mean any cars going out.'

'No, he didn't mention it,' Stan said. 'But, then again, any cars going out would be leavin' from the main car park on the other side o' the picket line, wouldn't they.'

Without another word Jim rushed out and over to the Control Room. He got a full account of what had happened from Jerry Malin, the control room attendant.

'I'm just a bit worried that Gabriella might have got caught up in it as she drove out,' Jim said. 'She dropped me off this morning 'cause my car's in for a service.'

'Yeah, I seen 'er arrive with you, through the window, Boss,' Jerry said. 'But she should be alright. I ain't heard owt.'

Jim was still anxious. Gabriella would either have been caught in the barrage or been close enough to be terrified by the onslaught.

He went back over to the dirty offices but remained preoccupied and stayed on the surface. He needed to wait until Gabriella got into her office, in Birmingham, so he could phone her and talk to her, to make sure she had got clear of the mine, safely.

After half an hour spent blaming himself for being so cavalier with her safety, he started to phone to try to contact her.

On his first call someone else picked up the phone to reveal that Gabriella had yet to arrive at work. A few minutes later, when he dialled her number, another colleague answered her phone.

On his third attempt, Gabriella answered.

'Hello. I've only just left you.' She said. 'Are you missing me already?'

'Yeah.' he laughed, reassured by how cheerful she sounded. 'Something like that.'

'So, what did you want?' She said.

'Oh, nothing. I just phoned to make sure you got out of the pit alright.'

'Yes, of course. Why wouldn't I?'

'Oh, no reason. Just worrying about you.'

'Ah, that's nice,' she said. 'See ya later!'

Occupied

In the distance, black, uniformed silhouettes moved, with purpose, against the glare from the generator-powered spotlights and the headlights of the battle-wagons, dominating the night with their numbers. There were no street lights lit; a few nights earlier, a couple of striking miners had taken them all out with ball bearings shot from catapults to make it easier for pickets to evade capture in the event of any night-time aggro.

The profusion and ferocity of the lights at the end of the village held Susan's gaze, robbing her of her night vision. She could see nothing in the pitch-blackness around her as she hurried on, weaving through the throngs of pitmen. She was only able to tell policeman from pitman by the height of the officers in their tall helmets, by the weapons they carried, their shields and batons, and by their organisation. The police had a chain of command and cohesion that came from training and their effective means of communication. The pickets were outnumbered, spontaneous, reactive and confused; a herd confronted by an indomitable machine.

Two ranks of police formed in a couple of seconds from several groups that had just bustled out of the pit-lane. With their clear plastic visors down on their round black helmets, they were, clearly, ready for action. It was impossible to tell what was going on. Susan looked around frantically; desperate to find Paul before the police attacked.

A tight grip locked onto her right arm.

'Quick! You don't wanna get caught up in this lot.'

Her legs forced into a jog, she allowed herself to be propelled away. At the edge of the road she stubbed her toe on the kerb, stumbled and would have gone down if not for Colin's firm grasp.

'Sorry Susan,' he said, pulling her back up and continuing to push her on to safety. 'Are y'alright?'

'Yes, I'm fine. Where's Paul?'

'I ain't seen 'im for a couple o' minutes. But 'e should be around 'ere somewhere.'

'What's going on? I've never seen so many police, there must be hundreds of them. And all those vans.' She said, nodding in the direction of the riot vehicles, parked down the pit lane.

'They managed to get Archie, from Barnsley, over into the pit lane. A load of them got hold of 'is arms and then a couple of 'em grabbed a leg each and chucked 'im into a van and drove off with 'im.' Colin said. 'We asked 'em why they were arrestin' 'im. They just said, "We don't have to tell you", then told us to get back or they'd run us in, as well.'

'Susan!'

She turned. Paul's lean form rushed back towards them.

'What're you doing 'ere? Come on, let's get you away.'

'I heard there was trouble. I was worried. Karen called round and said they were seizing people. What's going on? Why's all this police here?'

'Who knows,' Paul said. 'When they started arriving, on mass, we asked 'em what they were doin' 'ere and told 'em that no one was causing any trouble. The police said they'd had a report about a disturbance.'

'But you don't need fifteen riot vans full of police to check out a disturbance,' Colin said. 'I've counted a hundred and forty of 'em. All in riot gear.'

'And they've got another twenty vans full of reinforcements, round the bend on the road into the village,' Paul said. 'The police wanted to provoke all the Barnsley lads so they could round 'em all up and arrest them. They're like an army of occupation.'

'Come home with me, Paul,' Susan said. 'And you, Colin. You should go home and take care of Diane. This is out of control.'

'Alright, I'll walk you back,' Paul said, taking Susan's hand and stepping out to head for home at a brisk walk, with Colin tailing them.

'Who's looking after the children?' Paul said.

'I fetched Marion, from next door,' Susan said.

The noise of the confrontations had carried up to their house from the pit end of the village. Their neighbour had called round and told Susan that it looked like a battle was brewing, so Susan had fetched the fifteen-year old, who had acted as a baby-sitter for Martin before Jenny was born, to sit in the house while she came out to find her husband.

Susan unlocked the front door.

'Who is it?' Marion called, her voice quavering.

She snatched open the door to the lounge, wide-eyed and pale.

'What's wrong?' Susan said. 'Are the children alright?'

'Yes, they're fine. Martin's awake. 'E got a bit frightened so I fetched 'im down 'ere with me.'

Paul and Susan rushed into the lounge and found Martin sitting on the settee in his pyjamas, dressing gown and slippers. Marion and Colin followed them into the lounge.

'Four policemen came, hammerin' on the door,' Marion said. 'They woke Martin up, with all their noise and shouting. They were shoutin' at me, askin' all these questions: "Where's your dad? Where's your husband? Where's your brothers? Where are the men from Barnsley?"'

Susan sat down on the settee next to Martin and pulled the little boy onto her lap.

'It's alright now, Marion,' Susan said. 'Come on and sit down next to me. Paul and me are here, now. You're alright.'

'I thought they were going to arrest me and leave the kiddies without anyone to look after them,' Marion said, starting to cry.

'Paul, make us all a cup of tea, love, plenty of sugar,' Susan said.

'I should go back to the others,' Paul said, glancing at the front door.

Susan held her hand out and took his.

'Leave it tonight, darling. It's all out of control down there. Let's have a cup of tea and then you can see Marion home safe and come back and make sure me and the children are left alone. I reckon they'll have targeted us because we've got Ian staying with us.'

Thursday, 06:45

The police deployed a larger force the following morning. Ian wondered, aloud, where they were getting them all from.

'Criminals must be 'avin' a field day while this lot are all out here, playin' soldiers.'

More police battle-wagons had arrived, white Ford Transits with mechanically-operated, heavy-grade mesh shields that dropped down over the windscreen of each of the vehicles.

Some of the coppers in the front of the vans were taunting the miners, laying rows of tenners along the tops of the dashboards as they cruised past, to remind the impecunious pitmen that they were coining it in on overtime and subsistence allowances.

The police central command centre had hired large skips. When they had been delivered, in the night, the policemen had manoeuvred them into position to form a barrier, blocking the road to keep the pickets on the village side of the entrance to the pit lane. Eyeing up the lay of the land, Ian saw there were gaps between them, in readiness for the snatch squads and the cavalry to sally forth, onto the crowd of miners.

As Jim Greaves arrived and turned his car into the approach to the pit he could see from the massed miners and army of police that the day had the makings of a serious confrontation.

He saw a couple of police officers enhancing the defensive capabilities of their skips. They had lifted a couple of drums of the grease used to lubricate the guide ropes in the shaft. Jim could see what they were thinking; the front ranks of pitmen in any push would be reluctant to get that filthy, sticky, black muck all over their clothing. They knew that when it was plastered all over the edges of the skips it would cause the miners to hold back.

Jim pulled up, wound his window down and called over to a police inspector, standing with a sergeant kitted out in riot gear, tooled-up, ready for the fray, watching the officers greasing the skips.

'Excuse me.'

The inspector finished what he was saying to his subordinate, then strolled over with slow, steady steps until he was nearer to Jim's car.

'Yes?' He snapped.

'Did you ask anyone's permission before you took that grease off the colliery premises?' Jim said.

The inspector regarded the young colliery manager, allowing a smirk to form before giving Jim a silent snort.

'It's alright, they'll return it when they're ready.'

'Well, make sure they do,' Jim said. 'And, in future, ask permission before you take pit property.'

Back amongst the pickets, Ian saw the exchange between the young undermanager and the policeman and guessed correctly at its nature.

He wondered how Archie was faring. They had heard no more of him since the police had snatched him in the night. Ian would need to watch out for the snatch squad himself, they were going to be after him, the plain-clothes officers would have him marked out for today's fresh mass of coppers.

The first of the pit buses carrying working miners came into view, coming up the road on the other side of the police and their defensive wall of skips. The events of the previous night and the numbers of police involved had resulted in many more pickets being re-directed to Whitacre Heath that morning. As their booing and hollering started, the miners on the bus cowered low between the seats.

As the old bus approached, it was clear that several working miners, in cars, were intending it use it as a defensive screen. The crowd of pickets had all turned to face the police line. They surged forward, incensed by the approach of the scab-laden vehicles.

A small salvo of four or five brick-ends sailed over the heads of the men at the front of the picket line and the police in their ranks. One struck the side of the coach, another bounced off the wire mesh installed over its windows.

Before the next few coaches carrying working miners arrived, two dozen mounted police trotted out from behind the skips and the rows of riot police. They formed a line and then moved onto the crowd of pickets at the walk.

Riot police swarmed out behind them and formed up in three ranks, a couple of dozen men in each, as the horsemen advanced. The mounted men drove the crowd of pickets back, seventy yards up the road, into the village, then stopped.

When the shift start-time of seven o'clock arrived and all of the working miners on dayshift who would be coming were in, the ranks of police on foot opened to allow their horsemen to retire behind the line of skips and reserve ranks of police in conventional uniforms. The riot squads stayed forward of the skips, holding their line.

Police Constable Carl Bessemer was in the second rank, next to his colleague, Neil Bradford. The nearest pitmen were about thirty feet away.

Carl shouted over to them.

'Keep it up, lads. You're wasting your time but we don't want you giving up, I haven't finished paying for me holiday in the Bahamas yet.'

The pitmen scowled as a few of the officers in the front rank turned their heads to laugh with Carl. Neil said nothing, if he took issue with Carl it only made him worse.

Apart from feeling that the taunt was inappropriate, it was clear to Neil, from the way some of the miners were casting fierce looks and nodding in their direction, that the provocation had made Carl and those around him marked men, as far as the miners were concerned.

Ian had heard it, even though he felt it prudent to keep back in the crowd.

With the dayshift's scabs all in, there would be no action until lorries that had arrived earlier started to leave with their loads of black-leg coal.

In the absence of any activity, many of the pickets sat down on the road, like holiday-makers on the beach at Skegness. Others settled down in the characteristic crouch of the pitman. Being used to getting down in the confines of the mine, colliers were as comfortable crouching down as sitting on a chair. It was not unusual to see a pitman waiting, in relative comfort, at a bus stop in that position.

Ian had learnt the hard way; he knew it was better to be crouching like that or, better still, standing. You never knew when the police might charge, however quiet everything seemed in a lull.

About an hour later, he saw police horses moving back through the gap in line of skips. He knew this presaged the departure of the coal trucks. He told a few pitmen to pass the word round, "Get on yer feet"; the police would be getting ready for action.

As the first lorry roared out and sped away at the head of the convoy, the pickets surged forward. The police ranks parted and the mounted officers trotted out, breaking into a charge as they cleared their own lines. The foot police came running, pell mell, after the screen of horsemen.

The pickets turned and fled. Running with them, Ian glanced back to make sure he was staying clear of the marauding horsemen. He couldn't afford to go down to one of the blows they were administering with their long sticks, he needed to stay clear of the riot police and avoid being taken.

Being relatively fit and young, his sprint took him to an escape route, a dirt-floored alleyway between two rows of houses, well in front of the police. Darting into it, he saw that it led to some allotments. He ran until he could conceal himself on the footpath that ran up the ends of the back-gardens of the two terraces.

Back on the road, at the end of the alley, he saw a skinny, young miner, about twenty years old, bent double, trapped in a headlock by a massive policeman. Ian was just wondering why the policeman was using this unorthodox hold on a lad who was offering no resistance and appeared to have surrendered himself when he heard the officer holding him shout to a colleague.

'Go on, hit the fucker!'

The passing officer stopped and delivered a powerful blow to the back of the captured youngster's head with his truncheon. Ian saw the big policeman release the young miner, allowing him to fall, face-down, onto the pavement. With a single, surprisingly light skip, the policeman broke into a trot, and was gone, leaving the young pitman lying on the footpath, motionless except for the spasmodic twitching of his right leg.

Ian jogged up the narrow path, heading away from the pit entrance, glancing forward and back to make sure that he didn't get cut off by police emerging from the gaps, before and behind him, between the rows of houses. He needn't have worried; there were plenty of easier targets for the police on the pit village's main road.

Even that far away, as well as the screams and squeals of those who had failed to get away, Ian could hear the thwacking sound as the police laid on with their riot sticks, administering beatings to the miners they had overrun.

He knew it was a natural reaction but he still disgusted himself; he couldn't stop himself thinking, *"I'm glad that isn't me"*.

Neil and Carl were near the front when their unit reformed.

'There's going to be a lot off 'em wandering around with no shoes on.' Carl muttered. The whole road strewn with trainers.

Neil knew what had happened. In the crush of the retreating throng of strikers, pushing and bumping into each other in panic, many of them had had their pumps ripped off by the scrabbling feet of the men crushing up behind them, desperate.to escape. The Whitacre Heath miners weren't like the police and the Barnsley pickets, they'd been hopelessly ill-equipped for the skirmish.

As they marched back to the defensive line near the entrance to the pit lane, frantic scenes from their rout of the pickets played themselves back in Neil's head.

Back to Normality

Highbury Police Station. Monday, 30 April, 1984

'So, how was your little break in the country, then?'
Neil gave Bev a wry smile as picked up their two coffees from the canteen counter and let her lead the way to their usual table.

When they were both seated, Bev tilted her head and looked at him. Her friend was tight-lipped.

He needed to talk about it, preferably with someone who was sensitive, but he was wary of initiating a similar response to the one he had received from Julie, when he'd shared his thoughts about his time in the Midlands with her on Saturday evening. That had ended up with him having to stop himself, having sensed that his girlfriend was becoming as upset at hearing about various incidents as he had been in reliving them.

Bev studied Neil's face. He frowned.

'I dunno. You've seen all the stuff on telly. It's just – I don't know – I just end up thinking there must be a better way to police this strike. It all ends up being so – provocative, confrontational – aggressive, even.'

Bev said nothing and nodded slowly.

'To start off with, it was good to see all the preparations paying off – a relief. All the training, having the right kit and vehicles, good comm's, superior numbers, all that.' Neil paused and frowned again. 'But then there's the idiots, the usual suspects, taunting the miners, winding them up.'

Bev gave a little snort. 'Like Carl?'

'Yeah, like Carl – but there were plenty of others at it.'

'Then it started to feel like we're waging war,' Neil murmured. 'A war we can't lose.'

'And the enemy's our own people. Well, my people at least. Men from pit villages, like the ones where Julie and I grew up.'

'We're talking about the type of men who've been the backbone of Britain for centuries, the salt of the earth.'

He gazed away across the canteen, lost in his memories.

'One night, when we could see it was about to get lively, we were expecting it all to kick off or to be ordered to clear the street. Then it just went quiet. Everyone heard it; music playing, that warbling sound you get from the horns with any brass band. I thought, *"Hello, they're going to use the colliery band like a military band, they're about to muster and march on us"*. Everyone stopped and listened.'

'What happened?' Bev said.

'They must've come out of the village hall at the end of their weekly practice.'

'They only played one song. But it wasn't a stirring, military march or anything like that. They just played a hymn, you know, "The Lord's My Shepherd". It was gentle, soothing. Everyone relaxed a bit. It all went quiet.'

Neil paused and looking thoughtful and sad, Bev could see him calling the Psalm's refrain to mind, reliving the moment.

'That sounds lovely,' she said, with a gentle smile. It was so typical of Neil to have picked on an incident like that.

'Yeah, but what we're doing and making them do, it's so unnecessary. It just feels so over the top. I know what they're like; they're law-abiding, patriotic Englishmen. Most of them are the sort that we've always been able to depend on. But they're going to end up hating us forever, for this.'

Women Against Closures

Friday, 18 May, 1984: 18:10

T hey had established a weekly routine. Paul, Susan and the children would come round every Friday evening then Martin and Horace would take Sally for a walk round to Mario's chippie, in the row of three shops in the lay-bye on the village's main road, to pick up fish, chips and mushy peas for all of them. It gave them all something to look forward to.

That evening, when Pat cleared the plates away, Susan joined her in the kitchen and made four cups of coffee while Pat did the washing up. Susan had waited until they were alone to tell her mother-in-law all about her big event, that week.

She had arranged for Pat to look after Jenny all day, that Tuesday, and for Horace to meet Martin from school. This had freed her to be taken to Stafford, where she had joined two wives, from other mining areas, in attending a massive conference of NUPE on behalf of the Women Against Closures movement.

Pat had known how daunted Susan had been at the prospect of facing hundreds of members of the public services union and having to try to get a meaningful message across to them.

Susan had kept hoping that, on the day, she would be spared, that she wouldn't actually be called upon to speak. But Paul had helped her to prepare some thoughts then she had written up her notes and practiced her speech endlessly at home, on her own, just in case.

The NUM had sent a union official in a car to take her over to Stafford, where he'd introduced her to the other two wives, Kathy from Staffordshire and Maureen from South Yorkshire, and to Tony, one of NUPE's conference organisers. Tony had given them their name tags then escorted them through to their seats at the front of the stalls in the huge conference hall, with its packed tiers of circles stacked up behind them.

The first part of the morning was tedious for the three women. They looked up at the self-important representatives, sitting in four rows of chairs, behind the long table on the hall's high stage, listening to speeches and points from the floor relating to matters about which they knew nothing. But they remained alert, anxious, not knowing whether, or when, one or all of them might be called upon to speak.

At the mid-morning coffee-break, a senior delegate left the stage to join Tony.

In the concourse, Susan saw the two men emerge from one of the three pairs of doors into the hall, look round and, on seeing the three wives, make their way over to them to interrupt their forced, nervous conversation.

'Excuse me, ladies,' Tony said. 'Let me introduce you to Bill Sands. Bill's a member of our National Council.'

Tony introduced them individually, by name, to the senior man.

'Bill and I were just talking about giving you the chance to tell our members about your struggle in the coalfields and how the strike is going,' he said. 'We're keen to make sure you get whatever support we can give and we're thinking we could give you a few minutes when the meeting reconvenes. Then you could join the delegates for a bit of lunch before getting away. You'd be welcome to stay on after that but I expect you'd rather miss out on sitting through the rest of the day and get home to your families. You've all got a long way to go.'

The other two women glanced uncertainly at Tony and Susan then looked at each other and grimaced.

'What do you think, Susan?' Tony said.

It was suddenly apparent that he believed Susan to be the leader of the trio, probably because, despite her nervousness, she looked smarter and more intelligent than her two companions.

'I'll introduce you, Susan,' Bill said. 'Would you be able to keep it to twenty minutes or so?'

The relief of the other two wives was immediate and palpable.

'Er, yes. That's fine.' Susan said, managing to sound more positive and confident than she actually felt whilst glancing, in despair, at her new-found colleagues.

"Though, I don't think I'll last anything like as long as that, so you'll have to be ready to step in,' she murmured.

For the next twelve minutes, Kathy and Maureen relaxed and chatted and looked forward, with hungry anticipation, to a free meal, while Susan suffered the wait of the condemned woman, watching the minute hand on the clock over the middle set of doors back into the hall as it ground its way round towards the time when she would be called to mount the steps, at the front of the auditorium, like a prisoner going to the scaffold.

Back in the hall, she barely had time to sit down before Bill Sands started introducing her. In her mounting panic, she missed out on his dramatic preamble but couldn't fail to hear his conclusion.

'Please welcome Mrs Susan Wood, a miner's wife and member of Women Against Pit Closures.'

The unearned applause taunted and deafened her as she walked over to the steep stairs, to clamber up to arrive on the left-hand side of the stage.

Out of practice at walking in high-heels, the huge sea of staring eyes undressed her as she tottered across the stage, gripping the flimsy piece of folded cardboard containing the script she had crafted on some scraps of paper, to where Bill Sands waited with a beaming smile, to receive her, centre-stage, with a welcoming handshake.

She gave him a wide-eyed, crooked smile then turned and squared up to the podium, dry-mouthed and terrified. She placed her notes on it, under the goose-necked microphone stand that thrust towards her face, demanding her words.

She opened her folder, looked at the sheets of paper inside, leant forward, stooping a little too close towards the microphone. She cleared her throat, glanced up at the crowd and uttered her first few words.

'Good morning. Thank you for inviting us to come and talk to you today.'

It was the voice of a stranger; too quiet and hesitant, her Warwickshire accent exaggerated by the public address system, a croaking crackle on her last word.

She scanned her notes; it was all waffle.

She realised she was about to freeze.

Rousing herself reluctantly, she closed her cardboard folder.

In her desperation, she started telling them what the striking miners and their families were fighting for, of their loyalty to their communities and their way of life, how they believed in their industry and the contribution it made.

She told them that their industry and its union was under attack, how it might be the NUM and the pitmen and their families now but, if their oppressor was allowed to beat the miners' strong union, they would be coming for NUPE and the rest of the trade union movement next.

She spoke of the hardship in the coalfields, of levels of deprivation that were unbelievable in a first world country in the Twentieth Century.

Someone one rushed out and fetched buckets and the audience started passing them around, dropping money in for the strike funds.

Suddenly, Susan felt transformed. She was no longer just a miner's wife, she had become a person in her own right.

When the crowd started to interrupt Susan's speech with applause, she called over them in her clear voice.

'We don't want your applause. Give us your money if you like – but what we've come here for, today, is to call on you for real support; from you and all the unions. None of us – none of you can afford to see the NUM beaten. If they beat our union, they'll have beaten us all.'

'We need all of the trade union movement to come out on strike and join us.'

'We need to unite and we need to do it now.'

'This is the last chance for all of us!'

Susan had received a standing ovation and achieved far more than she and Paul had expected.

Pat could see the transformation this had wrought in her daughter-in-law.

'It feels like a real revolution, Pat. An opportunity to pull the unions together.' Susan said. 'And a chance to change the way we see ourselves as women – and the way people see us – forever.'

'We've always accepted that mining communities have been male-dominated. You know, it's the men that work and the women that "do", having babies, washing, getting the men's snap ready for them and worrying about them all the time their down there.'

'All that's changing now, the men are starting to recognise that we're in it as much as them, that we're vital for victory. What's happening is almost like the Suffragettes. We're making a change and a difference that'll be seen as historic.'

'We won't go back to how it was. We can't'.

'And it's not just the injustices against ourselves that we need to fight but others too. Having had all the propaganda in the papers and on the radio and telly thrown at us – all the lies – you have to ask yourself, how many lies they've told us about other things and other people? What's really going on in Ireland? What really happened in Toxteth and Brixton?

The black people have probably been on the receiving end of what we're getting for years.'

'Maureen and Kathy, the two other wives at Stafford, told me about what's going on in their coalfields. We have to get motivated down here, like them. They're doing workplace collections, door-to-door collections, street collections, pub collections, football ground collections; benefits, demonstrations, visits from miners to workplaces and other industries – and, down here, we've barely even started.'

When Susan paused to draw breath, Pat said nothing. Her son's lovely wife was starting to sound like a politician.

If Susan was right and this was the level of change that was happening for her daughter-in-law and son, Pat thought it sad; sad that Susan was ready and happy to turn away from the kind of steady, comfortable home-life that she and Horace had striven to build all their lives and that she'd hoped their son and his family would be able to enjoy together.

The Battle

It was easy to underestimate Sam Crowther; big, rough, basic, a bit of a clown – especially today when he'd turned out wearing his lad's toy, policeman's helmet. One of their mates on the coach down had told him it looked "like a tom-tit on a side of beef".

'Eighteenth o' June,' he grunted. 'Battle o' Waterloo. Un' 'ere we are, backs ag'in' the wall. Just like they wuz.'

Ian frowned, turned his head to look up at his mate then nodded, slowly. It still amazed him how certain pitmen who had shown no aptitude or interest, whatsoever, for learning anything at school could become a real authority on some specialist subject. But he had never asked Sammy what it was that had caused him to develop a particular fascination for the military history of the Napoleonic era.

'Makes you wonder, which of us is Wellin'ton's mob and which are Boney's,' Sammy murmured.

Either side could see themselves as holding the defensive position of the British and their allies; the police defending the coking plant, the miners cornered, on their ridge, in the stubble field, ready to fight for their very way of life.

They would know, soon enough, which side would meet its Waterloo, that day.

Six weeks earlier, in early May, a week after the skirmish at Whitacre Heath, Ian and his mates had been despatched to Nottinghamshire to picket Harworth Colliery. The police held off ten thousand miners on that occasion. The next day had seen Ian's team back down in the Notts Coalfield, joining an even greater horde intent on picketing out the Colliery at Cotgrave.

From the ninth of May, there were regional "Days of Action" all over the country with railwaymen, hospital staff, council workers and dockers striking on individual days. These were backed by mass demonstrations in London and Manchester, with over fifty thousand people at each.

The rail-workers' refusal to move fuel had escalated to a point where no more than ten of the scheduled three hundred and fifty-six coal trains were running each day. Then, on the seventh of June, the transport unions boycotting the transport of coal and coke agreed to block oil movements as well.

All this gave many miners the impression that the momentum was with them. But from what Ian was hearing, from his contacts at the Barnsley H.Q., there was a fundamental weakness to all of this; there was no coherent strategy and no overall organisation or coordination of the diverse, supportive unions. Even worse, Len Murray, the General Secretary of the Trades Union Congress, had denounced the Days of Action and dissociated his organisation from the NUM's campaign.

Then there was Orgreave.

At the heart of the nation's steel-producing area, its enormous coke works fuelled the great, steel foundries, factories and engineering works of Sheffield.

The NUM had agreed to allow a certain amount of coal to pass its picket lines and enter this plant. This was on the strict understanding that it would only be used to produce sufficient coke to keep the region's steel furnaces ticking-over, to prevent the insides of their great ovens from cooling and cracking and causing irreparable damage which would expose their foundries to the risk of closure.

In their wisdom, British Steel managers had abused this concession and been producing sufficient coke to continue a degree of steel production. As soon as the striking miners found out about this deceit from their comrades in the steel-workers' union, Orgreave became the NUM President's obsession; a battle to be won at all costs.

But like most of the officials at the NUM's Barnsley HQ, Ian saw the decision to target the coke plant as flawed. As far as he was concerned, their priority needed to be picketing out the working pits to stop their scab operations whilst blockading the ports to keep out the coal imports which threatened to break the stranglehold that the NUM and transport unions were trying to bring to bear upon the nation's power stations.

For Ian, it was the large-scale scabbing in Nottinghamshire that needed stopping first.

That view was immediately vindicated. Before the distraction of Orgreave, there hadn't been enough miners crossing picket-lines to man-up more than a shift a day at any pit in Nottinghamshire. The very day after Scargill shifted the focus of their action onto the South Yorkshire coking plant, the whole of the Nott's Coalfield was able to ramp up to two-shift production.

Ian was at Whitacre Heath on the Friday that the union first engaged in full-scale picketing of Ogreave's coke works. The following week it resumed with the level of confrontation escalating by the day. On the Tuesday of that week, coke workers leant legitimacy to the action by joining the miners on the picket line. Over two thousand riot police and officers on horseback mounted baton-charges to ensure the safe conduct of coke lorries out of the plant that day.

The next day there were even greater disturbances with more than sixty miners injured and eighty-two arrests, including that of Arthur Scargill himself.

Then, on Thursday, the thirty-first of May, the miners faced over three thousand policemen from thirteen different forces and suffered considerably more arrests and casualties in what was a convincing defeat.

Scargill decided to stay away from Orgreave until his union could muster sufficient numbers to overcome the escalating forces being deployed against them.

Orgreave, Monday, 18 June, 1984: 07:20

The miners occupied the cornfield, standing around in groups, basking in the midsummer sunshine. A group of younger lads started a kick-about with a black and yellow, plastic football that someone had brought along.

The coke works were a towering presence in the near distance, a crude imposition on the South Yorkshire valley. But their huge, blackened buildings and plant did not strike the pitmen as incongruous, most of them being accustomed to living with the invasive presence of their own heavy industry within the rural landscapes of their own pit villages.

Despite the early hour, Sammy unbuttoned his shirt to stand bare-chested, bare-bellied and belligerent, bottom-lip pouting. Ian looked up at his mate.

'It's gonna be an 'ot one,' he murmured.

'It's gerrin' that way.' Sammy growled. 'I bet them bloody coppers are sweatin' cobs, togged up in all that clobber.'

Ian looked thoughtful. He had spotted a senior policeman, seventy yards away, behind the ranks of heavily-dressed, sweltering police officers, dressed for comfort in shirt-sleeve order.

'Y'know, I thought that was odd, the coppers escorting us all down the road in us groups from village,' Ian said. 'They've spent weeks spyin' on us, usin' roadblocks and cordons; everythin' they could think of to stop us reachin' us targets. Then today they just escort us all the way down the road to the cokin' plant, happy as yer like, and 'erd us into this field.'

Sammy looked down at him.

'It all makes sense now.'

"Ow's that?' Sammy said.

'They've got that army down there, sealing us in.'

Ian waggled his head towards the gate through which they had allowed themselves to be directed, at the bottom of the field.

"Cruft's' over there,' he said, nodding at the police dog-handlers, thirty of them, in a long line, at the foot of the steep, high, scrub-covered bank, running down to their left. 'Then the Horse of the Year Show up top end of field.'

Sammy nodded.

'And you can see there's more of 'em with dogs movin' about in them woods,' Sammy muttered, pointing off to the right.

'Yeah,' Ian said. 'They've corralled us in on ground of their own choosin'. They're running this like a military operation, using that big, brick building they were throngin' around, on t'other side of the road from where they 'erded is in, as command centre. They're up t' somethin'. They've gorrus penned up in 'ere like lambs fer the slaughter.'

Ian gazed around at his own side. As ever, many of his comrades sported full, drooping moustaches, like those of the Saxon warriors who had made up the shield wall at Hastings. To his practiced eye there had to be six or even seven thousand of them. But it looked like the Government's intelligence organisation had breached the NUM's security again. The police had obviously obtained all the details of the union's plan to blockade Orgreave, *en masse,* that day. They'd made sure they'd be outnumbering the miners heavily. Ian had never seen so many mounted coppers. And the miners were starting to lose some of their number with some of the pitmen starting to sneak off through the undergrowth, brushing through the sooty branches of the densely-wooded area to the right of their position, playing cat and mouse with the police with their eager, baying Alsatians, to make their way back to the village where local miners' wives were serving simple refreshments.

'There's Arthur,' Sammy said, looking over the heads of the men in front of them, indicating down to the left of the slope with a nod of his head, shining an imaginary cap-lamp in their leader's direction.

The NUM President paced forward to the police, wearing the baseball cap he'd been sent by the leader of the American miners' union that had taken on Ian MacGregor and been beaten by him, some years earlier.

Starting from in front of the left flank of the body of police, Scargill walked along the full length of the front rank of officers with their long shields. He pretended to inspect them, shaking his head from side to side in disdain, berating them as he went. Ian could see the angry faces behind the plastic visors of the policemen in the line and hear some of the abuse they were hurling at him. It was apparent that Arthur was trying to provoke the police to arrest him again but this time they appeared determined to exercise restraint. The miners started to cheer their leader. When he had progressed along the whole of the police's front, he veered off and strolled back into the crowd of cheering pitmen.

Two of Ian's Barnsley men had been back to the village with a few of the others. Dave and his mate, Larry, had come back down the road to try to re-join Ian and the rest of them. Thwarted by the ranks of police at the bottom of the field, they were standing around chatting about that most clichéd of miners' topics, pigeon-fancying.

A flurry of movement across the road caught Dave's eye. The body of police parted to release an eight-man squad which burst out from its rear ranks. Their discarded full shields landed on the tarmac with a clatter as they sprinted down the road at Dave and his mate.

The two pitmen spun round to dash away, along the wide, grass verge, Dave managing only a few strides before tripping over a divot and tumbling to the ground.

He heard, rather than felt, the "thump, thump" on the side of his head then sensed a warm flood over his face. Barely conscious, heaved upright, his arms bent high behind his back, by two hefty policemen, it was only as they manhandled him back to their formation that Dave realised that it was his own blood, streaming from a deep gash in his torn scalp, that was flowing into his eyes. Vision blurred, he glimpsed the police horde, grinning under their visors at him as they drummed their truncheons on their shields, like Zulus.

He gazed back at them with glazed eyes, dazed.

As they frog-marched him through their ranks, hc thought. *'These aren't police. Police don't behave like this.'*

From the field, Sammy spotted him being hauled away.

'They've got Davey, the bastards,' he said, stalking off towards the police lines.

'Sammy, come back,' Ian called, taking a few steps after him. 'We can't take 'em on, on us own. You'll end up gerrin' snatched.'

'Yer silly, old fucker,' Ian muttered.

Sammy strode on down to the police lines.

No one moved from the phalanx of police, even when he closed with a big copper in the middle of the front rank, still wearing his toy helmet, to shove his moustached face forward to scowl within inches of that of his adversary. A sneer formed on the policeman's face as Sammy stared him out, challenging him to attack. After twenty seconds, Sammy swung round and spat on the ground to his right as he sauntered away. The miners cheered him.

As Sammy made his way back, Ian saw the stones start flying over, from behind. He looked back to see that they were being thrown by a tiny, isolated group of students, behind the main body of pickets. Some older pitmen in their vicinity castigated them.

'Pack that in, yer silly, young buggers.'

'Yeah, there's no need for that.'

Ian watched a large, round pebble arch through the air, towards the police ranks. An officer stepped out of the line, dropped his shield, moved into position and caught it, stylishly, in both hands as though fielding for Yorkshire.

"Howzat?" He shouted.

The pickets joined the police ranks in cheering and applauding his catch, as good-natured as any Headingly crowd.

Emboldened and inspired by Sammy's action, another smaller, fatter pitman walked down towards the police lines for a repeat performance. In a cruel irony, one of the stones flying over towards the police lines scored a direct hit on the back of his head. The pickets joined in with the raucous laughter from with the police ranks as he dropped to one knee, his head shrinking down, into his shoulders.

He stood back up and turned back round to face his own side.

'You dozy, fucking bastards!' He bawled.

He scurried back and disappeared into the crowd, holding his bleeding head, to a chorus of catcalls from the police ranks.

One of the older pitmen strode over and bellowed at the students.

'Now pack that in you stupid fuckers.'

A languid, summer atmosphere was restored.

Just after eight, a flurry of activity kicked up amongst the men furthest down the field; the coal lorries were arriving. The rest of the miners roused and consolidated themselves, ready to hurl themselves forward to join the ritual push against the police lines. The stone-throwing kicked off again.

As ever, as a known organiser, Ian was conscious of being a marked man, wary of being captured by one of the predatory snatch squads. But he knew from previous mass pickets that, if he stood back, he would suffer the indignity of other lads berating him and telling him to "stop skulking around back there", so he felt obliged to go forward in the front row of the push. As always, this was made up of a host of the biggest men who towered over him.

Ian was keeping a keen eye on the senior police officer he had spotted behind the great police phalanx, listening out for any orders he might issue to the forces deployed against them. Whilst the heavily-dressed officers in the ranks sweltered, the Assistant Chief Constable, was sipping from a cup in his shirt-sleeves. As soon as he had finished his tea and handed his cup and saucer to an underling, he started pacing up and down behind the rear rank, calling out encouragements to the policemen. Ian heard his bluff calls.

'Steady lads. Get ready for them.'

An anonymous voice in the police formation jeered back at him.

'Which fucker are you then, sir, Michael Caine or Stanley Baker?'

A chorus of laughter broke out amongst their ranks.

As often happened before a charge onto a police shield wall, the stone-throwing stopped as the miners started to psych themselves up with their chant.

"ERE WE GO!

'ERE WE GO!

'ERE WE GO!'

It surged to a roar when the police received the order "draw batons" and started hammering on their plastic shields, to fortify themselves and intimidate their opposition.

As always, a few, lone berserkers charged in first. Leaping into the air, they hurled themselves onto the police line in their desperate attempts at breaking open a gap in it. Ian was thrust up against Sammy as they clattered into the shield wall with the rest of the mass of miners, adrift in a cacophony of noise; grunting, heaving men, shouted police orders and exhortations and the fierce shouts and chants of the miners in accents from all over the country; Scots, Welshmen, Geordies and other Yorkshiremen.

The earthy taste caught in his throat as the dust rose from the uneven ground. He had to tread carefully to avoid going down. Then, suddenly, there was no danger of that; he was jammed fast, immobile and helpless, trapped between irresistible force and immovable object, his face squashed up against one of the long line of bucking, transparent shields, the smell of its hot plastic, the hotter bodies and the cloth of the police uniforms in his nose, hidden from the sun in the heat of the crush. Ian saw the furious, ferocious face grimacing on the other side of the shield. He could feel its owner's boots kicking out from under it as he raked at his shins.

The policeman had the leather thong of his truncheon wrapped around his wrist and started flicking the weapon, artfully, like a martial-arts rice-flail, over the top of his shield, eventually catching and nicking Ian's left ear with an inevitable, sharp, painful crack to the side of his head.

The police line seemed about to buckle then held. Some of the miners were trying to rip the shields from the policemen they were up against. A few feet away to his left, Ian saw someone had succeeded; a shield, torn from a copper in the front rank, was flung up high in the air. Ian heard the cheers, turned his head and saw it being passed back over the heads of the pickets, like some captured, battle standard. As he did so, another crafty flick of the policeman's truncheon nicked the edge of his ear again.

'Bastard!' He growled.

The miners surged forward to exploit the temporary break in the shield wall, trying to force a way through and break the police formation. A riot helmet, ripped from the head of an officer exposed by the gap in his side's defences, flew up high into the air.

The senior policeman in shirt-sleeves ordered mounted officers forward, to relieve the pressure on his lines. As the centre of the phalanx opened inwards, in the vicinity of the break, the horsemen cantered out. The miners broke and ran, in a mass, towards the rear.

The horsemen kept their order, hand-cantering without drawing their lengthy batons. They halted after little more than fifty yards, stood in a single line for a few minutes and then withdrew.

They had relieved the pressure on their foot colleagues, enabling them to reform, but, despite their relative restraint, as always, the use of horses had inflamed the striking miners.

The stone-throwing resumed with even greater intensity. Dust hung in the air as thousands of miners milled around on the dry stubble.

They heard the sounds of powerful engines and the heavily-laden coke lorries began to depart from the plant. As the miners rushed onto the police lines for a second, big push, Ian glanced at his watch as he jogged forward with them; 9:17.

Again both sides pushed and shoved as the short stream of trucks roared away. More of the police in the front rank had picked up the trick and were whipping their truncheons over the tops of their shields.

Being about twenty-deep by this time, the police ranks held the miners' second charge.

When it became clear that all of the coke lorries had managed to get away, the pitmen fell back a short way, once more. They stood around in groups, hot, panting and thirsty. The shirt-sleeved, senior officer seized the moment and sent out mounted officers and short-shield units to drive them further off, with considerably more than the "minimum force" proscribed in their police manual.

Having achieved their immediate aim of driving the pickets back, the short-shields fell back, still facing the miners, who trailed them, cautiously.

With the lorries gone, a lull settled over the scene.

Although a second lorry run was due later in the day, four hours into the mass picket, most of the miners drifted off towards Orgreave village, back to where the local women were still serving up water and to where their coaches were parked.

The few hundred left behind settled down in the sunshine, many of them removing their shirts and tee-shirts to bask in its heat, sitting or standing around in front of the police who were still fifteen rows deep, at the bottom of the field.

For half an hour, nothing happened. The police remained standing-to, crammed together tightly in their ranks, roasting and irritable in their heavy uniforms. Behind the shield wall, many of the police officers unclipped their ties without bothering to wait for the order, others lit cigarettes and passed them down the line.

Some young pitmen had found an old, tractor back-tyre at the edge of field. They got round it and carried it to the centre of the field then set it off to roll down the slope until it wobbled and toppled over, coming to rest some yards from the police lines.

The tightly-packed ranks eased apart and the shirt-sleeved senior officer stepped through. Standing in front of the body of police, he raised hid loudhailer.

'Move back. Move back one hundred paces.'

The remaining miners were now hopelessly outnumbered by the forces against them.

On a shout from the senior officer, the ranks behind him opened up. A squadron of galloping, mounted police hurtled out. Those miners sitting or lying in the sun scrambled to their feet then joined their mates in a frantic pelt up the hill, to avoid being ridden down and trampled by the charging cavalry.

A body of foot police in riot gear, with round short-shields and batons drawn, followed up the charge, at the double. One mob split off and chased down the nearest miners who had darted away to seek cover in a small clump of trees, bringing several of them down, as they cut and ran, with blows to the backs of their heads and legs before taking them and dragging them off. The captured pitmen were bare-chested, tee shirts having been ripped and buttons torn from shirts as they were seized then hauled to their feet by the police. Blood flowed freely from the broken crown of every captive.

The police continued to employ rapidly-repeated, coordinated cycles of mounted charges and snatch squads to gain ground up the field, fifty yards at a time. The huge, main body of foot police advanced relentlessly, with more Zulu-like drumming of shields, reinforcing them and holding the ground taken like a formation of infantry.

A few more miners started to lob stones at the advancing forces. Ian was keeping well back but, even so, he had to rely on being fleet-footed to evade capture by the dedicated snatch squad that kept eyeing him up, pointing him out and making runs for him.

On the fourth drive by the police, the squad targeting Ian ran at him again. He just managed to stay clear of them, as he darted away. He looked back and saw them overrun Sammy. On the receiving end of a fierce swipe to the side of his head, from a truncheon, his mate went down. Three policemen gave the big miner a brisk thrashing with their riot sticks then heaved him up, dashed the toy, police helmet from his head with a last swipe, then hauled him away, stupefied and bloodied.

As the main body of police advanced right up the field and swallowed up Sammy and his captors, Ian saw his mate receiving not just jibes but blows and kicks from the police as he was hauled through their massed ranks. He could see the police shouting at Sammy but couldn't catch their words. In his daze, Sammy heard them.

"Got yer, yer bastard!"

"You don't look so tough now you're not with your mates, do yer?"

Some miners still stood and fought or tried to help friends who had been downed or wounded.

Thousands of police swept forward, the long-shields and reserves still backing up the mounted and short-shield officers in the vanguard of their drive. Even more miners were seized, arrested and taken to the rear, blood streaming from head wounds. Overwhelmed, the last of the pitmen broke and ran.

Ian and some other miners scrabbled over the bank to their left flank, escaping onto the road and starting to dash away towards the village, pursued by more cavalry. Ian caught a glimpse of Arthur Scargill in the rout then saw him knocked to the ground by a blow from a mounted-policeman's stick. He darted over to him. Another passing horseman swerved at Ian and swiped at him with his baton. Ian stopped dead, ducked and weaved then dodged round behind the horse's clattering hooves to help his stunned leader back to his feet and drag him up the bank, by the side of the road, out of the way of the police horses, thundering past.

Arthur looked at Ian but, in his daze, he failed to recognise him as one of his most trusted Barnsley men.

Ian sat him down, turned away and scurried back down the bank and onto the railway bridge, intent on making his escape. Glancing left, he saw that the remaining miners in the cornfield were being driven out of it, up and over the slope at its top edge. As they reached the brow they pulled up, teetering at the top of an almost sheer cliff face, the ground falling away into a railway cutting with a drop of forty feet. The push of bodies, trying to escape the massed police, forced those who had stopped over the edge.

With no footholds and nothing to stop them, the pitmen hurtled down the side of the cutting, skinning their backs and buttocks, crashing on top of each other and onto the track ballast below.

Ian darted across the bridge to the other side of the road. He pulled himself up on the brick parapet and looked over, to see if there was a train coming. The line was clear, but from the shiny surfaces of the four sets of rails, it was apparent that the men below were staggering around on a heavily-used, operating railway.

The crowds of miners on the track couldn't afford to waste any time in getting across the tracks and out of that cutting. With most of them suffering from sprained ankles and bruised knees, their limping progress was agonising to watch. With another sheer face to ascend on the other side of the track as their only escape route, they were perilously exposed.

'The bastards,' Ian growled to another pitman, looking over the parapet at the scene in wide-eyed horror. 'They 'ad this planned, all along.'

The police had reached the bridge. Using a cordon of long-shields, they held the far side.

Ian carried on, running up the road. Behind him, the police were slowed by the wreck of a car that some of the miners had pushed out of a scrap yard on the village side of the bridge, before setting fire to it. Other miners had picked up bits of scrap in the yard and were using them as missiles, their volleys keeping the police at bay.

Another massed cavalry charge broke up this line of resistance, driving the miners off, up the road, past the first houses.

In the village, the routed pitmen tried, in vain, to stay ahead of their pursuers.

As half a dozen horsemen overran them, Ian swerved off to the left, round the back of a car on a second-hand car sales forecourt. A young miner, about eighteen years old, joined him.

Wide-eyed and chest heaving, the youth placed his hands on the roof of the car between them and the police on the road, to show that he was unarmed.

Two policeman darted round the car and flushed them out. As Ian weaved away to the left, to run deeper into the forecourt, the youngster ran off round to the front of the car to surrender.

A big, moustached policeman with a full shield advanced on the lad, his face covered by the plastic visor of his riot helmet, his long baton raised, ready to strike. The young miner held up both of his hands, in supplication, but forward and ready to try to fend off the inevitable blows. The policeman feinted a lunge then rushed the lad, knocking him over, onto his back on the bonnet of the car. He administered several clouts with his riot stick. With the young man stunned and winded, he transferred his weapon to his shield hand, grabbed the young miner's left foot, then pulled him off the car, holding the leg up at waist height and making the youngster hop along behind him. In less serious circumstances it would have appeared comical.

An enterprising ice cream man had parked up on a black, ash-covered parking area, next to the garage. It was only now that the battle was in full force, around his van, that he started to think about suspending his sale of ices. The whole day had become so insane Ian couldn't help thinking that he wouldn't be surprised to see the pickets, riot police and officers on horseback form an orderly queue to place their orders. Wary of succumbing to hysteria, Ian choked off his chuckles.

He threw himself down on the ground, next to the ice cream van, and wriggled sideways amongst the loose ashes, to seek refuge in the darkness under the vehicle. At the same time, the ice cream man, having realised the severity of the situation, shut up shop, jumped into his driver's seat, started the van and accelerated away, leaving Ian completely exposed and illuminated by bright sunshine.

He scrabbled up with a curse and was away, like a sprinter from the blocks.

With hordes of foot and mounted police milling around and harrying the fleeing miners, Ian cut across the road, aiming for an alley between two rows of houses. He jumped over a low wall to give himself a bit of protection.

As he trotted towards a slim woman who had found herself overrun by the melee, suddenly he recognised her. She was standing over the prone figure of a man who was laid out on the ground behind the low wall.

'Can you call an ambulance,' she called out, pleading to the passing police. 'There's a man here, he can't breathe. I think he may have broken ribs. Please, can you help him?'

Intent on obtaining assistance for the stricken miner, the woman was oblivious to the cantering horse and rider bearing down on her. Ian rushed her, just managing to pull her down as the mounted policeman's great, black baton swung down, missing her head and shoulder by the merest fraction of an inch.

Ian pulled her back up to her feet.

'Susan, what the hell are you doing here? We need to get you out of here. They've gone mad. They're completely out of control.'

'Ian! Where did you come from?' She said, peering at the blood dripping from his ear.

'Don't worry about that now.' He pointed to a house a little further on. 'Quick, go and stand in that doorway, out the way.'

He dropped down to crouch next to the injured miner, looking round to make sure that no other marauding policeman was about to strike.

'Are you alright, old lad?' He said, placing his hand onto the man's back.

'I'll be alreet,' he wheezed. 'I just couldn't breathe for a mo'. You get yersen an' that lass out of 'ere. I'll get over to that wall and hide mesen by it.'

Ian glanced around to make sure it was safe then jumped up. He jogged over to Susan.

'Come on,' he said. 'He'll be alright. Let's get out of here.'

They dashed to the alley way, ran between two terraces of houses and found respite in the path running up behind the back gardens of the houses.

For Ian it was *déjà vu*, so reminiscent of his escape from the police charge, down at Whitacre Heath. He just hoped the police would neglect the backways as they had on that occasion.

'How did you get here?' Ian said. 'Where's your Paul?'

'He's back home – picketing Whitacre. We came up by car with a woman from Women Against Closures, from down south. I was helping serve drinks with the local women. We didn't expect it to be anything like this.'

'Where's your friend parked?'

'If she's still there, it's just past the field where all the coaches that brought the men are parked.'

'Come on, then,' Ian said, setting off at a trot.

'You're not here on your own are you?' Susan called out as she jogged after him.

'No, I came down with a bus load of lads. But I got split up from them. The police grabbed a few. They got Sammy, big Sam Crowther – some mob as was tryin' to snatch me.'

Having found the woman and her car and seen Susan and her companions safely on their way, Ian ran back to the coach park.

Trotting through the rows of parked vehicles, he saw the bus on which he and the rest of the Railsmoor men had travelled down. It was going to be emptier on the way back, if they could get past the police and away. He slowed to a walk to catch his breath, it felt like they might enjoy a few minutes without further attack.

As he approached the coach, he saw one of their number sitting on the bottom step at the front door into their coach. A mate stood on either side of him, each with a hand on his shoulder. The stocky man looked up, three streams of blood running down through the grime on his face. Ian stopped in front of him, bent down and looked into his eyes. His pupils were like pinpricks.

'Fucking Gestapo,' the man groaned, in his daze.

'We should get 'im to hospital,' one of his mates said.

'Not 'round 'ere,' Ian said. 'The police'll be waiting to pick up anyone at the local 'ospital so 'e wouldn't get any treatment, they'd just bang 'im in't slammer. 'E'll have to wait 'til we get back to Barnsley. 'E's got concussion – but hopefully 'e'll be alreet, 'til we can get 'im closer to 'ome.'

Rotherham Police Station, 18:53

A surly, overweight police sergeant escorted the smart, young woman through to the big room being used as one of the extra, custodial holding areas. She had been called in to act as the defence lawyer for some of the scores of miners who had been arrested that day.

The sergeant held the door for her. As she stepped in, the reek of blood caught in her throat.

She looked around at the injured miners, some sitting on the chairs that lined the walls of the room in their torn rags, some lying, barely conscious, on the floor, others standing, holding obviously broken arms. There was blood everywhere.

With their shocked faces and the wounds on their heads, arms, legs and backs, they didn't look like defendants arrested by the police so much as prisoners of war, the wounded survivors of some terrible battle.

Sammy saw the look of shock on her face and heard the distress in her voice as she turned on the sergeant.

'These men don't need a lawyer, they don't need me,' she said. 'They need a doctor. They should all be in hospital.'

The Battle of Orgreave stopped the coke runs from that plant; a token victory for Scargill, at a terrible cost to his massed pickets.

The ninety-five miners arrested that day were charged with riot, unlawful assembly and similar offences, on the basis of evidence that the South Yorkshire police fabricated. From the advice they received from their lawyers, when charged, those pitmen knew they could all expect to be found guilty and sentenced to life imprisonment.

Removed

Tuesday, 19 June, 1984: 03:02

B*ang, Bang, Bang.*
    The shocks reverberated through the fabric of the house.
    Ian was awake and out of bed in an instant.
Who the hell was it? They couldn't mistake his house for…
BANG, BANG, BANG, BANG
… a bloody scab's.

It was louder. They weren't going away. Whoever it was, they clearly meant business.

Ian grabbed his jeans and started pulling them on.

'What is it?' Jackie gasped, sitting up and switching the bedside light on.

'Switch that off,' he whispered hoarsely.

'No, the baby.'

'You go to him. Quick.'

As Ian sprang down the stairs, three at a time, he could see a host of lights on the other side of the front door's frosted glass. They illuminated the hall enough for him to see the bottom of the door displacing as it was back-heeled again.

BANG, BANG, BANG, BANG, BANG.

He paused for a second at the door then fury took the place of alarm and confusion, as his protective instinct kicked in.

He stepped towards the door but, with an even bigger crash, the timber frame splintered as the Yale lock's socket was ripped from it and the door flew back at him.

Ian leapt forward, ready to defend his home. Bright lights dazzled him and two pairs of hands scrabbled at him from the glare. He patted them away, fending them off as he staggered backwards. A pair of giant mitts grabbed his legs, their owner's head smacked into his groin and he toppled over, backwards, as his feet were hoisted up. His bare back slapped down, hard, onto the wooden floor.

'Get off him, you bastards,' Jackie shouted, having appeared from behind him, soundlessly.

'*It must be serious.*' Ian thought, he had never heard her swear before.

'Back, back. Get back!' She shrieked.

The immediate threat to his wife made Ian rear up, only to find himself smashed back down by the knee of a man, three times his weight, on his chest.

He felt two cracks to his right forearm, still too dazzled to see that it was a truncheon that had broken his radius and ulna. His adrenalin and protective instinct kept him fighting, as two dark figures scrambled and stamped their way over him and his assailant to grapple with Jackie, naked except for her flimsy night-dress. The big man on his chest paused to grab something from his side then Ian felt a crack explode on his right temple as a truncheon blow swung in from one side and another heftier thwack to the left side of his head as the weapon swung back the other way.

He came round on the floor of a Black Maria, bare-foot, clad only in jeans. He could have been out for hours, except that he could hear Jackie, still screeching at them.

'Leave her alone,' he croaked. 'Bastards.'

A boot stamped down on his chest. He was pinned down by one of the squad of policemen sitting on the slatted, wooden seats down both sides of the vehicle's interior.

As the van lurched off, engine screaming, taking him away, the pain shot through his cuffed arm. Ian guessed, correctly, that the big sergeant with the size fourteen boot on his chest was probably the one who had flattened him in the entrance of his home.

There were a few jokes and some isolated chuckles.

Neil Bradford sat on the other side from the sergeant. He turned his head and looked at Carl Bessemer, who happened to be sitting next to him. Carl turned his head slightly to look back at him. Neither of them was smiling.

Neil looked at his watch; six minutes past three. It felt like it had taken half the night but they had been in, out, all done in less than five minutes.

Engineman

Graham Dent unlocked the light-blue door. He pushed it open, held it to allow Jim Greaves to enter then followed him in, stepping round to the lighting panel on the wall.

Golden light from the early morning sun spilt into the gloomy engine house, casting the shapes of the high windows, tall and narrow like those of a large chapel, across the floor of highly-polished, red tiles.

Despite the proximity of the upcast shaft and dusty coal prep' plant, surprisingly, the windows were far cleaner than those of the colliery's main office building.

As Graham flicked on the rows of switches, the bulb nearest the door came on and the strip lights, suspended from the high, pitched roof, began to flash then light up to reveal the building's aged but immaculate interior.

For a change, Jim did not need to wear his pit-black so he was still in his weekend attire of jeans, tee-shirt, trainers and Guernsey sweater.

'Will you be alright or do you want me to set you off?' Graham said.

'No, I should be alright, thanks,' Jim said.

'Okay,' Graham said, with a nod. 'See ya later. I'll be in the electric shop if you need me.'

'Okay, ta.'

Jim walked across the polished floor to the steel staircase then trotted down the wrought-iron steps into the cellar below. Graham's operation of the light switches had illuminated this great hall, beneath the building.

Jim strode over to the electrical circuit breaker for Number 2 Winding-Engine and heaved on its lever. With some effort, he forced it down and home, awakening in its electrical circuitry a powerful hum as the switchgear engaged. He did the same with the circuit breaker for the Number 1 Engine.

He clambered back up the steel stairs. With the great winding drums on his left, ready for action, he walked across to where the winding-engineman's operating cab for the Number 2 Engine stood, elevated above the floor of the engine house on its steel framework. Ten feet in diameter, the winding drums' great size gave the degree of curvature needed for the thick, stiff winding ropes and enabled the engine to provide them with the high speeds needed for winding the cages.

With the left-hand cage at the top of the shaft, the drum on that side was full of rope, whilst the shiny surface of the drum on the right was exposed, its rope extending half a mile down the shaft to where the other cage rested in the pit bottom.

The winding ropes were attached, one at each end of the drum barrel and arranged to coil on the drum in opposite directions, the underlap rope coiling underneath the left drum and the overlap rope over the top of the other, causing one cage to be raised as the other was lowered, as the winding engine turned the drums.

Five steel steps led up to the small, metal-grid platform outside the door into the engineman's cabin. The steel-panelled walls of the cabin were painted the same NCB light blue as the doors and windows of the building and the big electric winding-engine, itself.

Jim's grandfather had spoken, several times of Jacob Barnett, a close friend of his, who had been a winding-engineman at Whitacre Heath, over half a century earlier. It seemed strange to Jim to be settling down to work in the very personal space that man had occupied.

According to Jim's grandfather, Jacob had eventually gone blind and had relied upon a friend to lead him to this position, each day, to operate the winder by memory of the timings involved. Eventually, the relentless stress of doing a job in which so many men depended on him for their lives, whilst being unable to see the vital indicators, had proved too much for Jacob. Wrapping his arms around the tangled, submerged roots of a willow on the bank of the Anker, he had pulled his head and shoulders down, under the river's waters, and drowned himself.

Jim took off his sweater and hung it up on the hook on the cabin's steel wall. He settled himself in the engineman's seat, determined to treat the cabin in the engine house as his domain for that morning and for all the future occasions when he would be rostered for a shift on this particular job.

The responsibility was daunting; controlling a six thousand horsepower machine to lower a shift of two dozen officials at great speed and land them, safely, in the pit bottom without any incident that might damage the engine, shaft, winding ropes or cages.

Whilst having a mug of tea earlier with the deputies and overmen, as they had been gathering in the dirty offices, Jim had been forced to endure the joshing he had known he would receive. The officials had made references to his ability as a novice winding-engineman, requesting that he should try to avoid killing them all as he wound them down the pit. They had soon let the subject drop, sensing that it would be better, in the interests of his ability to concentrate on performing his forthcoming task safely, to avoid creating any more anxiety in their visibly apprehensive, young undermanager.

Jim was determined to give them, not just acceptable rides down and back up the shaft, but the best rides they had ever experienced.

In position and ready, he phoned the banksman to tell him that he would be preparing the cages, the upper of which had been left supported on its keps overnight, before running them through the shaft in both directions. Although the regular enginemen might not bother to do this, it was good practice and provided a quick check of whether anything had come away from the shaft wall, like a section of water-gathering garland or a downpipe, to stick out into the path of one of the speeding cages. It was also an opportunity for the engineman to observe both winding ropes, looking out for any broken strands on the ropes as one coiled onto its drum and the other flew off and out of the engine house, over the great, spinning wheel on top of the towering headstock and down the shaft.

As far as Jim was concerned, it also had the advantage of giving him a couple of practice runs before winding riders for the very first time.

A black pointer on a great, white dial indicated the cages' positions in the shaft. Jim placed his foot on the deadman, a pedal in front of his seat, which had to be depressed to allow operation of the winding-engine. This safety device addressed the risk of an engineman collapsing or dying suddenly on the job. If the engineman fell from his seat the pedal would no longer be depressed and this would cause the emergency brakes to be applied, bringing the cages to a stop. Whilst this would hopefully save every rider's life, if the engineman's foot came off the pedal whilst men were being wound at speed, it would be seriously uncomfortable because, whilst the drum would make a dead stop, the cage would continue to fly up the shaft until its momentum was overcome by gravity, then it would plummet back down until the winding rope became taught and stretched to its full consequent extension, whereupon it would stop and bounce back and fly upwards again. This desperately uncomfortable and hazardous cycle, for the riders, would tax the strength of the capel and continue until the cage settled in the shaft.

Immediately beyond the top right-hand corner of the window that looked out onto the winding-engine was the large, shaft signalling indicator panel. Jim was ready for the signal bell that sounded above his head, so he did not jump when it rang four times. The signal was confirmed on the signalling board with the "BANKSMAN" and "RAISE STEADILY" indicators lighting up.

In front of him were two large levers. Leaning forward and easing back the big steel lever on the right provided power to start the cages moving. As soon as he had started to ease the power handle on he started to push forward the other, similar lever on his left, thereby starting to release the massive brakes that clamped onto the ends of the winding drum.

With agonised creaks and groans, the huge, curved brake pads eased their grip for the first time that day and the drum started to turn in short snatches. Almost immediately the indicator lights went out to be followed by a single bell-ring from the banksman which illuminated the "STOP" light on the signal indictor board; Jim threw the left hand brake lever back and pushed the power off with the right lever.

After a pause the "KEPS UNDER" indicator light went out, to be replaced by "KEPS CLEAR" on the signal board, telling Jim that those restraints would not prevent him from lowering the cage from the top of the shaft.

With no signal from the banksman to indicate that they were opening gates to let men on, Jim could wind through the shaft with confidence, when he was ready.

He shifted in his seat then took a deep breath. It was time for his first run through at full winding speed, on his own.

He leant forward, depressed the deadman pedal and drew back on the big, steel lever on the right in front of him, providing power to start the cages moving. As soon as he had started to ease the power handle on he started to edge the brake lever on his left forward. The brakes started to ease and the great, electric winding motor started to hum; the pitch of its noise increasing as the winding drum accelerated.

Jim practiced getting a smooth acceleration, let it run at full speed until he reached the braking point then started to ease the right hand power lever forward then the left hand brake lever back in smooth, steady movements to slow the cages.

When the cage was three drum revolutions from the shaft's top a bell rang, reminding him to reduce the rope speed to less than three feet per second for the last couple of revolutions. This was to prevent the automatic overspeed devices from activating, bringing the cages to a jarring halt and making it necessary to reset the power.

Jim pulled the brake lever back, fully, and pushed the power lever the last few inches to land the cages smoothly as indicated on the big position indicator's dial.

The banksman rang a signal which illuminated the "FIRST MAN" sign, indicating that he needed to lower the member of Area staff, who had come to the pit to undertake the onsetter's role, down to the pit bottom, ready to let the officials off the cage when they were wound down to start their inspection shifts.

All Jim had to do was obey the signals and execute another smooth run through the shaft.

He landed the onsetter gently at the pit bottom. He knew the onsetter would be lifting the cage barrier and sliding the big, mesh gate aside, alighting from the cage, closing the gate, turning the compressor on to operate the gates for the duration of the shift and setting himself up, in charge at the shaft bottom.

Jim was expecting to have a few minutes to compose himself for winding the run of officials but the Banksman's bell rang three times. The three rings illuminated the signal marked "MANRIDING", indicating that men were to be lowered down the shaft immediately.

When the officials had boarded the top and bottom decks of the cage at the top of the shaft, the Banksman followed up with two more rings which caused the signals "BANKSMAN" and "LOWER" to show on the signalling board. The onsetter responded from the bottom of the shaft to confirm this with an immediate, single ring and the "PIT BOTTOM" and "RAISE" indicators lit up.

Jim eased them away, accelerating them hard but smoothly for twenty seconds until they were at the full manriding speed of over fifty miles an hour.

Seventy tonnes of rope, cages and men were flying through the shaft.

Jim forced himself to hold his nerve and keep them winding at full speed for another thirty seconds, resisting the temptation to restrain the cage's fearsome speed as he pictured it, full of men, hurtling down the shaft and the other cage flying back up towards it.

Perhaps a little eagerly, Jim started easing back on the power by pushing the right hand lever forward and then introducing a little brake with the big, left lever. With the officials' cage landed perfectly in the pit bottom, Jim settled back in his chair with the brake lever fully back and the power lever fully forward and released a quivering, sigh through pursed lips.

As soon as the officials were disembarked in the pit bottom, Mad Jack was ready with another engineer to examine the shafts, starting with this shaft.

Now Jim needed to demonstrate his skill in lowering the cage to the pit bottom at a safe and steady rate while Jack and his mate stood on the top of the cage to carry out their hazardous inspection.

After that he had to run the cages through several times while they did the usual examinations of the winding ropes from the top of the headstocks and observed the operation of the headstock wheels' bearings. He was glad of the additional practice this gave him, in anticipation of winding the officials back up at the end of their shift.

Having moved across to the operating cabin for the bigger and more powerful coal winding-engine, they repeated all of the inspection runs for Number 1 Shaft.

When the shaft examiners had inspected the sumps in both pit bottoms, the onsetter rang for "MANRIDING" and signalled to "RAISE", the banksman responded with "LOWER" and Jim, gaining in confidence, gave them a smooth, fast run through Number 2 Shaft.

With the shaft inspections completed, Jim had time for a cup of tea. As he came out of the winding-engine house to return to the dirty offices he met up with Jack and his partner for the day.

'That was borin', Jimmy,' Jack moaned at him.

'What do you mean?' Jim said.

'I was expecting a bit of fun with a novice on the controls, I was 'oping for a proper white-knuckle ride but that was perfect winding – no fun at all.'

'That's what I was aiming for,' Jim said, smiling in pleasure at Jack's backhanded compliment.

Jim was reading a few deputies' reports as he drank his tea. Les Parker was the senior overman in charge for that shift. As usual, he sat opposite Jim, smoking a cigarette.

Eventually he looked over at Jim.

'You'd better be goin', Boss. Those officials'll be waitin' for yer.'

'Is it that time already?' Jim said, glancing at his watch. 'Oh, you're right, I'd better be off.'

Jim jumped up, grabbed his jumper from the back of his chair and rushed out and off to the winding-engine house.

Back in the engineman's cabin, he had time to settle himself down and run through the sequence in his mind and get fully re-focused on the job.

The onsetter gave three rings, requesting "MANRIDING" then another one for "RAISE".

"Right, let's give them another smooth ride," Jim murmured to himself.

He accelerated the cages away smoothly, left them to rocket upwards at full speed for a full three quarters of a minute then started to slow the winder.

The winder started to surge, its digital rope speed indicator was all over the place. The pitch of the massive electric motor's note changed, whining high then groaning low. Jim controlled his rising panic and struggled to slow the cage's ascent smoothly. Looking out from the engineman's cab across the engine house he saw smoke rising from the lines of grout between the big, red floor-tiles. He blinked his eyes to clear what he hoped would prove to be an apparition. When he opened them wide and stared, the smoke had thickened.

With the winding speed slowed right down, the lack of control became even more alarming. Jim threw the power and brake levers, bringing the cage to a stop in the shaft.

"What the hell's going on?" He muttered. 'I hope to God I haven't knackered the winding-engine.'

The phone on the wall behind him rang. He jumped up and answered it.

'Banksman here. What's the problem, you haven't landed the cage yet?'

'I know. I can't control the winding-engine. There's smoke coming up out of the floor. It looks like there's something badly wrong over here.'

Jim put the phone down and darted out of the cabin, jumped down the five steel steps and ran over to the staircase down to below the engine room. In the big cellar below, he was relieved to see that it was not smoke billowing upwards but steam from the boiling, clear liquid in a great steel-framed, glass tank, the size of three household baths, that was elevated on a metal frame.

Jim went back to the cabin and phoned Graham Dent, to take advantage of the Deputy Electrical Engineer's invitation to call him if he had any problems. As he got there the phone rang. Jim answered. It was the banksmen again.

'The officials are moaning; they've shouted up to say can't you just try to land the cage? They say they're all prepared to give it a go if you are.'

'Tell 'em I'm not moving it again until I've got proper control of it. If it surges at the wrong time we could end up with the cage flying right up into the headstocks and the detaching gear operating. Then they'd be stuck up there in the cage for hours.'

The detaching gear connected the cage's four massive chains to the capel on the end of the winding rope. Without this device to disconnect the rope from the chains on the cage in the event of an overwind a catastrophe would result with the cage being wound right up to crash into the headgear at the top of the shaft. Unfortunately, no-one had managed to devise any device to prevent the lower cage heading for the bottom of the shaft from smashing into the boards across the shaft at the pit bottom and plunging into the sump in such an event.

Jim was particularly worried because there was always the risk that the emergency stays in the headstock might not operate fast enough to catch and hold the cage. When this had happened in similar incidents at other collieries in the past, the cage had fallen back down the shaft, taking its load of men plummeting to their deaths.

Even if that did not happen, an overwind would put the supply shaft out of action for the several shifts that would be needed to lower the cage out of the headstocks and reconnect the rope.

Whatever its outcome, an overwind was classed as a serious, hazardous incident by the Mines and Quarries Act Regulations and was reportable to Her Majesty's Inspector of Mines.

'I don't know what's going on here,' Jim said, when Graham arrived.

Graham followed Jim down into the cellar.

'What's this tank full of boiling water?' Jim said.

'It's the liquid controller,' Graham said.

'What's that?'

'It's like a variable resistance which controls the power delivered to the electric motor.'

255

'So that's why I couldn't control it then,' Jim said.

'Yep, that'd be it.'

'It couldn't be my driving that's knackered it, could it?'

'Nah,' Graham said. 'It happens occasionally if we don't remember to check the level in the tank and top it up.'

Jim was so relieved that he did not bother to comment on the Electrical Engineering Department's overly-relaxed attitude to their maintenance of this essential, control unit. He went to the phone and called the banksman.

'Shout down to the officials and tell them that we're going to be a few minutes while we sort out a problem with one of the controls,' Jim said.

Jim and Graham used a trolley to fetch two, large, plastic drums of distilled water from the colliery's electrical workshop.

'That should do it,' Graham said, when they had finished struggling together to lift and empty the containers into the liquid controller. 'If you land them, I'll bring some more barrels over and top it right up when you're done.'

When Jim returned to the cabin and sat down in the engineman's seat, with Graham looking over his shoulder, he was able to move the cage again and land it gently with perfect control restored.

'There you go,' Graham said. 'It'll be fine now.'

Jim drove the engine for the last run to bring the onsetter up from the pit bottom to the surface then closed down the power to the winding engines.

He headed back over to the dirty offices, thinking he had better go and get the inevitable piss-taking from the deputies and overmen over and done with. It would be better than having it keeping on cropping up as a point of ridicule as he encountered them through the coming week.

He walked in and was surprised to find none of the officials looking concerned or irritated by the delay and their wait in the shaft.

'Never mind, Bosso.' Joey Bondar called out. 'I've got used to you keepin' me 'angin' around.'

'You found out what it was then Boss?' Stan Townsend said, cheerfully

'Yeah, it was the liquid controller.'

Stan shook his head.

'Never 'eard of it.'

'No, nor had I, before today. Something electrical, Stan.'

'Anyway, better safe than sorry.' Stan said, quietly. 'The lads just said as they was glad it was you we 'ad on the engine, looking after us this morning.'

Come the Revolution

Monday, 16 July, 1984: 05:52

There were half a dozen pickets on the line, ready to confront the strike-breakers heading in to work the dayshift.

'I got a letter from the taxman on Saturday sayin' they were holding back money they owe us,' Alan said.

'Yeah,' Colin said. 'I got one o' them meself.'

'What's it say?' Paul said.

'Mine said I was entitled to a rebate but they won't be payin' it 'cause I'm a striking miner.'

'Mine said that, un all,' Alan said. 'How can they justify that, the robbin' bastards? They say it's our money and we're entitled to it and then they 'old onto it. What's it got to do with the fuckin' taxman?'

'It's like I say,' Paul said. 'They're turning the whole state machine against us.'

They saw the young Non-Statutory Undermanager's Escort turn into the pit lane. Jim drove up the pit lane, as always, winding his window down on the driver's side and pulling up to speak to them.

'Morning lads,' he called out. 'How's it going?'

'Still winning, Boss,' Paul said.

'Well, at least you've got a better day for it today,' Jim said. A cool breeze moderated the early heat of the sun.

'Yeah, we don't n-need the b-b-brazier today.' Brian said

'Who's that scruffy bunch of buggers, back there?' Jim said, with a backward jerk of his head towards five young males twenty metres back up the pit lane.

'They're fuckin' idiots,' Alan said.

'They reckon to be Socialist Workers Party from Leicester Poly',' Paul said.

'Oh, right,' Jim said.

'Huh, w-w-workers. They don't look like they'd kn-n-know a d-day's work if it got up and b-bit them.' Brian grumbled.

Jim looked in his mirror, slapped his car into reverse and shot back to halt alongside them.

They looked to be in their late twenties. Jim suspected they were probably career students. Extremely thin with long, unwashed hair and wispy beards, they were dressed in olive green, army surplus jackets, two of them having completed their outfits with Che Guevara berets.

'Good morning,' Jim said cheerfully, eliciting no response from any of them.

'Do any of you know how to strip, clean and shoot a self-loading rifle?' He called out.

'No,' one of them sneered, feigning aloof contempt.

'Well, I do,' Jim said.

'And, come the revolution, I'll be on the other side.'

The students scowled as the pitmen on the picket-line guffawed at them.

Jim stuck his car in first, drove up to the pitmen and coasted past them.

'See you later, lads.'

'See ya, Boss.' 'Good on yer, Boss!' The pitmen shouted.

17:14

At the start of the colliery's daily management meeting, Phil Grant, the Electrical Engineer identified an outcome from the ongoing overtime ban.

'There's one spin off to it,' he said. 'The loco batteries have been reported as being in the best condition ever recorded, in terms of how well they hold charge and the other readings.'

'What's that got to do with the ban?' Harry Everitt said.

'Well, that's the only thing I can think of that's changed. I reckon it's a result of them getting maintenance by the managers taking the weekend shifts in the underground loco battery charging station. What we're seeing is them doing a much better job at topping up the batteries with distilled water and on the other minor maintenance work. It's raised the batteries overall condition.'

'That's not a bad outcome then,' Harry Everitt conceded.

Jim thought about the implications.

'The men deployed in there come under you,' he said to Phil. 'You shouldn't have been allowing 'em to get away with not doing a proper job.'

'He's got a point, Phillip.' Harry said, before the stunned Electrical Engineer could respond to his young, management colleague. 'If one manager on two shifts at weekends can make more of an impression than the ruck of men you 'ave down there, round the clock during the week, we ought to be lookin' at makin' a few economies and sheddin' some o' their numbers.'

'I always thought it was too much like a fuckin' rest home in that place.'

Persuasion

Friday, 10 August, 1984: 05:48

As Paul stepped down, out of his back door, Brian unfolded a crumpled sheet of paper and thrust it up in front of his face.

'D-did you get one of these y-y-yesterday?' He said.

Paul raised a finger to lips.

'Oh,' Brian whispered. 'Sorry.'

Paul closed the door soundlessly and turned the key in the lock.

"Yeah, I got one. A message from the miners' friend, MacGregor the Axeman,' Paul muttered. *[Note 23]*

He read the letter's title.

'"Your Future in Danger".'

'What d'ye reckon?' Brian said when they were on the pavement and heading for the pit. 'When I saw "NCB" on the e-e-envelope and read that on it I thought they were s-sackin' me for b-b-bein' on strike.'

Paul snorted.

'I wouldn't put it past 'em. But what that letter shows is that MacGregor and Maggie and the rest of the Tories are worried.'

'So are we, er, w-w-winnin'?'

'Not yet.' Paul said. 'But we're not losing, neither. We need more support from the rest of the TUC, that's what Scargill and the rest of the leadership are trying to mobilise.'

They walked on in silence. Paul had no intention of worrying Brian by sharing his concerns at recent events.

At the start of the previous week, the news, on television, had reported the issuing of a High Court Order for the seizure of the NUM's South Wales Regions' funds of three quarters of a million pounds, in response to the region's defiance of an injunction, granted to two haulage firms, against the union's picketing of their movements, into and out of collieries. Paul had seen this as an attack on the whole labour movement. In his view, it would need all of the big unions to show solidarity by fighting back and facing down the Government that stood behind the Court.

He was not alone in this belief. When a horde of South Wales miners forced their way into their Wales' NUM Headquarters, the Region's NUM President, Emlyn Williams, addressed them.

He ended by saying, "We hope trade unions will show solidarity with the miners, and that, as of today, throughout the country, there will be a general strike."

Arthur Scargill had backed this up by calling on the TUC for "physical support".

But they had received no additional support, whatsoever.

Paul was struggling to hide his despair. In his view, the Tories were upping the stakes but, apart from the train drivers who were blacking coal trains, the rest of the unions were leaving the miners to fight off an assault on the entire labour movement, alone.

He had to crush his fears himself and fight on; it was vital that nothing was allowed to undermine the belief of the strikers and their wives in the cause and their prospect of victory.

Friday: 17:33

The last daily management meeting of the week was coming to its end in the General Manager's office. Having reviewed the week's production and discussed a few exceptional matters for the following week, Mr Everitt had one last item.

'I've had a request from Area for us to give some thought to getting out to go and try to persuade more of the men to come back to work.'

Ken Goodall's brow furrowed as he gave his gaffer a sidelong look.

'What have they got in mind?'

'Nationally, the Board is thinking about balloting the men over the NUM's heads,' Mr Everitt said. 'And you know about the Chairman's letter.'

Seeing the complete lack of comprehension around the table, apart from in Mr Hope, with whom he had discussed it earlier, the General Manager explained.

'Every striker's had a letter from MacGregor, in the last couple o' days, tellin' 'em their pits are at risk if they don't return to work. They're looking for us to build on that. That fuckin' idiot of an Area Director for North

Derbyshire has been tryin' to make a name for hisself. 'E's got 'is colliery management teams going out and doing home visits, door-steppin' the men and tryin' to persuade 'em to come back to work. Area wants us to think about tryin' it as well.'

'Well, they can fuck off as far as I'm concerned, Gaffer,' Jim said. 'The men know they can come back when they're ready. It's a matter for their own consciences. They know we'll do everything we can to support them and make it possible for them to come back but they'd see it as a monumental, fucking impertinence if we started turning up on their own doorsteps pestering them.'

'I agree with Jim,' Ken said. 'If anything, it's likely to be counter-productive – make 'em more likely to stay out.'

Mr Everitt smiled, with a snort of amusement, 'I think that's what matey's been finding in Derbyshire. The men have been telling his managers to fuck off.'

'We have to take a longer view. After all this is over, we still need to be able to lead the men when they're back off strike.' Jim said. 'You can't do that if you destroy your credibility.'

Mr Everitt considered their responses, looked around the table and nodded.

'Righto Gentlemen,' he said. 'I think we've got a consensus. You've said the same sorts of things as me and Mr Hope said earlier. Anyway, Area wanted me to test the level of enthusiasm amongst our team and I think I can convey our considered response to 'em now, thank you.'

A Forced Sale

Saturday, 11 August, 1984: 10:47

That was it; gone.

Despite the bright sunshine outside, the cool of the room chilled him as he sat at the end of the kitchen table.

Diane was round at her mum's, with Emily, keeping out of his way. She'd known he would want to be on his own when he did it.

At least he still had Diane. And his little girl. And the roof over their heads. For the moment. But what a price.

The baby Diane was expecting ought to have been a beacon of hope but, in the circumstances, it was just one more, unavoidable burden. Colin couldn't imagine how they would manage, how they'd cope with that huge, additional responsibility.

The pile of notes was a guilty secret, peeping out from the envelope on the kitchen table; five hundred and twenty pounds.

When he'd first started thinking about it he'd tried to fool himself that he might get as much as seven hundred, or even seven fifty, for her. He knew now, that optimism had only been a way of justifying, to himself, the placing of the advert in Motorcycle News, the first of a sequence of wilful betrayals.

As long as Paul kept his Norton, Colin had been determined to try to do the same with his Triumph. Perhaps being a couple of years older, having worked a couple of years longer had made it possible for Paul to put something away for a rainy day or, rather, this tempestuous season of five, straitened months.

He'd held out for as long as he could but he couldn't justify keeping the bike, even if they hadn't had their next child on the way, not when they were out of money and unable to make the monthly mortgage repayments on their home they had bought off the Coal Board.

Despite still being good, biking weather, Colin had known it was a poor time to be trying to sell. It would have been better back in the spring when there were people in the market for new bikes, with the prospect of a whole summer's riding ahead of them. But back then, despite all the pessimistic predictions, he had still been hoping that something might give in the struggle in which they'd become engaged and they'd be back at work.

There had been only one response to his advert, just one potential buyer. Apart from that, nothing. A rich kid from Solihull, whose dad had brought him over then hung around in his Merc' while the son took her off for a spin.

When the youngster had returned with Colin's pristine machine, the father had driven the hard bargain. Colin could tell he knew the score; Colin's house in sight of the pit headgear, obviously stuck with others of his kind in their fight for survival.

Colin saw now, when it's obvious you're broke and desperate, it's a bloody poor basis for negotiation.

The kid had smirked when Colin had asked him to give him chance to get inside his house before starting her up and riding her away. But it was clear, from the father's downward, mouth shrug and steady nod, that he had understood. Perhaps he knew the true value of things, perhaps been around long enough to have been on the receiving end of the odd kick in the knackers.

Five hundred and twenty quid.

When all was said and done it made no difference to their position. It was all accounted for. When they'd used it to pay off some of what they owed, they'd still be broke; still up to their necks in debt.

The hot, heavy tear surprised him, shooting down his cheek to fall onto the Formica table-top.

It wasn't just for his bike, it was for everything they'd suffered so far, with no end in sight. And for how much more they might have to bear.

It was for the hatred aimed at them, when all he'd ever done was work hard, with pride, for his family and his pit. It was despair.

He squeezed his eyes shut, opened them, wiped his face, brushed away the tell-tale droplet on the table, lifted his chin and took a slow, ragged breath through gritted teeth.

He'd soldier on and play the man. Like Paul.

Thieves

Whitacre Heath Colliery, dirty offices. Wednesday, 22 August, 1984: 03:58.

Jim pushed down on the big table as he got up from his old, captain's chair.

'Right! I'm off for a bath.'

Mad Jack had joined him for a mug of tea. Whilst Jim had read and signed off the afternoon-shift deputies' reports in his pit-black, Jack had been updating him on progress on the rectification of a minor defect on 56's Face's tailgate shearer.

Jim hooked his Davy lamp onto his belt, put the bundle of report books under his arm then slipped a finger through the handles of their empty, coffee mugs. He went out to the shelving on the wall of the communal office, put each of the report books back in its correct, marked slot then took the mugs to the sink to wash them out.

'See ya later, Jimmy,' Jack called, on his way out.

'Yeah, see you later,' Jim called back. 'Thanks, Jack.'

Jim left the mugs to drain on the side of the sink then strode out of the dirty offices, along the short length of red-tiled corridor and into the lamp cabin. As always, he slowed on entering, twisting sideways to unbuckle the narrow, leather strap that bound his self-rescuer to the left side of his belt, then turned left to walk along the long section of wall in the lamp cabin that housed the racks for the individual self-rescuers for all of the underground workforce. He whipped his rescuer off its strap and dropped it into its space in the racks.

The lamps and rescuers of the men and officials bore the user's tally number but a manager's lamp, rescuer, tallies and Davy lamp were always engraved with his name or initials.

Jim unhooked his Davy from his belt as he walked over to the entrance of the flame lamp room, where the valuable lamps were stored for security. The top half of the heavily-armoured, steel door was open. From the loud, whirring noise coming from inside the enclosure, Jim could tell that the nightshift lamp room attendant was hidden away inside, polishing flame-safety lamps with the big, rotary, wire brush driven by its bench-mounted motor.

'Good morning, Sidney,' Jim called out. 'Here y'are.'

He rattled his lamp on the brass-covered counter across the top of the door's closed, lower half to make sure he had been heard over the noisy machine, so that the attendant would know that his lamp was there and put it away, safely.

Having undone his belt and pulled it out of the fixings on his cap-lamp battery, Jim slotted the battery into its gap in the charging shelves. He stuck the lamp-fitting, on the end of its cable, into the charging position on the rack above the battery and turned it round so that its terminal pressed home, against the sprung-metal strip, making the electrical connection. He checked the ammeter above his battery slot, the needle flicked round then wavered in the central position on its dial; his lamp was on charge.

As Jim came round the end of his lamp rack, belt in hand, he encountered two, tall, relatively young policemen, drifting into the lamp cabin from the clean side. They ambled across the lamp cabin, side by side, in sharp, white shirts with dark blue epaulettes, both wearing their helmets. The slightly older of the two was slapping his truncheon, behind his back, on the palm of his left hand. Both seemed to have perfected the steady, pantomime policeman amble of Dixon of Dock Green.

'Good morning,' Jim said, eyeing them both, pointedly.

'Good morning.'

Only the younger of the two responded, the other gave the impression of having failed to register Jim's existence; he might as well have been a ghost.

Jim thought it not only incongruous but highly undesirable to have policemen in the lamp room, wandering around as if they owned the place. They were at the pit to police its entrance and perimeter and had no reason to be there, strolling around the surface buildings.

A few minutes later, Jim was in the undermanagers' bathroom, stripped off and adjusting the temperature setting on the shower when there was a knock on the door. It eased open. Sidney's skinny face and shoulders leant right in, seeking him out.

'Excuse me Boss, sorry to trouble you but did you bring yer lamp out the pit?'

'Yes, I shouted "here you are" or something as I put it down on the counter after I'd shouted "good morning" to you, a couple of minutes ago,' Jim said with a frown.

'I thought so,' Sidney said, with rising indignation. 'Those two fucking coppers 'ave 'ad it away.'

'What? The two walking through the lamp cabin?'

'Yeah. I wondered what they was doin' in there. They've nicked it. I bet they thought they'd take a souvenir of their time in the coalfields, the thievin' bastards. Sorry to disturb you, Boss.'

Jim stood still under the hot water of the shower, fuming at the policemen's casual removal of something so significant. Not only did you have to work hard to earn the right to carry an official's or manager's lamp, if you lost one it had to be reported to the Colliery General Manager and you were automatically fined for at least the value of its replacement, however senior your position. But, more importantly, whilst the lamp belonged to the pit, it was part of the pitman to whom it was entrusted. It carried some of his luck.

"Dixon of Dock Green wouldn't have done that," Jim thought bitterly.

'Bastards!'

Forty-five minutes later, bathed and dressed, Jim walked down the dark corridor of the main office building. He heard one of the policemen laughing at something he had just said to his colleague and Roger Dawkin in the Control Room.

'Good morning Roger,' Jim said, throwing the door open and bursting into the Control Room.

'Good morning, Boss.'

Jim breezed round to the other side of the control panel from Roger's position and leant on it.

'I tell you what, Roger,' Jim said, ignoring the two policemen. 'You don't want to leave anything lying around. If you do it's, likely to go missing.'

'Why, have you lost something?"

"Yeah, some fucker nicked my flame safety lamp from the lamp cabin when I came up the pit.'

'Never!' Roger said. 'Who could 'ave done that? There's no fucker about on the bank at this time of night.'

'I know,' Jim said. 'Strange ain't it? If I find out who's had it, I'll kill the fucking bastards.'

The two policemen exchanged quick glances then glanced towards the door but said nothing.

'Yeah,' Jim said, pensively. 'You saw me come through the lamp room, didn't you officers? I put my lamp down on the counter and a couple of seconds later it was gone, before the lamp room attendant had time to pick it up. Did you see anyone?"

'No, we just passed through and went round and back towards the canteen,' the older of the two said. 'Didn't see anyone, did we Neil?'

His partner looked miserable and said nothing.

'Hmmm, strange that,' Jim said, giving each of them a hard stare.

The two policemen drank up their coffees quickly and stood up leaving their mugs on the table.

'Er, see you mate.' the older copper said to Roger.

'Yeah, thanks for the coffee, Roger," the younger one said.

They went out and closed the door. Roger frowned quizzically at Jim who shook his head and raised a finger to his lips. They could hear mumbling in the corridor, on the other side of the Control Room door.

After a couple of minutes, the handle turned and younger of the two policemen eased the door ajar.

'Someone's taken our truncheons and helmets.'

'Never!' Jim declared, deliberately unconvincing in his mock disbelief. 'Where'd you left them?'

'We left them on this table, out here in the corridor.'

'What did I tell you? There's only a handful of us here, on the surface, but we've got thieves among us tonight. And I bet you'll be in trouble when your superior finds out your uniforms are incomplete and you aren't carrying your truncheons.'

'Yeah. it's a disciplinary,' the policeman said, looking hopefully at Jim.

'What a bugger, eh Roger?'

'Yeah Boss,' Roger said, a suspicious smile forming on his lips. 'It's too bad.'

'Will you let us know if they turn up?'

'Of course, officer,' Jim said. 'But I wouldn't hold out much hope, would you Roger?'

'Oh no, Boss, once owt's gone, it's gone for good 'round here.'

The door clicked shut behind the policeman. Jim winked in answer to Roger's quizzical expression and, in a loud voice, started covering a couple of points he had noted while down the pit. Roger replied in a similarly staged, professional manner.

When they heard the voices in the corridor stop as the door at the top of the corridor clicked shut, Roger broke off from what he was saying.

'So what's been goin' on, Boss?'

Jim described the policemen's suspicious presence in the lamp room as he had handed in his lamp and how the lamp room attendant had looked into the undermanagers' bathroom to alert him to his lamp's disappearance and what he had done next.

After he had finished bathing and dressing, as Jim left the dirty office building he looked down the main office building. Through the big windows of the brightly lit Control Room, he spotted the police uniforms of the two thieves, lounging behind Roger at the control room desk, drinking mugs of coffee.

At the side entrance of the main building, he looked through the glass pane in the door. He could see, by the light spilling from the Control Room door, that, for some reason, a small table had been left outside the Control Room, on the other side of the dark corridor. It was apparent that the two policemen had left their helmets and truncheons on it whilst they blagged a couple of coffees off Roger.

Jim eased the door open, stepped silently into the building then crept down the corridor to the table. He lifted the helmets and truncheons and moved soundlessly back up the corridor with them.

He wasn't worried about being caught; if the policemen came out, he'd say that he was just looking for their owners so he could restore the items to them, having already suffered a theft himself that night.

Jim opened and closed the side door without a sound as he made his exit and made off with the policemen's items.

Over in the dirty offices, Jim found Terry Harrison, one of the pre-shift deputies who did the statutory inspections required before men on the working shift could enter their districts, underground. Terry had finished his mug of tea and was hooking his flame lamp on his belt, ready to head for the shaft. Jim handed him a brown-paper packing bag.

'Do me a favour, Terry. Take this down the pit with you and leave it round the back of the transformer house in the pit bottom will you?'

'Okay, Boss. Can I ask what's in it?'

'Yeah, take a look.'

As Terry took a peek, Jim said. 'Two thieving bastard coppers nicked my lamp off the counter from the lamp room when I came out the pit this morning. Then they made the mistake of leaving those outside the Control Room while they were in there scrounging a coffee off Roger.'

'The bastards! Okay boss. Do you want 'em bringing out when I come back up?'

'No, bring 'em out on Friday when those fuckers are getting ready to go back to London and drop 'em into the undermanagers' baths will you?'

'Righto, will do.'

Ten minutes later, Jim was on his way back over to the Control Room again. He could see through the external window of the long, office building that the two policemen were still sitting comfortably in the Control Room, distracting Roger from his work, as yet, unaware that their helmets and truncheons were gone. Jim was looking forward to alerting them to their losses.

'That's the last cup of coffee they get off me, the bastards.' Roger said, when Jim had finished his explanation. 'Or if I do make 'em one, I shall piss in the fucker.'

On the Friday afternoon, when Jim walked into the management meeting, he had a clean, packing bag in one hand.

'I've got something for you Gaffer,' he said.

'What's this?' Harry Everitt said, looking into the bag. 'It's a copper's 'elmet – and stick.'

'I thought you might like to have them as a souvenir of having the boys in blue at your pit,' Jim said. 'It came from one of the bastards who nicked my Davy lamp.

Jim had already tossed the other helmet and truncheon into a rubbish skip. He wanted no reminder of the police presence at their pit, himself.

The Enemy Within

M argaret Thatcher was addressing the Conservative Party's 1922 Committee. She told its members that, in beating Galtieri's Argentinian forces in the recent Falklands Conflict, the country had defeated "the enemy without". She went on to say that the Government now needed to go on to defeat "the enemy within", referring to Arthur Scargill and the rest of the NUM leadership.

Paul Wood heard her statement reported on the radio. If there was an enemy within the country, as far as he and many other miners were concerned, it was her party that appeared to be hell-bent on destroying their union, their industry and, thereby, their jobs and their communities. He took her words to be the launch, by a lower-middle class, Conservative Prime Minister, of the class war that Scargill had warned them about; a clear indication of the Tories' hatred for the nation's colliers.

For Paul, the victory in the Falkland's had been a victory the miners had celebrated as patriotic Englishmen. Several Whitacre Heath strikers had sons who had seen active service down there; a Marine, a Para', a couple of Royal Navy seamen. Being compared to an alien enemy had outraged the miners and been taken as a vile insult.

For Jim Greaves and most of the NCB's management it struck a chord, as it did for many in the country who believed that the militant minority had been a major cause of Britain's slump to ruination in the 1970's. To many British people, outside the coal industry, it seemed reasonable to describe the NUM's leadership, as Mrs Thatcher had, as being "ill-motivated, ill-intentioned, politically-inspired and dangerous to liberty".

Jim could never believe that Arthur Scargill's motivation was the salvation of pits and jobs, the militant trade unionist had said too much in the past that was purely revolutionary.

Mrs Thatcher had not applied the term to the miners, only their union's senior officials. However, as other senior Conservative politicians embraced the concept, Jim sensed that some of them were starting to interpret it as applying more generally, not just to the NUM's leadership, not even just to that union's members but to everyone in the industry, including managers, like him, who were working hard to prevail against the NUM and its strike.

A Monthly Ritual

Monday, 3 September, 1984: 05:45

Brian trudged along behind Paul and Colin. Even in the early morning's darkness, his two mates could sense his misery.

Lorna had stirred and moaned as he had struggled out of bed that morning.

'You're not wastin' yer time down there again, are yer?' She'd sneered drowsily as she'd turned away and pulled the blankets over her head.

Brian knew it was obvious to anyone that he wasn't as clever as Paul and Colin. But he reckoned he was clever enough to know that it was because his wife wasn't as bright as Susan or Diane that she couldn't understand what they were all fighting for, couldn't support him in the ways that the wives of his two mates appeared to support them.

He sympathised with her anger and frustration. It was desperately tough for both of them. With no savings, they both had to forego breakfast and lunch every day, so they could provide breakfast and tea for their two sons; Barry, aged four, and Paul, six. And with money too scarce to feed the electricity meter, they had to rely on candle stubs to light their house at night, and the nights were getting longer and colder. And there was no end in sight.

But what could he do? He had to stick together, with Colin and Paul.

Jim Greaves pulled up next to their picket line, on his way in to start a week of days, after his latest spell on nights.

He had made a gesture, back in May. When they had not spurned it he had made it a routine. Around the first Monday of every month, when he stopped to talk to them he always said the same thing.

'A bit of corned dog for your snap, lads.'

'Thanks, Boss,' Paul said, accepting the plastic, shopping bag that Jim thrust out of his car window. 'Much appreciated.'

'Every little b-bit 'elps, Boss,' Brian said.

'You're welcome,' Jim smiled. 'I'd better get off and try and keep that pit open. Ready for when you lot come back to us.'

'Yer can c-come an' join us 'ere, if you like, Boss,' Brian said.

'It's tempting, it's going to be a nice day for it – and thanks for the offer – but the Gaffer'd sack me if I did.'

As Jim's car pulled back onto the left side of the pit lane to speed off past the pit offices and into the main car park, Paul lifted the bag and peered into it.

"Ey up! We're in luck today, lads.' Paul said. 'Four tins of corned dog – and – there's three Kit-Kats in here as well. We can have a couple o' fingers each.'

They had talked about it before. Mr Greaves always said it was a little something for their snap but Paul and the others were sure that their young boss knew that, each time he did it, the tins would be used to provide the basis for an evening meal for several of their families.

That evening, if they scrumped some spuds from the potato field, half a mile outside the village, four of them would be able to get their wives to supplement a tin of corned beef with a saucepan-full of mash.

Another Minority

Saturday, 22 September, 1984

Cyril's cell-mate usually said very little but, that evening, he had just spent the best part of an hour describing the battle for the coking plant and the subsequent attack on his home. Having listened to the whole story in silence, the Jamaican's dreadlocks swayed gently as he nodded, sagely.

'They took me down the nick and charged me with secondary picketing, resistin' arrest and a load of other public order stuff,' Ian muttered. 'Even though the main resistance was put up by me missus, layin' into 'em with the barometer off the 'all wall.'

'Then the next shift took me down to 'ospital to 'ave me arm set. They even posted a couple o' coppers to take turns keepin' guard on me, like I were a bank robber or an IRA bomber.'

'When they hauled me up before the judge, the coppers' statements were complete works of bloody fiction. Me defence lawyer reckoned the written accounts used to convict me and the rest of the lads were all the same, they even used all the same sentences. 'E reckoned it sounded like they'd all been sat down in a room together, somewhere, an' 'ad someone dictatin' it to 'em all.'

It had been during Jackie's visit, that afternoon, that Ian had decided to share it all with the enigmatic, Rastafarian.

It was only the second time in Ian's first three months inside that she had managed to leave Matthew with her sister, Linda, and make the one hundred and forty mile round trip from Railsmoor to see him. Her journey had been made considerably more difficult by Ian's sale of their car, a week before he was arrested, to pay off the most pressing of their rapidly-growing debts.

He was missing her and Matthew more and more each day as the sentence dragged out. He just hoped the smile he'd given her, on entering the visiting room, had fooled her, that she hadn't seen how close he'd been to breaking down at seeing her sitting there, at one of the little tables, amongst all the other visitors, waiting for him.

After a while, Jackie had given him a strange, intense look.

'How are you finding the other inmates?' She said.

Ian frowned.

"Ow d'ye mean?'

Jackie lowered her voice.

'You know. Being among so many black men – the way you've spoken about them, in the past.'

Ian thought about it.

'I suppose it was a bit of a shock at first, took a bit of gettin' used to. But Cyril, my cell-mate, 'e was set-up like me, snatched in London by the Met', the same mob as came and broke our door down. Reckons 'e 'ad a bit of pot on 'im, ganja as 'e calls it, but they produced a load more in court as evidence that it wasn't for 'is own use and did 'im for dealin'.'

'And then part of it's the way they've supported us, you know, the money we received in the strike funds, from the West Indians in the cities and the Sikh temples. It just feels like we 'ave a lot more in common with them now.'

'The main thing is the blacks used to be one of the minorities, like the Asians and now the Government and the media and police have turned on us, the miners and our families.'

Ian thought about the shout of a Met' police inspector in the police ranks before one of the picket line pushes at Whitacre Heath.

"We're the Met'. We hate spades – but we hate miners even more."

'They hate us and our communities and culture,' Ian said. 'We're English but to the police, the Tories – and even Labour – we're an unwelcome, ethnic minority now.'

'We're untouchables now, as far as they're concerned.'

Baptism

Sunday, 23 September, 1984

Neil sat quietly crushed up against Julie, in the front pew, waiting for the service to start. As he listened to the organ playing, he reflected that the glares he'd received, as they'd entered the church, had been fewer and not quite as hostile as he had feared.

They had come back up to Yorkshire, for the baptism of Julie's cousin Penny's baby daughter.

A large proportion of the congregation was going to be made up of miners and their families so Julie had tried to persuade Neil to give it a miss. But Neil had been determined to be there, determined to prevent barriers from becoming established between himself and Julie's extended family.

The ritual round the font, a little later, gave those present something else to think about and Neil had started to relax.

With the little baby welcomed into the Church and the service over, only the closest family members were required to pose for photographs in the gentle sunshine, just outside the church porch.

Neil slipped away, round to the south side of the church, pacing slowly over the grass between the gravestones, hands clasped behind his back, lost in thought. If he'd had his uniform on, instead of his trim, grey suit, he would have looked like a copper on his beat.

He paused under a huge cedar and gazed out, over the churchyard wall, down the valley to where the colliery's dormant headstocks towered over the old, pit village.

Julie peered round the corner of the church. Spotting him by the wall, she picked her way over to him, arms raised to balance, on her toes, to prevent her heels from penetrating the soft turf.

Neil turned round and waited for her, thinking how elegant and feminine she looked, like a bridesmaid at a wedding, in her blue, satin suit with her hair up.

'I think we'd better give the christening party a miss,' Julie murmured as she closed with him.

Neil looked quizzical.

'It's in the Miners' Welfare,' she said quietly, looking around to make sure that no one had followed her.

Neil shrugged.

'Well that's only a social club. When they let a room out for a function it's not closed to non-miners.'

'It's not that,' Julie said, with a frown. 'Well, I suppose it is. Anyway, you know what I mean, it's their place. It might be – provocative.'

'I'm not going to be provocative, I'm here with you, as part of the family.' Neil said. 'There's no point in being miserable with each other.'

'They need to grow up,' he muttered.

Julie continued to express her misgivings as they walked across the churchyard and out of its gate to follow her relatives down the road to the club.

In the function room, people had started to help themselves to the sparse buffet that was dwarfed by the white-clothed tables on which it had been laid.

The bar wasn't busy, hardly anyone had any money to buy a drink. Neil bought himself pint of bitter and a dry, white wine for Julie and paid for the drinks that Julie had ordered and taken over to her parents. Three men, sitting with plates of cheese and onion sandwiches at the nearest table to the bar, watched Neil turn with his and Julie's drinks.

He acknowledged them with a nod.

'It's alright for you,' one of them growled at him. 'Drinkin' yer blood money,'

Neil stopped by their table.

'I'm sorry,' he said quietly. 'But I don't like it any more than you do. The sooner it ends the better.'

'Likely bloody tale,' another of the miners said.

'Look, let me buy you lads a drink,' Neil said.

The three men scowled at him then the first miner leapt up.

'You arrogant twat!' He bellowed. 'Offerin' to buy us a drink like Lord Muck. You've no business bein' in 'ere. This ain't the fuckin' Police Federation, it's t'pit club. It's bad enough having you fucking Gestapo crawlin' all o'er our village like vermin wi'out 'avin' to put up wi' plain clothes pigs infiltratin' the Welfare to spy on us un' all.'

Neil watched him, bracing himself, expecting to be attacked.

'I'm not stayin' in here wi' a fuckin' pig,' the man snarled.

He spun round and stomped off to the door. His two mates glared at Neil, eased themselved up from their chairs, grabbed their sandwiches from their paper plates and followed his lead, casting ferocious looks back at Neil.

Julie appeared at her fiancé's side.

'Come on, let's go before there's any more trouble,' she murmured urgently.

She took the glasses from his hands and went over and placed them back on the bar. She returned to grab hold of his arm and steer him towards the exit. As they reached the door she waved at Penny, mouthing "sorry" at her scowling cousin before bustling him out of the building.

On the motorway, heading south, it was twenty minutes before Julie broke her silence.

'Can you honestly say that you were surprised at how that turned out?'

Neil carried on driving for a few seconds.

'I knew it wouldn't be easy. I just thought it was important to try, that's all.'

Julie knew that it had taken courage for Neil to go along to the party and that he'd been determined to do it for all the right reasons. But she was furious at his inability to anticipate the impact his presence would have on people there.

'The things they were saying outside the church, about what's been going on, back up there,' Julie said. 'You lot are building up a whole load of resentment. You – the police – you've turned yourselves into the enemy.'

'I know it's tough, but the main thing we're doing is protecting people's freedom,' Neil said. 'Making it possible for people who want to go to work to do it without being beaten up or set on by an army of thugs.

'I've seen what goes on. And I've been on the receiving end. There's a good few who enjoy having the opportunity to take part in some legitimised aggro and take a swing at a policeman.'

'Enjoying it?' Julie screeched. 'They're being starved out in those pit villages. And you say they're enjoying it? Rightly or wrongly their fighting for their families and their communities, now that no one's giving them any other option. There's no point going on to me about how disgusted you are by the behaviour of that idiot Carl Bessemer, taunting the pitmen and winding them up, and then saying that it's all the miners' fault.'

Neil could understand, even share his girlfriend's indignation.

"I'm not saying it's all their fault, I'm just saying they're not blameless, that's all.'

'They're utterly fed up with you. They say it feels like their village has been invaded,' Julie said. 'You just swan around the mining communities and the pits as if you own them. It's like your mate, Carl, taking that Davy lamp. That was nothing other than theft – by a policeman for God's sake.'

Having been disciplined for losing his helmet and truncheon, Neil had felt obliged to tell Julie about the incident. He wished, now, that he'd kept it to himself.

'I know,' he said. 'I told you, I told him not to do it. And I tried my best to put it right.'

'So what? Think how it looks to them. Police wandering around taking something, just 'cause it takes their fancy. Those things are important to miners. They're like symbols, they mean something more than any of your lot could ever understand, they're sacred to them. And he just walks off with it. Then you expect them to drink with you and wonder why they hate you.'

Neil let the silence hang.

'I don't expect them to like us,' he said gently. 'I just want to start rebuilding bridges.'

After her own pause for thought, Julie said, 'Well at the moment, you're not building bridges, you're burning them as fast as you can. And I'd be surprised if you can ever re-build what you lot are breaking down.'

It only served to frustrate Neil, even more, that what Julie said about the incident with the lamp echoed the sentiments that he had expressed to Carl at the time.

'Put it back before they miss it,' Neil whispered.

'Don't be so soft.' Carl drawled.

'That's essential safety equipment.'

'Bloody hell, it looks like something out of a museum. It's no wonder they're shutting pits, if they're this bloody antiquated.'

'Well, apart from that, a lamp like that is like a badge of rank,' Neil said. 'They're important objects, part of the whole pit culture. My grandfather was a deputy down the pit, he carried one. He was proud to have had a position where he was required to carry a 'silver lamp'.

'I shot the sheriff'.' Carl crooned softly, looking at the lamp in his hand, stamped with the marking MR J G on its brass plate, continuing their walk back over to the main offices, 'but I did not shoot the deputy.'

'A lamp like that costs a lot of money,' Neil said.

'That's even better then, a valuable souvenir of our time spent sorting this little lot out.'

'Well, you won't get away with it. The man you've taken it off is important, he holds a senior position.'

'How do you know?' Carl said. 'A young upstart like him.'

'Roger, the bloke in the Control Room, told me who he is. He's in on nightshift to act on behalf of the General Manager of the mine, he's an undermanager.'

'There you are then, I told you he wasn't important, he's only an "under"-manager.'

'An undermanager isn't like it sounds, it's a senior position. He'll be in charge of several hundred men, like a chief superintendent or a commander.'

Neil had found out about the career paths in the coal industry when he had looked into it as he was leaving school. He knew that he and Carl had nothing like the academic qualifications needed to be considered for a management grade job. Despite what he had said about the equivalence of the chief superintendent's rank, he knew that, unlike senior police officers, the young colliery manager would have needed to get an engineering degree and other serious qualifications to qualify for his job.

As they reached the side entrance at the top end of the main offices, Carl tried to conceal the lamp by removing his helmet and putting it inside.

'Well, we both know a chief superintendent or two that are wankers, so who cares?' He said.

Neil and Carl had been detailed to spend the end of their nightshift with other officers from their division, policing the picket line during the period immediately before the start of dayshift when working miners would be coming to cross the line to get in to work. Instead, they had had to suffer the embarrassment of confessing to the inspector that they would not be able to fulfil that duty because they'd both lost their truncheons and helmets then having to try to offer a credible explanation for how this had happened.

The furious inspector had dragged an admission out of Neil that they had been sitting down, drinking coffee, and that their helmets and truncheons had been left unattended when they were stolen. There was no plausible explanation for how they could have gone missing in the heavily-policed and relatively deserted office accommodation of the pit at night.

Neil could not reveal their inevitable suspicion that it had been an act of retaliation by a senior NCB employee, in response to his partner's theft of an important item of colliery equipment.

The young manager would have known that he had them, that they would not be able to confront him about it without admitting to Carl's earlier, inexcusable theft of his lamp.

After the undermanager's appearance in the Control Room following the removal of their truncheons and helmets, Neil had gone to the gents' toilets in the main offices. He had gone into one of the closets and lifted the lid on the cistern that Carl had used as a hiding place. Having recovered the lamp and dried it off with paper towels, he had got it back to the shelf on lower half of the door to the secure room of the lamp cabin.

Neil had thought he had managed to achieve this without being seen, but, as Sidney, the lamp room attendant had told Jim Greaves that morning, 'You was right; Boss. It were those coppers as 'ad your lamp. One of 'em snuck back into the lamp cabin and put it back on the counter. Thought I couldn't see 'im but I 'eard a noise, rushed to the counter, saw it there and saw the younger of that pair o' fuckers disappearing out the lamp room.'

Neil had known it was a faint hope; there had been no corresponding return of their truncheons and helmets. He could not know it but, by then, they were already hidden away, half a mile underground.

'I expect he was hopin' you'd put their stuff back for 'em,' Sidney said.

"Yeah, I expect he was. Fucking hard luck." Jim said, with bitter irony.

Democracy at Work

Friday, 12 October, 1984: 18:20

P aul and Diane arrived at Horace's and Pat's, with the two children, for the family's weekly treat of fish, chips and mushy peas. They were both in high spirits; victory appeared to be in sight, at last.

At the start of the month the Government had attempted to break the massive restriction on the movement of coal with NUR and ASLEF members being sent home for their unions' continued refusal to move coal trains. Paul had seen this as evidence that the Tory Government was rattled, that the strike was really beginning to bite at last. This impression was reinforced when it was reported on the radio and television that the High Court had imposed an injunction on the NUM to terminate the strike within five days.

When Coal Board directors had agreed to arbitration with the NUM, facilitated by the Advisory, Conciliation and Arbitration Service, ACAS, Paul had taken it as a real indication that the opposition was wavering.

The arbitration meetings had actually started the previous day but this day's events seemed far more momentous. The result of a ballot of the members of NACODS, the officials' union, on whether to take action against the pit closures had been announced. Eighty-two per cent of its members had voted to strike. The pressure on the Government, from this, would be enormous.

'It's good news, as long as they actually do come out on strike this time,' Horace said.

The striking miners had been praying for a similar outcome when NACODS had held its previous ballot six months earlier, in the first few weeks the NUM's strike. But, on that occasion, the NACODS leadership had committed their members to working, on the grounds that it was important to stop the pits from flooding and keep them safe and open.

'It's different now. They don't have to risk the pits; it could all be over in a matter o' days.' Paul said. 'They can see how unreasonable the Government's bein' – and they can see we've got 'em on the run.'

Seven months into the strike, the Prime Minister and her Cabinet had to face the fact; their Government was at risk. If allowed to proceed, the strike by the deputies and overmen of the NACODS union would stop production at all of the working pits in the country, just as winter was arriving, at a time when the country had little more than a month's coal stocks left.

The Central Electricity Generating Board predicted that victory for a combined NACODS and NUM action was certain and the Government would be forced to capitulate before Christmas.

Margaret Thatcher was so alarmed that she proposed using the Army to transport imported coal, from the docks, in order to break the strike. The Chairman of the CEGB dissuaded her, telling her that doing this would be certain to result in the immediate walk-out of all of his power workers.

Mrs. Thatcher summoned Ian MacGregor, the man she had appointed as Chairman of the NCB, to her office in 10 Downing Street.

'I'm very worried about this,' she said. 'You have to realise that the fate of this Government is in your hands, Mr. MacGregor. You have got to solve this problem.'

If Paul had known that the Prime Minister was actually inclined to settle the dispute on the miners' terms, following the result of the NACODS ballot, he would have been elated.

Thursday, 25 October, 1984: 19:37
Two weeks later, Keith Deeming turned the television down and answered the phone. As expected, it was Shuey McFadden.

'Have you heard what the stupid bastards 'ave gone and done?' Shuey roared down the phone.

'Yep,' Keith said, his tone resigned.

"There's nothing we can do about it, I was expecting something like this anyway," he thought to himself, *"And Shuey's pissed again, so there's no point in talking to him."*

ACAS had proposed an independent "colliery review process" for pits facing closure. Both NACODS and the NUM had accepted the concept but NACODS had used the NCB's agreement to the proposal as a reason to call off its strike, separately agreeing that the new independent review process for any proposed pit closure could be viewed as non-binding.

McFadden was still shouting down the phone.

'It's fucking un-fucking-believable. They keep goin' on at us for strikin' without a national ballot and then NACODS holds one, their members vote for an all-out and they subvert democracy by tellin' 'em to carry on workin'.'

'I know,' Keith murmured, not bothering to conceal his lack of interest in discussing the development with his drunken Lodge Secretary.

He let Shuey drone on a little while longer, for all the good it would do.

Their best chance of victory; gone, forever.

Worse was to come; with the threat of a NACODS strike averted, the National Coal Board reneged on its agreement and rejected the independent review proposed by ACAS. Its Chairman, Ian MacGregor, went even further.

'There is no basis for further talks with the NUM,' he announced.

The NUM was now even worse off than it had been before the NACODS vote.

The disaster was complete.

Intimidation

Tuesday, 30 October, 1984: 06:11

Jim stood in Ken Goodall's clean office, gazing out of the window, across the main carpark. Jerry Malin, on his way back from filling the Control Room kettle in the main office's kitchen, stopped, stepped back across the corridor from the Control Room door and looked in on the undermanager.

'Somethin' brewin' out there, Boss?'

'No, I'm just waiting for Warbler Worrell to turn up.'

'Oh, I 'eard about that. Mr Hope was telling the Gaffer about it, when 'e turned up yesterday.'

'Yeah, well, I'm going to make sure he doesn't pull that stunt again,' Jim said.

Jerry had been a competent and professional face deputy, before an injury had forced him to take the job in the Control Room.

'Mr Hope said it was some idiot of a general manager in South Derbyshire who'd agreed with NACODS that 'e'd pay any officials who didn't cross a picket line if they were intimidated by the pickets, so someone claimed that excuse and got paid,' he said.

'Yeah, bloody marvellous isn't it,' Jim grumbled. 'Set a precedent. He might've known what'd happen, when you've got wasters like Worrell.'

'And here he is now!'

Jim watched Worrell's car pull up and saw the official and his two companions, Tom Gibbons and another Grade II, outbye deputy, starting to get out.

He strode out of the room, down the corridor and out of the big double doors of the office building's main entrance. He turned right and marched out to meet the three officials.

Jim passed close by the five pitmen on the picket-line.

'Now remember what I told you lads,' he muttered.

Jerry put the kettle down on the desk in Mr Goodall's office and took up Jim's observation point to watch the scenario unfold.

He saw the undermanager make his comment to the pickets as he passed them.

A brief conversation ensued between Mr Greaves and the officials when he reached them then the undermanager turned and called out something that Jerry couldn't make out.

One of the pickets called something back and gesticulated.

Mr Greaves looked at the three officials and they started dragging their feet towards the pit. The undermanager moved in behind them and snapped something that made them step out, as he dogged their steps, shepherding them in to work.

Jerry came out of Mr Goodall's office when he heard the main door open and saw Mr Greaves striding up the corridor with a look of angry satisfaction.

'Well done, Boss,' Jerry. 'How'd yer manage that?'

A mischievous smile formed on Jim's face.

'I stopped and had a chat with the lads on the picket-line, as usual, on my way in this morning and told 'em Worrell was taking advantage of the men out on strike, as well as their pit. I told 'em stopping wankers like him from working wouldn't have any effect but they shouldn't let the idle bastard get a paid-holiday racket going, especially seeing as NACODS had had the opportunity to strike and then played the "safety and keeping the pit open card" instead.'

'When I went out I reminded the lads on the picket line how we'd play it. Then I went and asked Worrell what this intimidation nonsense was all about. He gave me some of his usual bollocks so I called to the pickets to confirm that they wanted the officials at work, keeping the mine safe and open for when the strike was over. They backed me up, so I told Warbler and his mates that if they went home today and claimed intimidation they'd get a verbal warning of dismissal from me for unauthorised absence and the next day it'd be written and the day after that'd be dismissal, so if they carried on doing it all week I could 'ave 'em all sacked by Friday.'

'Good on yer, Boss, they give officials a bad name, fuckers like them,' Jerry said. 'I noticed they started off a bit reluctant, what did you say to make 'em step out, if you don't mind me askin'?'

'I told 'em, if they didn't step out, they'd be late down the pit and I'd stop their time and put it on their records.'

'So d'ye reckon you're gonna 'ave to 'and-'old the fuckers in every mornin' now Boss?''

'No, I doubt it. I reckon they've got the message. And I've told 'em that the pickets will be welcoming them in every morning. But I'll keep an eye on the idle fuckers tomorrow, just to be sure.'

Warning

Saturday, 3 November, 1984

Dick answered the door in his slippers.
'He didn't look like someone who's on the box,' Keith would tell his wife later. "E looked like 'e'd settled into cosy retirement. 'E's put weight on, if anythin'.'

Dick closed the front door and showed Keith into his well lived-in lounge, where they both sat down and looked at each other.

For Keith, having been used to spending hours, each weekday, with Dick in the union office and the canteen over the years, it was strange to see him in his own home, without a jacket and tie. Dick's absence of almost a year had made him a stranger to him, their shared history, everything they had been through together seemed distant and unreal.

Dick shouted through to the kitchen.

"Ey up Shirl', Keith's here. Make us a cup o' coffee will yer love?'

Shirley peered round the door.

'Hello Keith, two sugars i'n't it, me duck?'

'Er, yeah, that'd be champion. Thanks Shirley.'

Keith would be uncomfortable until he had managed to pour out his reason for phoning his old friend and asking for the meeting but Dick prolonged his discomfort.

'A little bird tells me as Arthur's been cap in 'and to the Ruskies, askin' the Soviet Union to help fund the strike,' Dick said.

Keith nodded, too preoccupied to worry about Scargill's machinations.

'You don't look bothered,' Dick said.

'No, no. Well, yes, it's a bit of a rum job.' Keith said with a scowl and single, quick shake of his head.

'A bit of a rum job?' Dick said. 'I should cocoa. With the way the union's bin leaking like a sieve, there's no way the fuckin' idiot,' ('excuse my French, Shirl',' Dick shouted, in case she had heard the slip into pit language) '– there's no way he'll be able to keep that away from MI5.'

'Yes, I suppose that's a serious risk.'

'It's a bloody serious risk. If that comes out they'll be haulin' 'im off to the Tower for working with an enemy power to overthrow the country.' Dick said. 'He could discredit the whole union, if not the whole trade union movement. It'd be right grist to the mill for Thatcher and Heseltine.'

Shirley bustled in with a tray to deliver a tin of biscuits and the two coffees.

'There you go,' Shirley said. 'Are you warm enough, duck?'

'Yeah,' Dick said. 'I'm fine.'

As Shirley withdrew from the lounge, Keith shuffled forward onto the edge of his seat.

'Come on then, what's botherin' yer?' Dick said, stirring his coffee. 'You might as well get it off your chest.'

'I needed to tell you about some moves that are afoot.'

'Oh yeah, and what might these moves be, then?'

'Well, it's about you being on the box.'

'Oh yeah?'

'Er, yes,' Keith said. 'Well, you see, some of the men, some of the members have started takin' a dim view of it.'

'A dim view of it?' Dick screeched. 'I took a bloody dim view of it – I had me a bloody 'eart attack, a month in hospital, best part of a week in intensive care. Do they reckon that constitutes swingin' the lead or summat?'

'No, of course not,' Keith snapped. 'But they reckon you should be on strike, not 'on the box'.

'What are they worried about? I'm not bloody working, whatever they call it, so I'm not scabbing am I?'

'Well they don't see it like that. They say you should be showing solidarity. It's not necessarily my view, y'understand, but they're saying that by drawing your pay it means the management'll be recording you as not being on strike, so the Coal Board can claim you as adding to the number of working miners.'

'What they're bothered about is me getting paid.'

'No! Well, yes. Why shouldn't they be, Dick? We've been out for eight months and 'aven't seen a penny in all that time.'

'Well, you could try telling them that I was just lucky to 'ave 'ad an 'eart attack and nearly bloody died.

'They know all about that. But there's a lot of bad feeling about it. McFadden's saying that you should either retire on ill-health or come off the box and join us on strike.'

Dick snorted. 'And we both know that that's what this is all about, don't we? McFadden hopin' to depose me, as Lodge Secretary.'

Keith could not answer. Despite Dick's pretended lack of forewarning, he had already heard all about this latest intrigue from a couple of his spies on the Lodge Committee.

'So, what are they looking for me to do, to retain my post?' Dick said, playing dumb. 'Come off the box and declare myself on strike?'

Keith's awkwardness verged on distress.

'Go on then, what is it?' Dick snapped.

'No, they haven't decided yet.'

'"They haven't decided yet",' Dick mimicked.

'And how do they intend to decide?' He said, already well aware of the answer.

'Shuey's called an extraordinary meeting.'

'When?'

'Yesterday.'

'No, I mean, when's the fucking meeting?'

'A week tomorrow, in the Welfare.'

'And at that meeting, he'll no doubt be proposing that I am booted out of me job as Secretary for being a scab.'

'Well, I wouldn't put it like that. But, I think he'll be proposin' that you're stood down.'

'Sacked yer mean,' Dick snarled. 'Sacked for givin' years of service to the union, putting the members' interests before me own and flogging mesen' on their behalf until I collapse wi' a heart attack. No gold watch, not even a word of thanks, just the fucking sack. And you, as Lodge President, one of me oldest mates, you've decided to align yersen with that fuckin', Scotch weasel, have yer?'

'Look Dick, you know it's not me. If it were down to me I wouldn't start rocking the boat. But McFadden's managed to stir 'em up. And these are grim times.'

'McFadden again, is it? It was your mate Shuey, two minutes ago,' Keith snapped. 'Anyway, if that's the way you want it, get on with it, then. See if I care. C'mon, I'll see you out. I trust that's all yer came for.'

Keith walked down the road with his head hanging down. It was only then that he realised how much he'd been dreading that confrontation. He was glad it was done but saddened that, with it, his friendship with Dick was over.

He knew he shouldn't be surprised at how painful it had been. But it hadn't occurred to him that going to see Dick to raise this matter would involve the severing of their lengthy partnership; a successful double-act that had always provided Keith with the confidence and security he'd needed to operate at his level in the union.

A Dark Morning

Susan knew something terrible had happened as soon as Paul arrived home early. He dropped onto the chair at the end of the kitchen table without bothering to take off his donkey jacket, his face drained and expressionless.

'Paul?' She said, fearing news of a fatality or some other disaster underground.

His gaze remained locked, unfocussed downwards, to beyond the end of the table.

Susan lowered herself onto the chair, by the side of the table, opposite where Jenny sat in her high chair, chuckling at her dad and calling out. 'Dada!'

She touched his forearm with her fingertips to break his trance.

'Paul, what is it?'

Paul shook his head and snorted.

'Nothin' really,' he muttered.

Monday, 5 November, 1984: 06:11

There were no police at the pit that morning, no need to enforce the restriction on picketing numbers; eight months into the strike, there were frequently fewer than the statutory maximum of six on the line. That morning, it was just Paul and Brian with Steve Bircher and Dan Wainwright, an older pitman.

Paul recognised the familiar gait and slight figure of the man who had just walked into the yellow light of the first street-lamp in the pit lane. Peering harder, he confirmed it was his mate he had spotted.

Colin had never spent as much time picketing as Paul and Brian but he hadn't actually turned out on the line since September. The sight of his approach gave Paul's spirits a lift.

But then he saw the strap over his friend's shoulder. It could mean only one thing; Colin was carrying his snap bag.

But it made no sense; Colin couldn't be deserting them. If he'd been weakening Paul would have been the first to know.

A couple of long-standing scabs walked past and Steve and Dan jeered viciously at them.

'Black-leggin' bastards.'

'Fuckin' scabs.'

Like Paul, Brian's focus was on their mate, as he closed with them.

Pickets retained a good, mental record of those miners who had returned to work, reserving their maximum effort for discouraging those miners trying to make it in for their first shift back. Colin got to within twenty-five yards of them. Steve and Dan glanced at Paul then started to picket him. Dan sauntered a few paces and position himself in his path.

'What's this, Colin?' He said. 'Yer not about to try and do somethin' stupid are yer?

Steve joined in.

'Come on now, Colin. Come and join yer mates on the line.'

Neither Paul nor Brian said a word. Paul locked eyes with Colin and saw the defeat written in his eyes but his friend's gaze held firm as he kept coming.

The months of deprivation, going short himself in order to feed Martin, Jenny and Susan had weakened Paul's spirit; even if he'd been able to rouse any aggression towards Colin, he didn't actually have the strength to bring it to bear.

Dan was twice the size of Colin. As Colin reached him, Dan stepped sideways, barring his way.

'Come on Colin, you don't wanna do this,' he said. 'Just 'old out a bit longer, we've gorr'em on the run. Come and join Paul and Brian. You're not going to let yer mates down, are yer?"

Colin stopped and looked up at Dan. He looked Paul in the eye again, glanced at Brian and then back at Paul. He looked down, at the ground, shook his head then stepped round Dan.

'Don't do it, Colin!' Steve called out. 'You'll regret it.'

As Colin carried on walking into the pit Steve shouted after him.

'You fuckin', treacherous, scabby-backed bastard, Beighton!'

Colin trudged on, head down, for five more, miserable steps. He slowed and pulled up.

'Colin, you've got to stick with Paul and me,' Brian cried.

Colin spun round and stomped back towards them but Paul's brief moment of hope was washed by the tears that had started down his friend's face.

Colin pulled up in front of them, chest heaving.

'We've bin out for months! We ain't even made a dent in their stockpiles. They've got more coal now than they know what to do with. And they're getting more back to mine more for 'em every day. My kid's starving and I've got a pregnant wife who's ready to drop another and I can't support them. I ain't been able to for months. We're beaten! I'm beaten!'

His voice dropped, 'I've 'ad enough.'

He stood three paces from Paul, surrendering himself to whatever they were going to do to him, gazing at them; Paul's eyes fierce, Brian's face stricken.

Then he broke the silence.

'But if you say you want me to come back and join you there, now, then I will. If that's what you want me to do, then I'll do it.'

Colin's anger shone through his tears as he implored Paul with his eyes. He looked at Brian then back at Paul.

There was nothing left to say. Colin turned away and stumbled off, towards the pit, his snap bag an unbearable burden, as he heaved its strap onto his shoulder.

A group of five, working miners approached the picket line and hurried past, amazed at receiving an easy passage instead of the usual entreaties, threats and abuse.

'Fuck it. We've 'ad no money for months and there's nearly 'alf o' the fuckers gone back,' Steve muttered. 'We're wastin' us time 'ere. We're just fuckin' suckers now. I'm going back un' all.'

The three others watched as Steve staggered off, into the pit, as though drawn by some irresistible, mesmeric force.

'I can't take any more o' this,' Brian murmured.

Paul's eyes widened as his mate lurched away, in the opposite direction. With a moan and a shake of his head, Brian broke into a jog, down the pit lane. Paul just watched him go, saw him reach the T-junction with the main road, look both ways, cross the road then turn right and carry on striding, out of the village.

Every time a vehicle passed, Brian stuck out a thumb. The third car was a blue Cavalier. As it approached, Brian turned round and kept walking backwards as he raised his left thumb.

The car stopped. The driver pushed the passenger door open. Brian trotted up to it, bent down and looked in.

'Er, thanks a lot. W-which er w-way are you headed?'

'Coventry, if that's any good to you,' the driver said.

'Yeah, g-g-great.'

Taking one last glance back, across at the pit, he lowered himself into the car. At that moment, Brian knew he would never see the headstocks of Whitacre Heath again.

Paul had thought Brian was leaving the picket line to go home. Instead, he had just disappeared.

Colin sat on the long, thin, metal seat down the front of the long row of top and bottom lockers in the clean side of the pithead baths. Having drawn his tallies, he needed to take a couple of minutes to come to terms with where he was and what he had just done before starting to strip off his clean clothes.

For the first time in his life, a gulf existed between Paul and himself. He couldn't imagine life without his best mate's strength and inspiration but knew that was what he faced now, as a consequence of what he had just done; what he had had to do.

The other men let him be. They'd all had to undergo similar transitions, after crossing the line on their return to work.

'How ya, ganning?'

'Tom,' Colin murmured the name of each man in response to their gentle and quiet greetings as they came into the bay that housed his locker.

'Get on, Col'.'

'John.'

No one acknowledged that he was breaking the strike for his first day.

'Y'alright Colin?'

'Badger.'

It felt like he no longer belonged there.

Diane hadn't pressed him to go back. She hadn't said anything about it, there'd been no need. Colin had seen Diane's increasing despair in everything she said and did, every minute of every day that he had spent at home, in all those months on strike. It matched his own.

How would they manage when the new baby was born? What if the rumblings were true and the Government succeeded in persuading the building societies to foreclose on the mortgages of striking miners who were unable to make their monthly payments?

There had been no relief. Even when he'd decided to break the strike and return to work, he was condemned to more misery. He'd had no alternative but he'd known that he'd be demolishing his life, condemning himself to a future without happiness or self-respect.

He had known that Diane had feared that he'd allow Paul to dissuade him; he'd been afraid of that himself. She hadn't asked how he planned to resist that pressure and he hadn't told her. The only way he could do it was to cross that line without talking to him beforehand. That was why he'd waited until he was supposed to be on days; he had had to return to work when Paul would be picketing; he couldn't have his mate thinking that he'd snuck back, across the picket line, behind his back.

Colin stood up, turned round, placed his two tallies on the seat and slipped his key into the little round lock on his clean locker's door. He twisted the handle, stiff from lack of use, opened the door and gazed in with glazed eyes.

The used bar of white, PHB soap had dried, in the months of heat in the locker, to set like a pebble in his metal, soap dish.

It had been his locker for years but he didn't recognise its interior.

It felt at least as bad as he had anticipated; his return gave him no relief from anything, he was trapped; he had lost everything.

Colin was one of thirteen miners who travelled the one-way street back to work at Whitacre Heath that week.

The clerk in the deployment centre recorded that Colin's tallies had been issued for the first time in months. His was just another name and number on the list of returnees that updated the Personnel Manager's strike statistics.

07:13

'Come on, love,' Susan said, placing her hands over her husband's. 'What is it?'

Paul turned his head slowly.

'It's all falling apart. We can't stop it. We can't stop 'em. Soon there'll be over fifty per cent back at work at Whitacre.'

Susan looked into his despairing eyes and saw the hurt.

'Colin's gone back,' he murmured, his hoarse words almost a whisper.

'What?' Susan's exclamation silenced Jenny's chatter, turning her face wide-eyed and serious. 'To work?'

'Yes, to work,' Paul growled thickly.

'When? When did he go back?'

'This mornin'. He crossed the line while me and Brian and a couple of others were picketin' this mornin'.'

'Oh, Paul,' she groaned. 'Hadn't he said anything. Before I mean?'

'No,' Paul said. 'But I suppose I should've seen it coming, shoulda done somethin' to help keep 'im strong, now I think about it. I knew it was gettin' 'im down. But then it gets us all down. But why didn't 'e tell me 'e was gonna do it? I can't believe 'e's gone and done it without talkin' it over first.'

Susan said nothing. She could imagine what Colin had been thinking and what Diane would have thought. She knew that if Colin had seen no alternative he wouldn't have wanted to be deflected from his decision, as he most surely would have been had he subjected himself to Paul's judgement before attempting to go back.

'And Brian's gone,' Paul said.

'What? Back to work?'

'No. Just gone. Walked off the picket line, out the village. Gone!' Paul said. 'He's weak, Colin is – always 'as been. Got no bottle.'

'If it's left to people like him we'll have no industry, no pits, no jobs, no nothing.' Paul snarled, his lip lifting.

'They're just desperate, love. Another baby on the way.'

'"Another baby on the way",' Paul sneered. 'That was the best 'e could come up with. If he was bothered about his children 'e should've stayed out and fought for his job. Did she say anything to you?'

He looked at Susan, 'Has Diane said anything to you?'

'No,' Susan said, affronted, shaking her head and standing up to go round to Jenny who was snuffling, about to cry.

'Well, I tell you one thing; we're finished with that pair, going behind us backs. They'll have plotted this together and 'e can't be straight and tell me, like a man, what 'e was planning. We have nothin' to do with 'em – 'im or 'er. All these years I've treated 'im like a brother and 'e goes and does this across us.'

Susan picked Jenny up to hush her, holding her baby tight and close, her face disconsolate.

The Fight is On

Thursday, 8 November, 1984: 12:51

Jim had spent the dayshift coal-chasing on 22's retreat face. There were only ten minutes of production left before the facemen would start making their way out of the pit, when the panzer driver's voice rang out urgently, over the face's tannoy system.

'Come in, Mr Greaves. Come in, the Undermanager.'

'That's for you, Boss,' one of the men called out to him over the tremendous din of the panzer and the maingate shearer as it cut its way down the face.

'Okay, thanks Alfie,' Jim shouted back as he crawled past him to get to a tannoy. Reaching it, he pressed in the rubber-covered, speak button and shouted.

'Hello, it's the Undermanager. Whadya want?'

The panzer driver shouted more quietly through the tannoy, 'Can you get to a phone and call Mr Hope in the Control Room? He says 'e needs to speak to you.'

The maingate was nearer, so Jim scrambled past Alfie again. Crawling quickly, Jim caught up with the shearer and squeezed past its two drivers, urging each of them on with a few words of encouragement as he passed.

In the maingate, he straddled the panzer in order to cross it then strode the thirty yards to the electrical panels suspended from their mono-rail over the conveyor belt. The panzer driver grabbed the phone hand-set from where it hung on its hooks in the white, metal box of the phone, dialled the Control Room number and handed it to Jim as he arrived.

Jerry Malin, the dayshift Control Room attendant, answered the phone immediately.

'Hello, Jerry,' Jim said.

'Oh, thanks for getting back to us, Boss,' Jerry said. 'I've got Mr Hope 'ere, 'e needs to speak to yer.'

Jim heard Jerry mutter to the Operations Manager.

'Mr Greaves, Boss.'

'Hi Jim,' Mr Hope said, when he had taken the phone. 'Are you alright?'

'Yeah, fine Boss,' Jim said, impatient to hear what he had to say.

'Sorry to be a pain, Jim, but we've been watching the Emcor on 56's. It looks like it's been easing up for the last couple of hours. We can't be sure yet, but it looks like it's on a rising trend, it appears to have risen to twelve parts per million.'

Jim knew that any optimism about the reading being an aberration was likely to prove unjustified. The Emcor for 56's had been showing the normal background reading of seven or eight parts per million of carbon monoxide when Jim had arrived at the pit, seven hours earlier. It was unlikely to be a false alarm, it never was; a rise in carbon monoxide of four parts indicated it was almost certain that a fire was brewing. If they didn't catch it, the "CO" would continue to rise and the fire would take hold of the district.

'I'll get round there. What does the deputy say?'

'There's no fire-stink in the tailgate,' Mr Hope said.

'We've got a couple of Drager pumps down here. I'll go and pick one up from the tailgate and a box of tubes and see what we've got up there. I'll let you know what I find.'

'Thanks Jim. Give me a call as soon as you've checked it out, if you would.'

Jim came out of 22's Tailgate, onto the East Main Return airway. and made his way through the heavy, wooden air-doors to the East Main Intake. He crossed the conveyor that was running, empty of coal, now that 56's and 22's dayshifts had finished turning and their facemen were heading out of the pit.

Jim turned a blind eye to the small groups of men, sailing past him in the opposite direction, riding illegally on the belt, lying, one after another, face down on the speeding conveyor to get a quick and easy ride out to the Loco Road, instead of walking, an offence for which Jim could fine them and which would leave them with disciplinary black marks on their records. The men played the game and pretended that they hadn't seen the undermanager walking back up the belt road until one passing faceman slipped up and greeted him with a cheery call as he sailed past.

"Ey up, Boss.'

'I didn't see you,' Jim called out to him.

'Arr!' The man shouted back, with a dawning realisation. 'Thanks Boss!'

A quarter of a mile further on, a belt road ran into the Main Intake from the other side of the belt. The conveyor belt in this road served 56's Face, carrying its coal from 56's Maingate, down to the Main Intake belt. Jim clambered over the little, metal bridge over the main conveyor and made his way up the steep, the rise that connected the deeper Bench seams level with that of the Seven Feet seam, which 56's Face was working.

After a fast climb up the steep belt road, Jim met 56's Face Deputy, Will Sharp, at the outbye end of his district's maingate, which headed off to the right, towards the face, at ninety degrees to the steep road.

'Hello, Will. Found anything?' Jim said, breathing a little heavier from his rush up the incline.

'Nothing yet Boss. Mr Hope said you were coming over so I stayed back to come round the district with you.'

'Okay, thanks Will. I'll give Control a call to let them know where I am and then we'll get going.'

Within thirty yards Jim saw a bougee pump, at what had been the face start-line, where the face had been headed out and installed. This was used for injecting large volumes of hardstop, a pink, plaster-like mixture of gypsum and water, to seal cavities underground. Will confirmed that it was in working order and could be powered-up easily. If there was a fire and it was deep-seated in the waste, they would need this to try to find and seal the air path that was feeding it.

Jim and Will enjoyed a relaxed chat as they marched up the maingate but both remembered to keep sniffing, periodically, to see if they could detect any smoke emerging into the intake air. When they reached the inbye end of the maingate, Jim took a Drager tube from the black plastic, panatela-sized box, broke off both ends. He squeezed the Drager hand-pump, evacuating it completely, then shoved the sampling tube into the rubber seal in the hand-pump and released it to draw a measured sample of air through the crystals in the tube.

When he looked at the colouration of the crystals it indicated a typical background level of around seven parts.

'Nothing showing in the intake,' Jim said.

Even though 56's was in the so-called Seven Feet seam, with the band of coal left in the roof to stop it from breaking down and the beams of the chocks above them and the chock bases under them, it was still a crawl through the face.

Jim let Will lead the way.

After two hundred yards they reached the other end of the face. They crawled out and got to their feet in the tailgate then Jim took another Drager reading to measure the concentration of carbon monoxide at the start of the return airway. He looked at the crystals in the tube then held it out for Will. There was still nothing more showing than in the intake, at the other end of the face.

Jim and the face deputy started to work out along the tailgate, sniffing with more diligence and with the intention of taking a Drager reading every two hundred yards. Their second reading, four hundred yards from the face, showed eighteen parts per million. The two men sniffed the air.

'Ooh yeah. I can smell it!' Will said.

'Yeah. It's there.' Jim said, catching the characteristic, coarse, heavy taste of the aromatic, oily smoke of an underground heating on the back of his tongue.

'Let's walk back and do a few tests on the way back to the phone at the face,' Jim said. 'Then I can let Control know what we've found.'

The Emcor readings in the Control Room wouldn't show the level they had just taken on the Drager tube until the continuous gas sample reached the monitors on the surface, about three hours later.

Jim phoned the Control Room and told Tim Hope the bad news.

'The carbon monoxide's emerging in the tailgate, just over three hundred yards outbye of the face-line. There's no sign of smoke and the stink isn't very strong yet, so it's probably deep-seated, and, given where we picked up the reading, it's probably well inbye, not far back from the face-line.' Jim said. 'I'll get the afternoon-shift to move the bougee pump up from the face start-line to nearer the face and start trying to improve the roadside seal on the maingate. There's not too much hardstop in the maingate, only about fifty or sixty bags so we'll need some more loads sending in this afternoon and tonight.'

'I need men deploying to start ripping the tins out at the outbye end of both gate roads so we can start keying in the stoppings,' Jim said, referring to thick walls that would be formed by pumping hardstop into the space between two walls, four metres apart, built with packing bags of dirt from the floor. These would create the substantial seal to stop the airflow around the district if the fire beat them and they lost 56's District.

'Okay Jim, I'll get all that sorted and on the move for you,' Tim Hope said.

With the manpower reduced by the strike, they had been placing priority on working 22's. Being a retreat face, this had a faster turnaround each time the machines cut into the roads at the end of the face, enabling them to maximise the colliery's output in the short-term.

There were disadvantages to this strategy; it meant they were working their most productive face faster, towards an earlier exhaustion date than originally planned. Also, it meant that the other faces were only ticking over. A consequence of this was that the airflows through the wastes were at more consistent levels, with more constant paths for longer, on those face districts. And that was an excellent way to fan a fire in Warwickshire.

On 56's, the face was afflicted by some areas of bad roof. Timbering the cavities that opened up above the face after each cut had slowed the face's rate of advance even further and resulted in timber being left behind in the waste to feed any fire that was brewing.

'I reckon we'd better man-up to work the face on three shifts, to get it moving and try to bury the fire,' Jim said.

'Yes, you're right,' Tim Hope said. 'Jim, can you stay down there in charge for the afternoon-shift and I'll get Ken Goodall to relieve you on nights. Bill Metcalfe can keep an eye on it on days tomorrow and you can settle onto nights thereafter for us.'

'Okay, Boss,' Jim said.

He hung the hand-set back in the phone's steel box. He had been confident as he had left 22's, to head over to 56's, that he would be kissing the coming weekend goodbye and that he would have a few seven-night weeks ahead of him.

'You get off and up the pit, Will,' Jim said.

'What are you doin' Boss?'

'I'm staying on, with the afternoon-shift.'

Jim pulled aside the grubby, narrow band of flexible bandage he was wearing to protect his wristwatch. The afternoon men would be arriving on the district fairly soon.

'I'm going to go back through to the maingate and see if I can spot any likely areas of leakage to tackle.'

'I'll crawl through with you, keep you company. I might as well do that. I can walk out in the fresh air then.' Will said. 'And, any road, you shouldn't be down 'ere on your own when there's CO about.'

'Will you be alright then, Boss?' Will said, when they reached the maingate.

'Yeah, thanks Will. I'll be switching over to nights for a while so I'll be with you next week.'

When Will had strode off and left him in the maingate, Jim used some smoke tubes to test for leakage. These were the same size as the thin Drager tubes but instead of drawing air through them they were used in the Drager handpump to emit puffs of harmless, white smoke. By releasing little clouds near the roof and waste-side of the road, it was sometimes possible, with a bit of craft and intuition, to find where air was being drawn into the waste.

Jim started at a point in the road about one hundred metres outbye from the corresponding point in the maingate to the point in the Tailgate where the higher carbon monoxide readings started. He found several points at which the smoke was whipped away, drawn into the waste. He marked these places for attention by the afternoon-shift by chalking arrows onto the rusty, steel arches.

The fight for 56's was on; another fight for the very future of Whitacre Heath Colliery, itself.

Extraordinary

For the third time in little over a year, Whitacre Heath NUM members were in the Miners' Welfare to be balloted. With all those working having stayed away, there were considerably fewer in the room on this occasion.

Keith Deeming sat next to Shuey McFadden, at the long, top table at the end of the ballroom, with the rest of the Lodge Committee. He rose from his seat, shouted out to get the men to take their seats then called them to order.

'Good morning, brothers,' he shouted.

'Right, we all know the purpose of this extraordinary meeting of the Whitacre Heath Lodge. But just to be clear, the Committee has called this meeting to consider a matter that's been raised by some of the membership.'

'As you all know, our Lodge Secretary, Mr Dick Fleming, has been unable to fulfil his duties since the serious heart attack he suffered, nearly a year ago. Obviously, we were all sorry to hear about this at the time and have offered our support. Unfortunately, at this time of industrial action by the NUM, Dick's absence is a matter of concern. In particular, many of you have said that it is not right for the Lodge Secretary to be accepting his pay, throughout the dispute, whilst the rest of the right-thinking members 'ave been on strike. The purpose of this meeting is to discuss this matter and decide what – er, if any – action is to be taken.'

'Without further ado, I open it up to the floor. Who wants to speak first?'

Shuey McFadden's right-hand man, Malcolm Kennedy, sprang to his feet.

'Mr President, this is all pretty simple. We have someone who has held onto his post as Secretary of the Union whilst not coming out on strike. I say we should kick the cunt oot and put Shuey in the job.'

'Come on now, this is a formal meeting of the Colliery Lodge. I'm not putting up with language like that,' Keith said. 'So be warned, any more of that from anyone and I shall have them evicted from the meeting.'

Muttering broke out around the room, McFadden remained seated next to Keith but turned in his seat, a little, towards him.

'Yeah, but exceptin' the bad language, Malckie's got a point,' Shuey called out. 'No one wishes Dick any ill-will, but by drawin' his money instead o' strikin' he's being recorded as a scab – a strike-breaker – throughout this phase of our struggle. As the number of people scabbin' is gettin' close to fifty percent we need every one we can get shown as a striker. Keith, here, gave Flavel the chance to come out on strike and he refused, so effectively he's prepared to continue black-leggin', so he's a scab.'

Before Keith could clarify that Dick had never been given the option of making his participation in the strike apparent by refusing his pay, Shuey continued.

'We've all gone without pay for eight months and he shoulda done as weel.'

'You 'aven't, you robbing bastard,' a voice called out from the rows of seated pitmen. 'You've been drinking the strike funds.'

A number of the pitmen guffawed at this indiscretion. McFadden glared at his accuser but made no attempt to refute the allegation.

'And we've never seen you on the picket line, not since the first morning,' Paul Wood called out. 'Someone said you'd gone on holiday.'

The majority of the room was laughing now.

The meeting was not going the way Shuey had planned.

He jumped to his feet.

'I've done me fuckin' bit,' he snarled. 'Apologies Mr President, er, Mr Chairman. And, what's more, I'm still on strike. The question is, are we prepared to put up with hae-in' a Lodge Secretary who's drawin' his money – a Lodge Secretary who's fuckin' black-leggin' – or are we gonna dae the only decent thing and remove the scab from that key union position.

'What if we don't want you as Secretary?' Paul called out, staring at him intently.

McFadden shook his head in exasperation.

'Then you vote me out next time.'

'That's ages away,' Paul shouted.

'Yeah, but the point o' the Deputy Secretary position is to take up the Secretary post when he canna dae it.'

'No it wasn't, according to you it was supposed to be to provide more muscle, given what was going on in the coal industry.' Paul shouted back. 'How would this provide more muscle?'

'It would give me a firm mandate, instead of being understudy to a fuckin' scab, that's how.'

'Mandate?' Paul sneered. 'You don't know the meaning of the word.'

'Brothers, brothers!' Keith called out. 'Let's conduct this meeting in the spirit of brotherhood. We can't be havin' it becomin' acrimonious.'

'Right,' Malcolm Kennedy called out, slumped in his seat. 'I propose that we kick the scab Flavel out of 'is job and make his deputy, Shuey, Secretary.'

'I second that,' Ivan Wykes, another of McFadden's supporters shouted, with a raised hand.

'Hold on a minute,' Keith said. 'We haven't made sure we've given everyone a chance to speak their minds.'

'Well, has anyone got anything else to say?' Kennedy said, looking around.

After a heavy silence, Keith said, 'Alright, the proposal is that Dick Flavel should be removed from the office of Lodge Secretary and that Shuey McFadden should be appointed instead. All those in favour?'

Keith and another Committee man both counted the raised hands.

'And against?'

They both counted again then conferred.

'That was forty-five in favour and twenty-nine against,' Keith called out, then sighed.

'The motion is carried.'

As Keith closed the meeting, Paul's chair slid back noisily as he stood up in the body of the room then stalked out.

Outside the building, Paul reflected that they had been in a no-win position; they couldn't carry on with someone as Lodge Secretary who was vulnerable to accusations of scabbing. But that had meant that they were bound to end up with McFadden as their leader.

Shuey and his two lackies emerged from the front door of the Miners' Welfare; Kennedy and Wykes smirking at being in the company of the new man on the throne, McFadden wearing his characteristic, sour expression. McFadden spotted Paul and cast him a look of pure enmity.

'Hardly a ringing endorsement for yer, wore it?' Paul called after the trio, as they made their way across the car park. 'I counted one 'undred and thirty-eight men in there and you could only muster forty-five. You couldn't even manage a token majority. Not much of a mandate, that, fer yer, McFadden.'

Bribery

Monday, 12 November, 1984

Paul couldn't afford to buy a Sunday paper so he'd missed all of the full-page adverts. The first he knew of the inducements was when he heard them reported on the news on Radio One, that morning.

The National Coal Board had offered every striking miner who went back to work by the nineteenth of November a Christmas bonus of six hundred and fifty pounds. The equivalent of a miner's monthly wage, it would sound like a king's ransom to an impoverished striker.

It wouldn't affect his resolve but he considered the implications; he could only hope that most of those who had held out this long would view it with contempt and be hard-line enough to keep the faith.

Thursday, 15 November, 1984

Paul had long seen the BBC and newspapers as propagandists for the other side in this war. He'd relied on this perspective to dismiss the reports he'd heard on BBC's Radio 4 News, that morning, that the advertisements had caused huge numbers of striking miners to return to work.

By the end of that day, they were reporting that over nineteen thousand men had succumbed to the temptation. After the months of hardship with the prospect of an otherwise cold, hungry, impecunious Christmas, they'd gone back to work to claim the promised bonus. Paul was forced to accept it; a return to work by more than a twelfth of the union's membership in one week was an appalling defeat.

The entrance to Whitacre Heath Colliery. Friday, 16 November, 1984: 06:08.

'It's a betrayal, everyone of 'em as bad as Judas acceptin' 'is pieces of silver.' Paul snarled at his comrades on the picket line. 'And I hope their proud of 'emselves.'

But, in his heart, he knew that none of them would be drawing any comfort from what they had done.

There wasn't even any satisfaction for him in knowing that, by returning to work less than a week before the announcement of the inducement, Colin had missed out on his share of the largesse by a matter of days.

The Earl of Stockton

November, 1984

"It breaks my heart to see – and I cannot interfere – what is happening in our country today.

This terrible strike, by the best men in the world, who beat the Kaiser's and Hitler's armies and never gave in.

It is pointless and we cannot afford that kind of thing.

Then there is the growing division of Conservative prosperity in the south and the ailing north and Midlands.

We used to have battles and rows but they were quarrels. Now there is a new kind of wicked hatred that has been brought in by different types of people."

From the maiden speech to the House of Lords by Harold MacMillan, the aged, former Conservative Prime Minister. The Earl of Stockton, Viscount Macmillan of Ovenden, had taken his title from his former parliamentary seat, on the edge of the Durham Coalfield.

His speech received an unprecedented standing ovation.

Fire Fighting

Monday, 19 November, 1984

For the first week of the fire on 56's, the emphasis was on starving it of oxygen by stopping the air from leaking from the maingate into the waste. This involved shuttering the metre-spacings between the legs of the maingate's steel, arch supports with lagging boards then hand-plastering them with hardstop to stop as much leakage as. Thereafter, they'd used the bougee pump to fill the gap, between the lagging boards and the corrugated, steel sheets that lined the road, with hardstop and to pump the same, plaster-like mixture into steel pipes, driven through the corrugated sheets, to seal deeper into the waste.

Pitmen's hands tended to be strong and work-hardened, their rough skin permanently blackened by ingrained coal dust. Hand-plastering aggravated this and when Jim lent a hand with it, he worked like the rest of the men, without gloves. The result was always the same, the hardstop leached the moisture from their skin, causing their hands to coarsen even more and their finger ends to break open with painful, sensitive splits.

On the face, an ancient method of continuous carbon monoxide detection was used to mitigate the possibility of a failure in the more modern, higher-tech' means of monitoring. Each shift of men brought in one of the colliery's canaries, siting its little wooden cage just outbye of the rip, at the tailgate end of the face.

Early in this shift, Jim heard the distinctive birdsong at the other end of the face, where the facemen had hung the cage, so the bird would be safe, in the untainted, intake air of the maingate.

'What's the canary doing down here?' Jim said to the men at that end of the face.

"E's alright there, Boss,' one of the men replied.

'Yeah, 'e's in the fresh air there,' another faceman called out.

'You silly buggers,' Jim said, struggling to suppress a chuckle. 'There's no point in the poor thing being down here if he can't do his job.

'Henry,' he said to one of the rippers. 'Take that cage up the face and hang him up in the tailgate.'

The faceman he had nominated unhooked the bird-cage and dawdled over towards the face with it.

'It's a shame,' another of the men said. 'The poor little thing might die.'

'Well, yes,' Jim agreed. 'But then that is why we have 'em, so they can save men's lives.'

'I'd rather this little bird lived than those tailgate-end wankers,' Henry said before ducking down and crawling off up the face to avoid any further argument on the matter.

Saturday, 24 November, 1984: 07:49

As the youngest and most junior member of the senior management team, Jim was always expected to take charge of the nightshift in the event of a fire.

Each morning he would get home after Gabriella had left for work, only seeing her, briefly, in the evenings before he left for the pit, except at the weekends when she would still be in bed when he got home.

That morning, Jim came out of the pit after working, through the night, with the team lagging the road and pumping hardstop into the waste from the tailgate to seal off the points at which air was being drawn air into the return airway.

Even after the lengthy, rigorous scrubbing and soaping that Jim had given himself in the undermanagers' bathroom, Gabriella could still smell the taint of the thick, heavy smoke of the underground fire on him when he got into bed with her. She could only wonder at what kind of environment it was that he and his men were inhabiting down there.

When he woke at noon, the dull headache from his tiredness had a sharper edge from the onset of progressive, carbon monoxide poisoning. That would only get worse over the coming days, as the "CO" concentration in his blood continued to increase.

Ten days later, it appeared that the sealing work underway in both gate roads and the consolidation of the waste from the face's faster rate of advance were proving effective at reducing the amount of air leaking through the waste. It was clear from the line on the Emcor graph in the

313

Control Room that the rate of increase in the level of carbon monoxide had slowed.

Although Jim continued to deploy teams to seal and plaster the gate roads, he shifted his main focus onto driving the face forward as fast as possible.

Trespass

'**B**oss!' Joss Kemp shouted across the communal room of the dirty offices, waving the handset of the telephone in Jim's room. 'It's fer you.'

Jim broke off from the officials he was briefing and walked over to the face deputy.

'Thanks,' Jim said, as Joss handed him the receiver. 'Who is it?'

'Roger, in Control.'

'Hello, Roger,' Jim said, into the phone.

'Boss, I just thought you'd want to know, them coppers've gone and shifted 'emselves into the Gaffer's office.'

'What? How've they got in there.'

'I dunno, Boss. I 'eard 'em movin' about, at the bottom of the corridor, so I went down to see what the fuckers were up to and found they'd got in there and were shifting all o' their kit in. They must've got into Angela's office, some'ow, and got the key out o' there. They're makin' a right mess. When the Gaffer sees it 'e'll go blue. An' Angela Barker's gonna kill us!'

'Right, thanks Roger, I'll be right over,' Jim said.

Will Sharp, that week's nightshift face deputy on 56's, saw Jim's black look.

'What's up, Boss?'

'I've got to go back over to the main offices, the police are fucking about again. I'll have to see you down there.'

'It's the bloody Met' again this week, ain't it?' Will said. 'They're a fucking, 'orrible, arrogant bunch, that lot. Cops from other areas who've been 'ere call 'em the bananas, 'cause they're all yellow and bent.'

Jim strode through the lamp cabin and out into the cold night air to cross over to the main offices, furious at being interrupted and delayed from getting underground to get to work on the face but even more irritated by the implications of the police's abuse of the colliery's hospitality.

Everyone knew you treated the Colliery General Manager's Office with reverence. No one ever went in there without the permission of Angie or the Gaffer, himself. Even when Jim covered for the Gaffer, in his absence, he didn't use his office, or even enter it; to do so would have been a sacrilege.

When the Met' had returned to Whitacre Heath, to take over its policing for that week, they'd been given the conference room as their local control room, as usual. Every force that had provided support there had found that arrangement perfectly workable, including the Met' on its previous visits. In some ways, the conference room was even better appointed for the purpose than the General Manager's office, so what they were doing was worse than pointless.

Importantly, the next shift of police would have to vacate the office again in time for Mr Everitt's arrival in the morning, anyway. And that relocation was going to make their control room inoperable at the worst possible time, a pit shift changeover, as the men on nights left and the dayshift arrived; the busiest time for policing the pit and picket line.

Jim entered the door at the top end of the main block, stalked down the corridor and looked in at the colliery's Control Room.

'Hey-up Roger,' he said, standing in the doorway.

Roger swung round on his chair.

'Oh, hello Boss. Sorry to have to trouble you. I told 'em they shouldn't be in there but the ignorant bastards just said, "we're commandeerin' it".'

'They did, did they? Well, I'll sort the fuckers out.'

Jim marched down the corridor, his Davy lamp swinging from his belt; the memory of its theft, three months earlier, only serving to fuel his indignation.

As Jim rounded the corner into the bottom corridor, a policeman, carrying a box of files, was entering the Gaffer's office by the door, off to the right, at the end of that passage-way. He paused, looked back at Jim then jumped into the room and shoved the door shut.

Jim reached the end of the corridor. He seized the hefty, brass handle, turned it, shoved the heavy, oak door open and stepped into the office.

He looked around the room. Its great table was already littered with equipment.

He was surprised to recognise the two coppers who had taken his Davy

316

lamp among the gang of eight policemen busying themselves in the room and to see the tall, chief superintendent who had arrived in the patrol car, wearing his parade ground uniform, on that very first morning of picketing, back in March. On this occasion the senior man was in shirt sleeves.

For a few seconds, Jim didn't speak while he wondered why the chief superintendent would be the one setting up their local radio console, while the junior ranks looked on. He obviously intended to make himself comfortable at the radio because he was establishing the unit on the General Manager's desk, at the end of the long table.

The chief superintendent stopped fiddling with the cables out of the radio console, drew himself up to his full height then scowled at Jim.

'Can I help you?'

'Yes, you can,' Jim said, from the other end of the table. 'Get out!'

'I beg your pardon?' The policeman barked, looking contemptuously over Jim in his filthy pit-black. 'Who are you – and who the hell do you think you're talking to?'

'I'll tell you who I am, I'm the Undermanager and tonight I'm acting on behalf of the Colliery General Manager with all his statutory powers. And I know exactly who the hell I'm talking to; I'm talking to a bunch of trespassers.'

'You've been given a perfectly good room to use whilst you're here. This is the Colliery General Manager's office, so that makes it my office tonight. If you don't get out of it, immediately, I'll take your name, rank and number and get straight on the phone to your national Control Room to make a formal complaint on behalf of the National Coal Board.'

The chief superintendent fumed.

'Right lads,' he muttered, after a few seconds. 'Let's move everything back to the other room.'

'Thank you,' Jim snapped.

Before he moved to leave the room he glared at Neil Bradford who looked away quickly in embarrassment, then he looked at the older of the two involved in the theft of his lamp. Carl Bessemer considered staring the young manager out but let his eyes drop to the floor.

Jim waited and watched a little longer, until he was satisfied that the policemen were making serious moves to transfer their control centre back to the conference room, then stalked out and back to the collier's Control Room.

'Roger, can you go and check that those fuckers are out of there and

317

make sure they've left it tidy. Then get the key back off 'em and lock it up again.'

'If they leave any mess, tell 'em to sort it out or the Gaffer'll fuck 'em when he turns up in the morning.'

'With pleasure, Boss.'

'Right, now I'll try to get underground and get on with some fucking work,' Jim said. 'Thanks for alerting me to it, Roger,'

'You're welcome, Boss. Off to 56's, boss?'

'Yep,' Jim said as he left the Control Room to head up the corridor.

On the face that night, Jim was making sure they kept the shearers cutting coal for as much of the time as possible. His efforts were still hampered by the poor roof conditions in two main areas on the face, one about a third of the way up the face, around sixty metres from the maingate, the other a hundred metres further up, about three quarters of the way through the face.

They were running the split bars they needed for timbering-up down the face as the machines were cutting coal. When a shearer was cutting coal on the tailgate side of the face from a fall area, the machine drivers would hold the machines as the timber came through, to avoid burying it under a load of coal.

The Number 1 shearer was on stop, having to wait down at the maingate while the rippers at that end of the face set two parts of a steel arch, a bow in the roof and a 'ribside leg', against the solid, on the other side of the road from the face.

Two teams of four took advantage to put a lockout on the armoured face conveyor and crawl out under the exposed face-side to timber up the two areas of weak roof, filling the holes in the roof where there had been small falls of ground from above the chocks.

The cavities were two to three feet high over the chock beams, in the ugly, light-coloured, grey and yellow bands of stone above the face's, otherwise, firm roof of shale. It was vital to catch these areas quickly and support them securely, to prevent them opening up to release massive slabs of stone to collapse in on the face team.

Jim was working with three men, to make up the group of four working at the area of bad ground nearer to the maingate. Jim and one faceman timbered the fall while another used a bow-saw to cut the split bars to the lengths they needed and his mate held the pieces of timber while they were sawn then passed them to the timberers.

They just needed a last three or four lengths of split bar to finish their neat and tight securing of the cavity before drawing in the chocks. Jim used the tannoy to call the men at the tailgate to get them to throw on the extra pieces of timber.

A minute later the voice of one of the men at the tailgate called back over the tannoy. 'The timber's on the panzer, ready for yer to run through.'

'Righto,' Jim called back. 'Let the panzer go.'

'Take the lock-out off and I'll let it go,' the panzer-driver called up from the maingate, over the tannoy.

'Can you stop timbering up, up there, for a couple of minutes so we can run this timber through so we can all be working?' Jim called over the tannoy up to the men on the other fall area, further up the face.

Receiving no reply, he called again.

A voice responded on the tannoy, 'Who rattled your fucking cage?'

'Take that lockout off so we can run that timber down,' Jim shouted into the tannoy.

'Fuck off.' the voice called back. 'We'll take it off when we're ready.'

'Right,' Jim snarled to himself before starting a furious, crawling sprint through the low face.

The incident with the trespassing police and the permanent, sharp headache he'd had for days from the cumulative effects of carbon monoxide poisoning had shortened his fuse. He was intent on sorting out the impertinent collier further up the face.

He was wearing knee-pads but when his left knee banged, heavily, into an unusual bracket sticking out from the front leg of one of the chocks it provided no protection.

'Aargh,' Jim growled.

It felt like he had completely displaced his kneecap round the side of his leg for a moment. He grabbed his knee, expecting the pain to kick in, amazed that it felt intact with its kneecap still in position. When he checked it, he was amazed; he could still move his leg.

He decided not to tell anyone about his mishap in order to avoid appearing as if he didn't know what he was doing on a low face.

His knee felt extremely tender as he tested his weight on it. As he carried on crawling, gingerly, up towards the other timbering team, the lockout went off the conveyor and the panzer driver started it up.

As Jim approached the other timberers, the split bars he had requested drifted past on the flights of the conveyor.

He got up to the men with his anger dissipated by the distraction of the knock to his knee. He still challenged them about the insubordination.

'Which one of you fuckers was being mouthy when I was trying to get that timber run through?'

'It wasn't us, Boss.'

'No. it wasn't us,' a second man said.

The third said, 'I reckon it was probably someone up the tailgate buggerin' about.'

'Yeah,' the last said. 'Tryin' to drop us in it.'

Jim realised that none of their voices matched that of whoever had responded to him over the tannoy.

Having had a wasted trip and an unnecessary bump on his knee, Jim inspected their work, complimented them on it then crawled off back down the face.

He was going to regret that knock on the knee for his next few shifts, crawling around on 56's.

Madness

It had worried Paul when his mother had phoned to ask him to call round so his dad could talk to him. Having noticed a change in Horace recently, he couldn't help fearing that his old dad was going to be telling him that he'd been diagnosed as suffering from something serious or even fatal.

Saturday, 1 December, 1984: 10:15

'Thanks, pet,' Horace murmured, as Pat handed him his coffee.

She moved round and settled, gently, onto the settee, next to his armchair.

Sally, the Highland Terrier, roused herself and deserted Horace's feet to settle, leaning back, expectantly, against Pat's shins, desperate for some fuss.

'Come on, then,' Pat said. 'What's on your mind?'

Horace shook his head, his mouth tightening into a sad line.

Pat waited. Horace shook his head.

'It's madness,' he muttered.

They sat quietly, Pat ignoring the continuing, silent but eloquent demands from the little dog at her feet.

'It makes no sense,' Horace said. 'I can't go on with it anymore.'

Pat sighed.

'It's only ever been you who could make that decision, me duck,' she said.

She sat and looked at him.

'Go on, then. Tell me.'

'That taxi driver,' Horace said.

He shook his head again.

Pat nodded.

'I knew it'd be that,' she said.

It had been on the news. The previous morning, David Wilkie's fare had been a working miner, going in for the dayshift at Merthyr Vale Colliery in South Wales. Two strikers had lain in wait for his car.

Whether their intention had been to kill the miner or just deter him, when they had dropped a concrete block from an over-bridge onto the taxi travelling along the dual carriageway below, the miner, although injured, had survived. It had been the taxi driver who had lost his life.

'He had a little girl the same age as our Jenny, a son and daughter like our Paul,' Pat murmured. 'What sort of Christmas will those little kiddies have, without their dad?'

'Paul's said it was a civil war, all along,' Horace said. 'Well, 'e's right now. It's a war with miners on both sides.'

He looked up at the painting of the tranquil, Warwickshire landscape above the coal-fired boiler.

'I've supported you for nine months,' Pat said. 'And I'll go on supporting you if that's what you want. But what you say's right; it's wrong now. It's been too political, all along. It's been too much about bringing down the Tories. It's made me wonder, at times, whether it's because we've got a woman for a Prime Minister, if that's what they can't live with. They think she's the enemy.'

'Since the IRA blew up the hotel at the Conference in Brighton with Norman Tebbit's wife crippled and all those people killed and injured, they've stopped shouting, "Maggie, Maggie, Maggie – out, out, out". It's "boom, boom, boom" they shout now.'

'It's like they're siding with the enemies of this country. And all that talk of the union accepting money from Gadaffi, when the Libyans shot that poor policewoman in London, just a few months back – it's a disgrace.'

Pat stopped.

'Well anyway.' She lowered her voice. 'It's wrong. Scargill has refused to have a ballot. And if they are trying to bring down the Government, it's not democratic, it's a revolution.'

'I know our Paul thinks himself a Labour man but I voted for Maggie in both of the last elections and I've got more time for her than that madman Scargill. One of the reasons the Tories got elected was 'cause they promised to stop the unions from trying to run the country, so no one should be surprised that that's what she's doing.'

Pat stopped and looked at her husband again. He was a picture of misery.

She knew what he'd be wondering. What had been the point of these past, wasted months on strike? And worse than that, how was he going to tell Paul and deal with his disappointment if he were to go back?

There would be no workable way to tell their son, Pat knew that. But it felt like they were on the wrong side now and there was no longer any point in suffering the consequences, if there ever had been any.

'You're the one who'll have to cross that picket line so you're the one who has to decide,' she said. 'But if you want my opinion, I think you should go back to work on Monday.'

Saturday, 15:35

Paul sat forward on the settee, eager to find out what it was that Horace had to tell him. He had disappointed Pat by bringing Susan and the children. But Susan, sensitive to the need for Horace to be left alone with Paul, had taken Martin and Jenny with her to join Pat in the kitchen for a cup of tea.

'I'm sorry, son, I know this will be a disappointment to you, I know it'll come hard,' Horace said. 'But I can't go on.'

'What d'ye mean?'

'The killing of that poor taxi driver, I can't go on anymore. With the strike.'

'Why? It's not the NUM or the strike that killed 'im, that was just a couple o' nutters. This is the point in the year when we always knew we'd have to dig in to beat them.'

'To beat who? We're fighting amongst us-selves now.' Horace said. 'If that communist, McFadden, hadn't phoned up Barnsley and got 'em to send a picket line we'd 'ave been workin', like Baddesley, all along.'

'So what're you saying?' Paul said.

Horace closed his eyes.

'I've had enough. I'm going back to work.'

Paul said nothing.

The silence hung heavily for a few seconds.

'What's the real reason?'

Horace opened his eyes and looked at his son.

'What do you mean? Isn't it enough that they've started killing people in our name.'

'There's been people killed before now, miners on the picket line and hundreds, thousands, injured by the police. What's the real reason?'

'There's nothing else. It's just I can't strike for what Scargill wants us to fight for; to refuse to allow any pit to be shut, however much money it's losing.' Horace shook his head. 'When the Kent pits are workin' they produce coal at three times the cost they can sell it for. I can't defend that, let alone fight for it. I can't support it anymore.'

Paul glowered.

'It seems a hell of a coincidence to me that this comes a couple o' weeks after the Coal Board've offered everyone on strike that massive Christmas bonus to scab,' Paul said.

'You can't think like that, son, not when you've seen your mum and me struggling like everyone else for nine months. We've still got a mortgage to settle on this place and we've spent all our life savings when I'm down to retire in a year. But we've still 'eld out 'til now. Now, when it's just too mad.'

'Well, there you are! You're the one who's talkin' about money.' Paul snapped. 'Well, let me tell you, if you do go back, you're no dad of mine.'

Paul sprang to his feet, stomped round to the kitchen door and yanked it open.

'Come on Susan,' he snapped. 'Bring the kids.'

'We're going!'

Pat had been quietly priming her daughter-in-law on what Horace had to say, so Susan knew better than to inflame the situation by trying to reason with Paul while his blood was up. Paul darted into the kitchen and snatched up Jenny from her chair.

Susan shook her head then took Martin's hand and led him out, after her husband.

Horace stepped into Paul's path.

'Don't rush off like this, son.'

Paul laid his spare hand onto his father's chest and shoved. The old man lost his balance and slumped back into his armchair, as Paul stormed on.

'Sorry,' Susan mouthed to Horace as she bent down quickly to touch him on his hand before hurrying out after Paul with Martin.

Pat came out of the kitchen, rushing towards the hall to go out to shout after her son but Horace pushed himself back up and took a gentle hold of her arm.

'Don't, love. I'll go round an' see 'im tomorrow. When 'e's had time to calm down a bit.'

Sunday, 2 December, 1984: 09:45

Horace went round to Paul and Susan's house, on his own.

Unusually, their back door was locked to him. He went back round to the front door. Nobody answered when he knocked.

He walked back the long way, going all round the village, stopping to have a sit down on his tool-box at his allotment as he tried to find the words to tell Pat that their son's door was closed to them.

It was a disaster.

But, as heaved himself up to head for home, Horace was resolved, he had to do what he had agreed with Pat; he would be going back to work the next day.

Saved

Jim came up the pit at the end of nightshift. He'd arrived at the pit, the previous evening, to find that Ken Goodall hadn't been in on afternoons but had already returned to his usual, early morning start-time. Whilst Jim had been out of the loop on nights, still in fire-fighting mode, the rest of the management team had decided the crisis was past; that they could expect the fire on 56's to die, if they just kept up the rate of advance and consequent, continued compaction of the waste.

Jim wasn't surprised. That was typical of the way fires tended to end, if they ended positively.

When they ended with the more dramatic loss of a district, you knew it was over when you'd finished building the stoppings, the two airtight walls that sealed off the entrance to the main and tail gates of an abandoned district. When they ended in success, they tended to just run down to a point where the management felt confident and then they just petered out, anti-climactically, like this.

'You might as well finish the week on nights, Jim,' Ken said to him, the next morning, in the dirty offices, when the nightshift officials had gone to the pithead baths and the dayshift men were all underground.

'Sod that, I'll double back on afternoons today and then do days tomorrow, then I can come to the end of fire party.'

The colliery's undermanagers and officials always got together at the end of the weeks of fire-fighting, and it was much more of a party when they had actually managed to save a vital district.

Jim had just heard that the officials had chosen to meet up in the Miners' Welfare, on Wednesday evening, for the latest celebratory session. And he wasn't going to miss out on that.

Good News

Pat had known that Diane had gone into labour the previous evening, so she was delighted when Doreen called round to tell her that her daughter had given birth to a second, little girl.

When Horace got back from the pit, that afternoon, she gave him the good news.

'Are they both alright?' He said. 'The babe got all 'er fingers and toes?'

'Doreen says she's perfect, a little beauty,' Pat said. 'Diane's fine but worn out. They're just keeping them in the maternity unit at the Eliot for another day. Doreen's looking after Emily and Colin'll fetch them both home by taxi, tomorrow.'

'It's always a relief when the mother comes through alright and the babe's 'ealthy,' Horace said

'There's nothing better than a new arrival, born safe and sound.' Pat said. 'It's nice to have something to give you a bit of hope for the future for a change.'

Neither Horace nor Pat would mention their sadness at Susan not being able to see Diane and welcome her new baby into the world.

Celebration

Wednesday, 12 December, 1984: 21:06

Gabriella was the only woman amongst the raucous group in the bar of the Welfare. Jim had invited her because they had seen so little of each other for the previous six weeks but mainly because she always enjoyed the pitmen's company almost as much as he. It was one of the reasons they had bought a house in a neighbouring pit village, rather than living more remotely from the mining communities, like most of the rest of the Area's colliery managers, in one of the villages around Market Bosworth.

The numbers at their party were inflated, suddenly, by the arrival of the afternoon-shift's officials.

For Gabriella, it was a celebration that her man and his colleagues had all made it safely through their latest, hazardous incident; yet another worrying period made worse by her inability to imagine the true nature and risks of what was actually involved.

For Jim, Ken Goodall and the officials, it was a recognition that, having saved a vitally important production unit, they'd be able to get back to normal, insofar as their lives down the pit could ever be described as such.

As they had arrived, Jim had spotted Paul Wood by the dart board at the end of the room, chalking-up for another couple of strikers, Steve Bircher and Wes Hands. It became apparent that the impoverished trio were only able to drink glasses of tap water to refresh themselves as they threw.

Jim knew that being cut off from the pit throughout another fight to keep it alive and excluded from joining in their celebrations would be hard for a pitman like Paul. He walked over and chatted to Paul and his companions for a while. He saw the discomfort in their eyes as they asked about 56's fire, how Jim and the rest of the team had tackled it, how their triumph had been achieved.

Out of the corner of his eye, Jim glimpsed Norman Bramble, one of his working miners, by the bar, bristling at the sight of his Undermanager fraternising with strikers. Jim could tell that Paul and his mates, having noticed him register Norman's agitation, were growing concerned that their young undermanager could be causing serious trouble for himself by associating with them. Despite this, Jim refused to leave them without making a gesture.

As a keen beer drinker, Jim knew how much the three men would be missing the pleasure of a pint.

'Let me buy you a drink, lads,' he said, ignoring Norman's evident and rising anger.

The three men exchanged a flurry of uncomfortable glances but Wes looked inclined to accept.

'Best not thanks, Boss,' Paul interjected, giving Jim a serious look. 'Best not cause any trouble.'

'Let me worry about that,' Jim said. 'Are you all on bitter?'

'Yes please, Boss,' Steve said.

Jim hurried away to the bar. Passing Norman on his way, he wished him a cheerful "good evening", receiving only a furious, pointed scowl in return.

Later on, back amongst the officials, Jim paid the barmaid, as discretely as possible, for another round for Paul and his mates and asked her to deliver it to their table on a tray.

Jim had just satisfied himself that the three men were being served with their next round when Gabriella nudged him, helpless with mirth.

'What?' Jim chuckled, delighted by her hilarity.

'Them,' she said, pointing at two afternoon-shift deputies, seated next to each other at one of the tables.

'What's so funny?' Jim said. 'They're only singing.'

'I know, but look how they're singing to each other.'

Jim looked and saw the humour for himself. The two officials, one big and one tiny were serenading each other with a duet of "I'd Like to Take You on a Slow Boat to China", looking deep into each other's eyes, like a pair of love-struck lovers.

Jim laughed with Gabriella, delighted to have her observation. Being immersed in the pit culture, it would never have struck him as unusual so, if left to himself, he would never have noticed it.

Who Do You Love?

Horace had been dreading this. He was glad to be further down the face, away from the four facemen lying in wait, ready to confront their Undermanager, as he crawled through from the tailgate end of the face.

"Morning, men. How are you all?' Mr Greaves called to them, over the rumble of the face conveyor.

Before anyone else could reply, Norman, furthest away from him, jumped in.

'Here he is!' He shouted. 'I saw him, last night, in the Welfare, buying beer for those bloody strikers.'

Normally, the men would dismiss their chock-fitter's attempts at agitation but, on their way in that morning, what Norman had revealed had unsettled them, leaving them all pondering its significance.

He had told them of how he had seen their boss at the gathering in the Miners' Welfare, the previous evening, going over and fraternising with three strikers and, even worse, buying all three of them a couple of pints each.

Having spat out a stream of tobacco juice, Norman had called out, in front of all the men on the train.

'An' that militant, bloody striker of a son of yourn were amongst 'em an' all, 'Orace Wood.'

Paul would have been without his regular darts partners, Brian and Colin. Horace knew that his son and his companions would have appreciated that bit of contact and fellowship from their Undermanager as much as the rare treat of a decent pint.

But, however kind an act, it could only be seen as inexcusable in a colliery manager, in the circumstances.

'He thinks more of those bloody strikers than 'e does us workers,' Norman bellowed, jabbing a finger at Mr Greaves and glaring up the face at him.

The Undermanager looked serious for a few seconds. Then he shouted back, so they could all hear over the racket on the face.

'It's my money and I don't have to explain how I spend it to anyone.'

'But, as it happens, Norman's right; I did buy some strikers a beer last night.'

'Now, I can see why that might upset you. But the thing to remember is this: All of you had some time on strike and I never stopped caring about you, just 'cause you were out on strike.'

'So, they may be strikers – but they're still my men!'

Horace saw the change in his workmates. The facemen hadn't known what they'd been looking for from their Undermanager but his answer had done more than address their concerns; they understood it and approved of it. Even Norman nodded, quietened.

Mr Greaves carried on his passage through the face. As he came to each man, they struggled to ease into the gaps between the front chock legs, to make way for him. He placed a hand on their shoulder or a knee-pad-covered knee as he climbed over them, each one saying a cheerful "'Morning, Boss," to him.

When he got to Norman, he gave him a gentle smile.

'Are you alright, yer grumpy, old bugger?'

Norman dropped his eyes and smiled, sadly.

'Yeah, fine, Boss. Never better.'

'I'll make sure I buy you a pint, next time I see you out,' Jim said.

'You know yer don't 'ave to do that, Boss,' Norman said.

Jim crawled on down to Horace.

'Good morning, Horace. Are y'alright?'

'Champion, Boss,' Horace said, scrambling to let his Undermanager pass.

Mr Greaves winked at him then eased himself past with a smile.

He stopped and half-turned back to Horace.

'Anyway, how are you getting on with that "bloody striker" of a son of yours?' Jim said.

Horace's face fell.

'Best not ask, Boss. 'E's cut me off. Won't speak to his black-leggin' dad.'

'Yeah, I'd heard that,' Jim said.

'It's tough, Horace. It's bloody tough. It's wrong, the way the leaders on both sides have ended up setting us all against each other.'

'Anyway, hopefully we'll sort it all out in the end.'

'Yes. I hope so, Boss.'

But Horace didn't sound convinced.

Jim had crawled away, a couple of chocks down the face when he stopped and twisted round again, to call back.

'What? No aniseed balls today, Horace?'

Horace scrabbled round to face him, crawled down to him and wrestled a relatively fresh-looking, paper bag from his pocket.

'Sorry Boss, I wasn't leavin' y'out. It's just a bit earlier in the shift than I usually start flashin' 'em.'

Jim thrust his blackened fingers into the bag and drew out one of the rust-coloured balls and stuck it in his mouth.

'Thanks Horace,' Jim said as he turned to continue his crawl through the face.

Horace would never know that, as always, as soon as the Undermanager had moved far enough down the face and into the dust haze to be sure of not being seen he had spat the aniseed ball that Horace had given him into the waste, where the roof caved behind the chocks.

Jim had discovered he had lost his taste for these sweets a couple of years earlier, when working at Canley Colliery. A pitman had given him an aniseed ball and it had made him gag. But an incident in his grandfather's pub, when he had been four years old, had meant that he had never spurned Horace's ritual kindness.

Early one Saturday lunchtime, Jim's dad had been helping behind the bar as a younger Jim stood on its threshold. An ancient, retired miner had been leaning over the bar to talk to Jim.

The old man smiled and dug into his pocket, pulled out an apple and reached over to hold it out to the four-year old. The old miner's coat was shabby and the apple misshapen and covered with unsightly, scabby markings.

The boy looked at the apple, shyly.

'No, thank you,' he mumbled.

Later, Jim's dad had explained gently that the old gentleman would have thought he was giving Jim a treat so it would have been better if he had accepted the kindness, even if he had only pretended to save the apple for later and not actually eaten it.

Visitation

Friday, 14 December, 1984: 15.13

Horace walked down the side of the house and opened the back door, his hair still damp and well-combed from the pithead baths. As he stepped into the kitchen Pat turned away, kettle in hand. She did not greet him.

She put the kettle down on the work surface and plugged it in. Horace hovered, worried and waiting, ignoring Sally whimpering and pawing at his shins.

Pat turned back to look at him.

'He's been here today,' she said

'Who?' Horace said, unnecessarily.

'Our Paul. While I was out.'

Horace's spirits soared for an instant.

'He left his garage key,' Pat said. 'He's been in and taken his bike.'

Horace knew, as well as Pat, what that meant; Paul was having to sell his Commando.

Pat knew it would hit Horace hard but she had to tell him. And she had needed to get it over with as quickly as possible. Her heart ached as she watched her husband's face fall as he lowered his maroon, leather, snap bag onto one of the kitchen chairs with care, as though frightened of causing any more upset.

'The poor lad,' he murmured.

It was the most stupid time of year to try to get a decent price for it; the wrong season for motor-biking and, anyway, any money people were spending would be on Christmas. But Paul was determined to provide some sort of Christmas for Susan and the children, after all their months of going without. And he needed to pay off some of their growing debts and outstanding mortgage repayments, to keep the wolf of house re-possession from their door.

However little he got for his Norton, he'd be able to get Susan to buy a small bird for their Christmas dinner and get a present each for her and Martin and Jenny.

He wanted to make sure that Martin didn't miss out at Christmas. With working miners' children at school outnumbering those of the strikers now, being small for his age, Martin had started being bullied for the first time. And it would amaze Martin. The sensitive, intelligent, little chap knew that they had no money and that, somehow, that would affect what he could expect from Santa. But he had let enough slip for Paul to know that sixteen pounds invested in a 'Transformer Action Figure' set would serve to transform his son's Christmas.

Paul never doubted that the strike was worth it. But Susan said it felt like life had stopped for a year. Things that he and Susan had taken for granted, a modest, family holiday on the coast, an occasional baby-sitter and a night out for the pair of them, even a couple of beers in the Miners' Welfare, had stopped being memories and become things of which they could only dream.

Martin ripping the wrapping paper off that Transformer would brighten the hard, winter's gloom, if only for a few hours.

A Chance Encounter

Friday, 21 December, 1984: 15:43

'It's Sally.' Martin shouted.

Susan froze then twisted round to look back, still holding onto the handle of Jenny's pushchair with both hands.

The Scottie was circling the boy, whimpering and jumping up at him. In his excitement, Martin was oblivious to the white lines the little white dog was scoring on the skin of his bare legs, as she scrabbled at him with her claws.

Before Martin could get hold of her, she'd darted off to Susan, jumped up at her, bounced off her thigh, dodged away as Susan bent down to catch her lead then shot round to the front of the pushchair to leap up and lick Jenny's fingers.

Martin darted round after her and grabbed hold of her lead.

As he crouched down, stroking her and beaming, the little terrier wriggled her stocky body against his knee.

Horace had just emerged with Sally from the alley-way between the two rows of houses, on his way back from the allotments. Having a fork in one hand and a bucket in the other had made it easy for the little dog to snatch her lead from his loose grasp and dart after the boy.

He had been taking care to avoid a chance encounter like this, knowing his daughter-in-law would be doing the same. The last thing he had wanted was to put her in an awkward position.

Before Susan could stop him, Martin had leapt up and was away.

'Come on Sal',' he cried, as they ran back together to his grandfather.

It had only been three weeks but it seemed like an age since Horace had seen his grandson up close.

'Hiya, Grandad,' Martin shouted as he ran back to him.

Horace smiled gently.

335

'Hello Martin, lad,' he murmured, as the boy reached him.

He felt the electric touch of boy's fingers on his own as he placed the lead in his hand and beamed up at him.

'Come on Martin,' Susan said, looking away, across the road.

Martin's face dropped.

'Off you go to your mum, lad,' Horace murmured.

For a moment, the boy hesitated, about to speak. He looked, hard, into his grandfather's eyes then turned away, looking back over his shoulder as he broke into a trot to catch up with his mother, as she strode away.

Released

Being a good pitman, Ian Flint knew how to behave himself and
get by in a rules-based regime. With his broken arm mended and
his pitman's work ethic he had wielded his spade in the prison's
market garden with more enthusiasm and to better effect than any convict
the warders had ever known. Being a model prisoner had resulted in his
early release, on good behaviour, six months before the end of his year-long
sentence.

He'd returned, the previous Wednesday, to a hero's welcome; the
dramatic nature of his capture, his 'war-wound' and false imprisonment
having turned him into a celebrity within the community.

The strike funds, still coming in from northern cities, had made it
possible for the union to reward him for his contribution and for what
he had endured for their cause. So, despite the general lack of money
in Railsmoor, Ian hadn't gone short of beer during a long weekend of
celebration.

Sammy hadn't been able to join him. He was still having to honour the
year-long curfew, imposed on him by the Court, which required him to
make daily trips to Barnsley to report to the police station, at twelve thirty,
and banned him from stepping outside his house after five in the evening.

Ian had always been respected in the community but now everyone
wanted to be associated with him. He'd boasted to a couple of mates over a
pint, the previous night, that he'd had more sex with more, different women,
in those first few days back home, than he'd had in the rest of his life.

'It's like being a rock musician or a film star or international footballer,
the way the young bits o' stuff – and some o' the wives – 'ave bin throwin'
'emselves at me,' he had confided.

They wouldn't have the money for a lavish celebration but, with the communal meal that the strike kitchen would be providing in the village hall, the amazing community spirit and his enhanced standing, it looked to Ian like being his best Christmas ever.

That was, until Jackie heard the tales being told out of school, while taking her turn in the strike kitchen.

Ducking too late, the teacup scored a direct hit on the side of his head then dropped to smash on the floor.

'Who the hell do you think you are?' Jackie screamed at him, from the other side of the kitchen table, the broken cup still spinning and ringing on the floor.

'Just because you've been inside for a few months doesn't give you the right to behave any way you please.'

Ian raised both hands.

'Please Jacqueline, Matthew'll 'ear you.'

'What do you care about him – or me? If you cared anything at all about your family you wouldn't have been fucking about with those stupid slappers, dopey Vicky bloody Bennett and that gormless Tina Hobson and whoever else.'

'Give me strength. What a pitiful, scruffy bunch of bitches to fuck about with.'

She shook her head.

'You're pathetic!'

He felt pathetic, at how shocked he was at hearing her swear at all, let alone in their own house. He hung his bleeding head.

'I've had enough of your nonsense,' she shouted. 'You keep going on about loyalty – to your mates, to the bloody union and the pit, you talk about the treachery of the scabs but what loyalty have you got to those closest to you? – To your own wife and son!'

'Jacqueline, I'm sorry.'

'"Sorry",' she snorted. 'All that time you were away, playing at jailbirds, you knew you could trust me to stay faithful to you. The only time I went out was to go and do my bit with the bloody Women Against Closures and to cook stupid meals at the stupid, bloody strike kitchen and travelling all over the place to get up in front of people and speak to them about "the struggle"; the bloody struggle, your stupid, bloody strike, the strike that's made you a bloody hero.'

'You bloody love it, don't you; playing soldiers or bandits or cops and bleeding robbers or whatever it is you think you are.'

'Well you can bloody well stay in tonight,' she snarled as she darted out of the kitchen, opened the door of the cupboard under the stairs, snatched her coat from the rack inside then slammed the door closed.

'I'm going out. I'll drink your bloody beer for you. And I'll see who wants to shag me.'

'Don't Jackie,' Ian said, reaching for her arm as she lunged for the backdoor.

'Get off me,' she growled, waving her arm high and wild to evade his grasp. 'I've had enough of you for one day.'

She pulled the back-door open and skipped out over its step. Ian kept hold of its handle to stop her from pulling it closed behind her, so she just let go of it, spun away and stalked off down the path between the house and the side fence, doing up the buttons of her coat with furious fingers.

Ian jumped out and called after her.

'Jackie, I'm sorry, I've been an idiot. Come back, let's talk about it.'

He rushed after her then stopped as the garden gate slammed shut, rattling its posts in the ground behind her.

'I've done all the talking I'm going to do with you,' she shouted, over her shoulder. 'You're right, you *are* a bloody idiot. Now you go and babysit for a change.'

Ian stood and watched her strut off down the pavement.

Two women on the other side of the road watched him with heads together and pretended shock and murmured "Oooh's" before cackling in delight at the indiscretion. Closing his ears to them, Ian turned, shaking his bleeding head, and shuffled back into the house.

'Mind your own fucking business, yer nosey old bags,' he muttered.

He dawdled back down the path.

He shut the back door behind himself and looked at the kitchen clock; three thirty-five. He had only meant to go out for a quick two or three but he'd ended up staying for an inevitable eight or nine.

Crapping on his own doorstep; before they'd put him inside he would never have believed he could be so stupid. Not only had everything Jackie said been true, the cruel irony was that it had only been that morning, as he'd headed for the Miners' Welfare, counting his blessings and looking forward to their Christmas together, that he had chastised himself for being an idiot, for letting her down, and resolved to behave himself, to stop playing with fire.

He would have an evening in, without any tea, to reflect on her words.

He tried to take heart from one phrase she had used. He could only hope that when she'd said, "I've had enough of you for one day" that meant that she hadn't gone or given up on him for good.

"What a complete prat," he thought.

He had always known she was too good for him. She'd been a teacher until Matthew was born and would be again, when he started at school. Despite her anger and threats, he had no fears about her messing about with anyone else that night; she wasn't an idiot or an overgrown kid like him.

For the second time that day, he resolved to start behaving himself and try to make it right with the only woman he had ever loved.

Christmas and New Year

Christmas Day, 25 December, 1984

For the second year running, Jim Greaves had volunteered to take charge of the pit on Christmas Day, in order to be sure of being off on the first of January with his New Year's hangover.

Remembering the impatience of the deputies and overmen to get away, as soon as they were up the pit the previous year, Jim had only brought in enough of his mother's mince pies for himself, for Roger in the Control Room, Les Parker who would be staying on the surface for that shift, as senior overman and Mad Jack who was in the fitting shop, in charge of the Mechanical side for the day.

Jim finished reading and signing all of the statutory reports for the dayshift when the officials had gone and then called in to the Control Room.

'All done, Boss?'

'Yep, I'm getting off.'

'Have you got family round for Christmas dinner?' Roger said.

''Yeah, we have this year, we've got Gabriella's parents joining us for the traditional dinner. But I'm going for a pint on the way home. I've just given Gabriella a ring to tell her to drive up and meet me there. She can leave the turkey roasting for a bit.'

'We've got a goose this year, when I get to it.'

'Are you getting relieved early again, this year?'

'Yeah, good old Jerry's coming in early again, so I'll have time for a quick couple on the way home for dinner, meself. I'll be ready for it.'

'Right, say "Happy Christmas" to Jerry for me, I'll give him a ring later, anyway. And I'll see you tomorrow, then. Happy Christmas and a Happy New Year to you and the family.'

'And the same to you. Boss,' Roger called after Jim has he headed off up the corridor. 'And, let's hope it's a better one.'

'Amen to that,' Jim shouted back.

Half an hour later, Jim was warming himself by the fire in bar of the Horse and Jockey.

The pub was noticeably quieter than it had been on the previous year's Christmas Day, reflecting the number of men still out on strike at Whitacre Heath and the effect of the enduring overtime ban on Baddesley Colliery's miners. He was on his second pint, enjoying the joviality of those pitmen who were there and who had had more time than he sinking pints.

Gabriella had still not arrived. He guessed that she had probably been delayed by the great volume of work involved in preparing everything that went with the turkey. He had a half of Bass sitting on the narrow, tiled mantelpiece, ready for her when she did turn up.

Kath called out in her croaky voice.

'Drink up! It's time you lot went home.'

Jim finished his pint, stepped over to the bar and handed his glass to the Landlady who pulled him a last pint. As Jim handed her a pound note a pitman passed her three empty pint glasses.

'Can we have a quick three, please Kath?' The pitman said.

'No you can't. It's Christmas Day, you should all be at 'ome, with your families.'

'But you're servin' 'im.'

'Mind yer mouth!' She scolded. ''E's the Undermanager. And 'e's been at work this mornin'.

'And 'e don't take 'alf an hour to drink a pint, he'll see his'n off promptly,' she snapped, a smile playing around her tight lips and a hint of a wink at Jim belying her fierce demeanour.

'And you'll have to get a move on,' she muttered to Jim, with a nod in the direction of the window. 'She's come to find yer.'

Jim swung round and there was Gabriella coming through the door. She had found time to do her eyes and apply some red lipstick. Before he could hand her half pint to her, two of his pitmen had intercepted her to seize a Christmas kiss.

Christmas Day was a sombre occasion for Horace and Pat. They had agreed that it would be impossible to celebrate it without Paul and his family, impossible to forget the young family's straitened circumstances.

It was a particularly cruel blow, to be deprived of being able to give presents to their two grandchildren and see them enjoying the magic of Christmas.

Horace had been relieved when Pat had said that she couldn't bring herself to prepare and eat a turkey dinner without them all. They would be having gammon and chips in the evening, instead.

The lounge was decorated, festively, with Christmas cards on various surfaces.

Deprived of his traditional Christmas morning trip with Paul, Colin and Brian to sample the draught Bass at the Horse and Jockey, Horace had poured Pat a sherry and a tumbler of light ale for himself.

They sat quietly; Pat on the settee, Horace in his armchair, with Sally at this feet, a programme of carols from a church choir on television, in the background.

The house was warm. Horace's concessionary coal delivery had been surprisingly prompt after his recent return to work; their comfort another source of sadness for them, only causing them both to imagine how chill Paul's and Susan's house would be for them and their young family.

Paul's beautiful landscape on the chimney breast was a constant reminder of their son, his family and their unbearable estrangement. Horace's desire to use Christmas as an opportunity to go round to see his son and appeal to him to re-establish contact had been a recurring inclination for days.

But he had known that doing that would only have inflamed things, that there was no possibility of reconciliation, particularly since it seemed that Pat was set on being as uncompromising as her son. Early after Paul's outburst, when Horace had spoken about it, Pat had expressed her view.

'If that's how he's determined to treat us, putting that man, Scargill, and this strike nonsense before his family then that's how it must be.'

Pat had chosen not to tell Horace, at the time, that she had already had more success in getting to speak to their son.

The day after Horace had encountered locked doors, she had gone round and told Paul that none of them should let outside influences come between them as a family. She could not tell Horace about the meeting or its outcome. Paul had stuck to his view that Horace had betrayed everything they both believed in, saying he was a disgrace to the memory of his own father, the Jarrow marcher.

In his fury, Paul had repeated his terrible sentiment, *"He's no father of mine".'*

She couldn't tell her poor husband that.

Pat and Horace said little to each other through the day, knowing that if they did talk it was likely to be to wonder what sort of day Paul, Susan,

Martin and Jenny would be having. And to wish that they could all be together.

Like Colin, when he had sold his Triumph, Paul had been appalled by the amount he had received in return for parting with his beloved motorbike, although not as surprised.

But it had given them the funds to help them to survive as a family for a little longer and Paul had put some aside to pay for a few things for Christmas. Susan had used some of this to buy a cockerel and a small, Christmas pudding.

Since the end of the summer, Paul had taken to roaming over the fields and into the woods, when he wasn't picketing, to find fallen boughs he could drag back, saw up and burn in their lounge's fireplace, in the absence of his coal allowance.

Most importantly, for Paul and Susan, they had been able to buy presents for the children. Martin found a large box at the foot of his bed and unwrapping it, not daring to hope, he was amazed to find it housed the Transformer Action Figure he'd thought he could only dream of receiving. Even though he still believed in him, he was old enough to know that Father Christmas's magic was dependent, in some mysterious way, on his parents making purchases.

To Martin, it seemed his heroic father and angelic mother had worked a Christmas miracle for them.

Whilst he wanted to play with the transformer anyway, Martin made sure that he did so, unceasingly, throughout the day to make his delight in it evident to his mum and dad. He took the toy with him when they went, as a family, to the village hall for Christmas tea, paid for by the strike funds, where Paul and Susan were able to enjoy a couple of drinks in the company of other strikers and their families.

Neil and Julie had managed to align their respective duties so that they were both off on Christmas Day and Boxing Day. They had driven up to Yorkshire from Luton on the morning of Christmas Eve and called in at Julie's parents' home for a cup of tea before going on to Neil's parents, in the next village, where they would be staying.

After having breakfast and opening presents with Neil's parents and his sisters, they went to Julie's for Christmas lunch. By tacit agreement, Neil and Julie timed their arrival to make sure that it would be after the time at which Julie's dad and brother would have left for a pre-lunch pint at the pub in the pit village.

Railsmoor, Barnsley Area. Monday, 31 December, 1984.

As soon as Ian had heard that he had a chance of being released from prison in December, Christmas had been his landmark, the subject of his daydreams and all his thoughts, as he'd lain awake listening to the coughing and crazed shouting of other, sleepless prisoners throughout the grim, seemingly-endless nights of his remaining time in prison.

He still couldn't believe how he could have been so stupid as to jeopardise not just his Christmas but his marriage with his idiocy in those first few days of freedom. He'd never been interested in screwing around, he'd just felt a kind of obligation to enjoy the recognition and what so many had seemed to think were his just rewards, for what he'd been through for the sake of the strike.

It hadn't even been particularly enjoyable. A few meaningless quickies in grubby corners; outside, behind the Chapel, up against the wall in the crate store behind 'the Welfare', in the bus shelter. He knew he'd shagged them but it might as well have been someone else who had enjoyed whatever pleasure was actually involved; he'd been so drunk that the encounters had barely registered in his consciousness.

Christmas couldn't have been more awkward or embarrassing. He had not held the position he should have; not been the man of the house enjoying the love and respect of his woman. Instead, he'd had to endure Jackie treating him like a stupid, juvenile delinquent, for whom she had neither time nor interest.

Before Christmas, on the morning of his third day back, he had caught the bus to Barnsley, with Sammy, pawned his grandfather's fob watch and bought a large bottle of Jackie's favourite perfume. The shop assistant had wrapped the gift in classy, silver, wrapping paper with fancy ribbons for him.

Opening it, in the lounge, on Christmas morning had been a chore for Jackie.

'Thank you,' she had said, tartly, as she placed the unopened box on the coffee table. 'Very nice.'

Ian had suddenly feared that she would say, "And what did you buy your girlfriends?"

But she hadn't needed to, she had known it would be in his mind as much as her own.

Jackie had insisted that they stick to their plan and go down to Railsmoor's strike kitchen for Christmas lunch.

From the time they walked into the village hall to the time they left, three and a half hours later, she had made her disdain for him evident and, inevitably, everyone knew what it was all about.

He got no sympathy from the other men's wives. His role as the returning hero was over; they treated him like a teenaged delinquent, even those who had been happy to flirt with him before.

Always assertive in their own ways, it seemed to Ian that the women of the pit village were becoming even more so. He had been aware of it starting to happen in all of the mining communities in which he had spent time as a picket. For some women, like Jackie, this was most evident in their self-belief.

After Boxing Day, the atmosphere had started to thaw. Mercifully, Jackie had allowed him to move off the settee in the lounge and return to their bed. It seemed like she was starting to accept his contrition and recognise that his foolishness had been a way of working through all he had endured.

When he went out to meet his mates on New Year's Eve, he had not needed her pointed instruction "don't do anything daft" as he had stepped out of the back door; he was determined there'd be no repeat of his recent foolishness.

When he promised to be back to see the New Year in at home with her, Jackie had allowed a little smile to let a glimmer of encouragement show through as she had shrugged with her mouth and shoulders.

The men had a sub' from the strike funds as a reward for their hard year of picketing. They started off at the Falcon, one of Railsmoor's two pubs. Sammy was complying with his curfew, so Ian led the rest of them in remembering their absent comrade by raising his glass to toast him on their second round.

Just after seven, after several in the Falcon and quick couple in the White Horse, they marched purposefully down the middle of the empty street to the Miners' Welfare, shouting and laughing.

Before following his mates through the double doors into the Welfare, Ian stopped and turned to look at the headgear of his pit across the road, its wheels stilled, not just by the holiday but by their strike.

Inside, as well as all the pints they were consuming, the barmaids kept slipping Ian the occasional double whiskey. By just after ten o'clock, when Sammy's wife came over and sat down next to him to say hello, Ian knew that he was plastered.

Vanessa was seven years older than Ian. Despite the guilt that he had always felt at fancying his mate's wife, he had been unable to prevent her from being the subject of some occasional, erotic and explicit fantasies when he was younger.

Freckle-faced and foxy with long, straight, dark brown hair, a curvaceous figure, husky voice and filthy laugh, Ian always thought of Vanessa as the type of woman they would have burnt at the stake for bewitching the menfolk in earlier centuries. Now, as she sat chatting and giggling with him, her hot, firm thigh brushed and pressed against his.

When she stood up to go back to join her girlfriends, she bent back down and gave him a moist kiss on the cheek. As she leant over him, Ian caught her hot, musky scent as his bleary gaze ogled her breasts' heavy motion in the plunging neckline of her blue, satin, party frock, inches away from his face,

It was a relief when she straightened up, with a light touch on his shoulder, and tottered off, wiggling away on her high heels, so he could clear his head.

Twenty minutes later, when he came out of the gents', Vanessa was lying in wait for him. With a dangerous smirk, she grabbed his hand.

'Quick, come with me,' she murmured.

Reactions and judgement dulled by drink, he stumbled after her as she drew him down the corridor.

She heaved him round into the dimness of the stairwell up to the Welfare's upstairs meeting room. Out of view from the corridor, she leant back against the wall, grabbed his belt with both hands and pulled him to her. In her heels, she stood taller than him. Her breasts pressed onto the top of his chest.

They were both breathing heavily. With practiced skill, Vanessa had his belt undone and his trousers unfastened and unzipped in a moment. He felt the scratch of her nails as her long fingers reached into his briefs. He tried to pull away from her open mouth and tongue to say "No".

They heard the noise from the bar increase as its door flew open, then a pair of heels came clacking down the corridor.

Vanessa stiffened.

The footsteps stopped.

'Don't move, they won't find us,' she whispered with a saucy smile.

The footsteps clattered on again.

Vanessa was wrong.

As Ian stepped back, he just managed to stop his trousers from dropping to his feet. He was struggling to pull them back up his legs as Jackie's older sister, Linda, appeared, standing on one foot, to lean her head and shoulders round the corner.

Ian saw her eyes drop to watch as he finished pulling his trousers up and tucking his shirt back in. She moved into the centre of the opening, put her hands on her hips and continued watching until he had finished hurriedly fastening his trousers and belt and pulling up his zip.

He didn't waste his breath saying anything. Linda had always hated him, always thought he wasn't good enough for Jackie and had always made those feelings evident.

She spat out the single word, 'Bastard,' then flounced off back down the corridor and clattered her way through the door back into the bar.

'Don't worry, Vanessa said. 'She wouldn't dare tell your Jackie.'

'Not much!' Ian snorted. 'She'll be round our place like a shot. This'll make 'er Christmas for 'er.'

Ian knew there was no way of ducking the repercussions. He darted out of the pit-club and went straight home to confess to Jackie that he'd let her down and managed to make a fool of himself again, this time in front of her older sister.

After a night relegated back onto the settee, Ian had his hangover interrupted when Linda called round, early on a miserable New Year's morning, as he had known she would.

As she described the scene to Jackie in more detail than Ian had given, the previous night, Jackie stood silently, her folded arms wrapped around herself as she stared at the floor.

'And I expect you already know that Sam Crowther's missus isn't the only woman he's bin playin' around with,' Linda said, as she concluded her account.

'I wasn't playing around with her,' Ian started to protest.

'No, I suppose you weren't,' Linda sneered. 'With your trousers round yer ankles you were deadly serious about what you were up to.'

Jackie drew a deep breath and looked at Linda.

'Okay, leave it with me.'

Linda drifted out through the kitchen, the satisfaction evident in the hard line of her mouth. When she was out of the back door she dawdled down the garden path. Before she reached the gate, she was rewarded by Jackie's muffled shout.

'Who the hell do you think you are?'

'Not only do you cheat on me, your wife, and your own son, now you've gone and cheated on your best friend.'

'It wasn't like that, Jackie. I told you, I didn't know what was happening. I'd just been to the gents' and Vanessa grabbed me. Before I could fight her off, Linda was there. I was trying to get away when she turned up.'

'So it was all her fault was it?' Jackie shouted. 'What about all the others, was that their fault as well. I suppose you just can't do anything about the irresistible charm you have for all these women.'

Jackie's voice dropped.

'All the misery of the last year. It was all supposed to be about the union and your bloody brotherhood. But all that's forgotten now and you start screwing your best mate's wife. It was all just empty posturing. You've changed, Ian Flint; there's no honour or decency left in you.'

'Get out,' she spat the words at him.

'I'm not going,' Ian said. 'I'm not going to give up on us. It was all a load of nonsense.'

'You've had your chance. Either you get out or I'm going. If that's what you want; to see me and Matthew turned out of our own home. In the middle of winter.'

She went out to the hall, grabbed his coat from the cupboard and returned to the lounge.

'Here!' She snapped, throwing it at him. 'Get out!'

'Jackie, don't do this. I'm really sorry.'

Jackie fumed for a few seconds, nostrils flaring, chest heaving.

'Get out,' she growled. 'You disgust me. I'm finished with you, I tell you.'

Ian walked down the street. His head was clearer now as he faced his circumstances: No money and, having been sacked for the offences for which he'd been framed, no job. An ex-con' who'd lost his wife and family, he'd lost his position in the community and all the respect they'd so recently had for him and he would lose his mates as soon as they heard what Linda had caught him doing with his pit-dad's wife.

He thought back to the previous New Year's Day, twelve months earlier. If anyone had told him then that 1984 would be so fateful, he wouldn't have believed them. It was as though he had conspired with numerous other diverse parties to see just how much they could bugger up a man's life. He was starting the New Year having lost everything, his ruination was complete.

Ian Flint was finished.

Bullies

D iane had phoned three times in the last hour. She'd been desperate to get Doreen to run the errand for her but, in the end, had been forced to face up to it; it was obvious that her mum was going to be out for some time.

So here they were, braving the cold village, for the first time since Sophie's birth, three weeks earlier. It was also the first outing for the double-buggy that Colin had bought for their two girls.

Diane kept sucking the blood from the nasty, little nick on the finger she'd skinned when she'd caught it, wrestling the buggy open, outside the backdoor. Then Emily had been working up to a tantrum as Diane had fetched Baby Sophie out. It was daunting enough, having to struggle out, without all the little difficulties that just seemed to build up.

She felt guilty for feeling so drained and down all the time, at her inability to experience the elation that everyone seemed to expect her to be displaying at having been blessed with another beautiful, healthy, little girl. But there was no escaping the weakness that gripped her slight frame.

Doreen kept telling her not to worry. Being a mum herself she reckoned it was probably "just a touch of the baby blues". She said everything would seem brighter when the weather improved and Diane's hormones settled down.

The hormones; perhaps that was what it was all about.

At first, it was when the girls were asleep. Now it happened even when they were awake, so Emily would catch her. She kept on having a little cry.

Actually, not so little; sometimes, when she was alone, she couldn't stop sobbing and blubbing and then it all seemed so black, with no prospect of ever getting better, and she could end up howling. And there was no one she could talk to about it.

If only she had Susan. It kept hitting her now; how much she had always relied on her friend, how lost she was without her.

She had hoped some fresh air might blow away the darkness that gripped her but the steady, freezing wind chilled her, only serving to make it worse.

She plodded on, shoving the pram, taking care to avoid losing her footing on the glassy, black scabs of ice on the pavement, praying, with every step, that they'd be able to escape the estate and get round to the shops and back without encountering anyone, whether friend or foe as a consequence of this awful, endless strike.

At least Emily had quietened down, once they'd got going and she had the changing scenery to occupy her. And, fortunately, it looked like the bitter weather was deterring anyone else from venturing out that morning.

Diane turned the pram out of the road from the estate and stepped out, alongside the main road, to the shops.

It was her own stupid fault. She only needed some nappies. If she hadn't been feeling so sorry for herself all the time she'd have realised she was running out. She'd been sure there was another pack in the cupboard. Even then, she would have made it until Colin got home, if Sophie hadn't managed to fill three in quick succession that morning.

It was just her luck – but, hopefully, she could be in and out of the shop in a couple of minutes. And, as long as she got a trouble-free march home, she could be back, safely inside, in the warm, in ten minutes.

She pressed on.

At the VG Store, she struggled to get the new pram in through the door. Irene Mundy, who ran the shop with her husband, darted out from behind the little Formica-clad pay-point, where she was manning the till, to greet her and hold the door for her.

As Diane parked the pram, out of the way, by the ice cream chest-freezer and started to move into the shop, Emily cried out, 'Mama', holding her arms out, to be picked up.

Diane twirled round, crouched down and placed her fingers on Emily's shoulders.

'Just wait there a second, sweetheart,' she murmured, imploring her with her eyes. 'Mummy won't be a minute.'

Miraculously, this actually seemed to pacify the toddler, so Diane could jump up and make it to the aisle with the shelf of disposable nappies.

By the time she came back round with a small pack, Sheryl Skinner and Alice Parkin had drifted towards the till like a pair of Dreadnoughts. Both women were married to striking Whitacre Heath men. Despite the privations of the strike, neither showed any sign of losing any of their considerable bulks.

''Ere she is, Ali.' Sheryl said. 'Lady Fucking-Muck. Come 'ere to grace us peasants with a visit.'

'That's enough of that language in here, Sheryl Skinner,' Irene shouted, as she rushed off to the backroom to grab a new till receipt roll.

'That's right, Sher', we don't want to go upsettin' 'er Ladyship, do we?' Alice sneered as they both continued edging forward to bar Diane's path to the till.

Diane slowed to a stop. She'd been afraid of these two when they'd been the fattest girls in the year above her, at school, and they'd been using the strike as an excuse to regress.

'Look at that fancy pram she's got, Alice,' Sheryl sang out.

'I know,' Alice snorted. 'Proper Rolls Royce job, i'n't it?'

Diane gulped then sniffed.

"Oh no," she thought, *"I'm going to cry again."*

'Yeah, well that's what you get when you're the filthy whore of a scabby-backed traitor.' Alice sneered.

'That's enough!'

Fierce as the roar of a lioness, the familiar voice rang out from behind where Diane was cowering.

'Leave her alone,' Susan snarled, marching out from another aisle to position herself between the two heavy-weights and their victim.

'Irene,' she called out in her clear voice, towards the back of the shop. 'Could you come and take for something, please?'

As Irene bustled out and around the rows of shelves towards the till, Susan turned to Diane.

'Here,' she said, holding out her hands.

Susan took the little pack of Pampers from Diane. Ignoring the hefty women, she strode past them and put the pack down in the space for the wire shopping-baskets.

'How much is that, please, Irene?'

'What's it got to do wi' you? Your 'usband gonna join his black-leggin' bum-chum is 'e?' Sheryl sneered, as Irene rang up the amount.

Susan spun round. Sheryl flinched as Susan thrust her face towards her.

'Don't you dare talk to me like that,' Susan growled. 'Now, clear off.'

Susan swung her back on Sheryl and Alice, looked at Diane and nodded towards the till. Diane stepped forward and opened her purse.

Alice sneered, 'Just look at all that scab money she's got, Sher.'

'I've told you. Just leave her alone. Now shut up and go away!' Susan roared, stomping towards Sheryl and Alice.

The two bullies skulked back like a pair of gross hyenas.

Diane handed over the money and held out a shaking hand for her change. Back alongside her, Susan picked up the pack of nappies.

'Here, she muttered. 'I'll help you out.'

She held the door open. As Diane rushed to waggle the pushchair wheels over the door-ledge and out of the shop, Emily swung a hand, trying to pat her Auntie Susan's leg, to get her attention.

Susan had only rushed round to grab a plastic pack of cheap, grated cheddar to make macaroni cheese for Martin's tea. She'd actually needed a pack of tampons as well but she'd decided she couldn't justify the expense. She'd left Jenny asleep with Martin keeping watch for the few minutes she'd be away.

Susan saw Diane's new baby girl for the first time and turned her head away, lips pursed.

When she'd seen Diane struggling into the shop with her pram, Susan had hoped to be able to lie low at the back of the building and avoid her until she left or to make a rapid exit herself. The two fat women had thwarted that plan.

When Diane had managed to reverse out, drawing the pram behind her, Susan stepped out and handed her the pack of nappies.

'Susan,' Diane moaned, shaking her head.

'It's no good, Diane,' Susan snapped. 'I can't talk.'

She knew she'd regret it later but her blood was still up from the nastiness in the shop.

'It was you and Colin that stopped talking, not us.'

Susan pushed her way through the shop door. It slammed closed behind her. Diane was left outside in the cold, alone.

She tried to start back up the road. After a couple of shuffling steps, the nappies tumbled out onto the pavement, making a small tear in their plastic pack. She bent down and struggled to get hold of them and pick them up. She saw the black dirt on the edge of the nappy exposed by the hole.

She shoved the nappies more securely into the pram's wire basket then sucked her still-stinging finger.

As she stood up, Emily started to grizzle again.

Diane pushed on, struggling against the weight of her despair.

It was like nothing she had ever known.

UDM

Monday, 7 January, 1985: 16:15

'**M**y word!' Shirley exclaimed with a chuckle, as she walked into the lounge when she got back from the pit, that afternoon. 'You look like the cat that got the bloomin' cream.'

Dick had known that his delight at finding a solution to the awful prospect that had been looming over him since before Christmas would be written all over his face.

It had been about three weeks ago, a year since his sudden heart attack. He'd been at the doctor's surgery for his monthly check-up when the nurse had told him that, as soon she had finished examining him and taking his blood pressure, the doctor wanted to see him.

Dick's heart had sunk; he had been making a practice of avoiding their G.P., Dr Simmonds. He had suspected for years that the old dragon was in cahoots with the colliery's Personnel Manager. On innumerable occasions, she had given what had seemed, to Dick, like harsh decisions on the readiness of some of the men he represented to return to work, sending them back on light duties to help Bentinck hit his workforce attendance targets.

When Dick had gone into her surgery she had told him, brusquely, that she was pleased to say that his latest ECG looked positive and healthy. It was no surprise that, despite his quick, Oscar-winning performance of breathlessness, she had seen no reason why he shouldn't make a return to work after Christmas, even though he wouldn't be able to go underground.

The prospect of work on the surface, loading supply vehicles in the cold, sleet and snow or shovelling up coal spillages from under the conveyor belts in the heavy dust of the draughty, freezing cold, coal prep' plant was unthinkable.

Actually, he couldn't deny that the prospect of any manual work, after years of effectively being his own boss in his capacity as a full-time union rep', had filled him with dread.

A former NUM delegate he knew had emerged as Dick's potential saviour less than a week later. Roy Gently had phoned from Nottinghamshire to find out whether he would be interested in taking the lead in forming a Whitacre Heath branch of the budding, new Union of Democratic Mineworkers. Dick hadn't just been interested, he'd nearly bitten Roy's hand off. Not only did it offer a potential way of evading physical labour, if elected as UDM Branch Secretary, it would mean a return to his steady life on the bank and the opportunity to poach members from the very union that had ejected him from what he saw as his rightful position.

Roy had promised to discuss it with the rest of the UDM's new leadership and to try to get them to raise it with the South Midlands Area's senior management and explore what level of support they could expect in establishing a presence in Warwickshire.

Dick had called him several times, thereafter.

'Leave it wi' me, Dick,' Roy had said, on every occasion.

Dick had confided in Shirley that he was starting to get the distinct impression that the UDM delegate was "all mouth and trousers".

But then he'd got the call from him, that morning.

'It's good news, Dickie lad,' Roy had said. 'The senior bod's 'ave got a guarantee of management support from the Deputy Area Director for any men wanting to have a branch of the UDM. I couldn't tell yer, but we expected this 'cause the top boys had already been tipped the wink by a couple o' Labour MPs that the Government 'ad told the Coal Board to do everything they can to establish as much of a UDM presence as they can in every area. So the Board's been tellin' the Area Directors to get on wi' it.'

They both knew the score, the Tories would be hoping to use the introduction of the emergent alternative to the NUM as a way of splitting the workforce and breaking the NUM's power base.

"Serves the bastards right," Dick thought.

Dick gave Shirley a crafty grin.

'I'll be working with Bentinck as the local management support to get the union going. I'd already started making a few calls, casting me net to see who I might be able to bring o'er but I gave Bentinck a quick call and he says the Board's position was that it would be a goer however small a number of members I manage to drum up, initially.'

'So,' he said, rubbing his hands and beaming. 'It looks like it's all systems go.'

356

Snow-Picket

S trikers tended to experience more lows than highs. That morning, Paul experienced both.

He was on the line with five other lads, ready to picket the afternoon-shift.

They'd enjoyed no success, whatsoever, that morning, in their efforts at persuading truck drivers, making deliveries or arriving to pick up loads of coal, to show solidarity and turn away.

It was a bitterly cold day, with the certainty of black ice under the light dusting of snow on the roads. Despite this, on a day when Paul would never have dreamt of riding his own bike, a police motorcycle turned into the pit lane. Paul watched the gleaming, white machine cruise past them, the sound of its engine the unmistakable "crop" of a Norton Commando.

It hit Paul harder than even he would have expected, it felt like the cruellest taunt he had received from the police, so far.

'A Norton Commando,' Alan Walker said. 'Like yours eh, Paul?'

'Nope, not like mine. I ain't got mine no more. 'Ad to sell it. I'm gonna have to sell me Allegro next.'

'Well, don't look so down about it,' Alan said. 'When this lots o'er you can buy yersen a decent motor, instead o' that Leyland shite.'

Half an hour later, they noticed two police constables appear from the office car park, round the corner of the colliery's main office. P.C.s Bradford and Bessemer were returning to the police's temporary control room in the conference room in the main office building. They stopped and looked over at the picket line.

There was an exchange of words, then the older of the two started striding over towards their position. The pickets heard the other officer raise his voice and call after him.

'Just leave it, can't you?'

Ignoring his colleague's impatience, Carl pressed on, strutting towards them.

'Ay-ay!' Alan muttered. 'It looks like somethin's wound this fucker up.'

As it happened, there were two things irritating P.C. Bessemer. The first was being out in the coalfields again. The overtime and subsistence allowances were great compensations but, sometimes, Carl got fed up with being stuck out in the depths of the countryside, away from the City, as the strike dragged on. Particularly this time, when he had been hauled away just he was starting to pursue this cracking bit of stuff he'd reckoned he'd been in with a chance with.

The second irritant was the snowman the pickets had built. As tall as a real man, they had made a good job of him, kitting him out with a broken, plastic, pit helmet, filched from a skip on the pit yard, a scarf and a "Coal Not Dole" protester's placard. It seemed to Carl that the strikers must be having things too easy, if they had the time and inclination for such childish pranks.

'Just ignore 'im,' Paul said quietly.

The pickets turned their backs on the approaching policeman.

'Yeah,' Alan said. "E's a nasty bastard, this one.'

'Oi, you!' Carl shouted, when he was ten paces from the picket line.

'Ignore 'im,' Paul muttered to the others, trying to keep them facing down the pit lane, acting the outward-looking picket line.

Despite Paul's instruction, Alan turned round.

'What's your fuckin' problem?' He growled.

'You are,' Carl called out to him. 'You and that snowman. Take it down.'

'Don't talk so soft.' Alan sneered.

'Take it down, before I knock it down.'

'Why, what's wrong wi' a bloody snowman?'

'You're only allowed six men on a picket line and there's six of you already.'

The rest of the pickets turned and stared at the policeman.

Alan cackled.

'He thinks the snowman's a real person,' he chortled.

'It's obvious what you're trying to do, you're trying to make the picket line look bigger,' Carl snapped. 'It's intimidation.'

'Fuck off! You've got a screw loose.' Alan said. 'It's pitmen we're picketing – not fuckin', idiot coppers like you. They can tell the difference between a human bein' and fuckin' snowman.'

Carl glared at Alan and saw the contempt in the expressions of the pitmen staring back at him. He spun on his heel, the extent of his fury evident in the stiff, jerky motion of his body as he strutted off.

Barely two minutes later, a police Range Rover roared round, out of the colliery's main car park, its engine gunning.

Carl had found its driver sitting in the passenger seat of the patrol car next to it, keeping warm whilst chatting to a colleague and tapped on his window.

As he'd opened it, Carl had said, 'Can I borrow your vehicle, pal, just to go up the pit lane?'

'What? Oh, alright. But you'd better have it back here in ten.' The Range Rover's driver had said, looking at Carl uncomprehendingly as he handed him the keys. 'And take care of it, Bessie. I don't want you pulling one of your stupid stunts in it and bumpin' it.'

Carl pulled up in the Range Rover next to the pickets, lowered the passenger window and leant across.

'Are you going to knock that snowman down, or do I have to do it?' He shouted.

'You do it, if you want, yer silly twat,' Alan called back. 'You're the one who wants to see it knocked down.'

The Range Rover shot away, sped down to the entrance of the main car park, reversed in and swung back round towards them, the vehicle's four-wheel-drive preventing any wheel spin. As the vehicle carved its path across the snow on the pit's access lane, the pickets realised that the suspicion that had formed in most of their minds was correct; he was going to use the Range Rover to ram their snowman.

'Quick! Scatter!' Paul shouted, pushing the two nearest him off to the right whilst the others slithered away to the left, just before the big vehicle careened into their snowman.

To his cost, Carl found out that, just like real-life pitmen, this snowman had real backbone.

As the compacted snow of the figure's body and head flew off, the Range Rover's rear wheels left the ground then hit the ground with a bounce as it came to a complete and violent halt, a cloud of steam billowing, instantly, from its caved-in radiator.

Paul shook his head in disgust at the policeman slumped in the Range Rover's driving seat, blood starting from the gash across his nose where it had struck the steering wheel with force.

The other pickets hooted in delight, two of them dancing over towards the wreck.

The younger policeman reappeared from the main entrance of the office building and strode over to the wrecked vehicle, ignoring the laughing miners as he made for the passenger door. It creaked and groaned as he heaved it open.

'What the hell do you think you're doing?' Neil shouted.

'Worr 'appened?' Carl moaned, slumping over to the passenger side.

Neil shook his head as he looked at his colleague; nose broken, vehicle wrecked, career over.

'How many times have we been here? And still you manage to forget that damn great, steel gate post they'd built their snowman round. You idiot!'

Retribution

Shuey had decided it was time that Malcolm Kennedy got involved in taking direct action against the scabs. He had disclosed his plan to his henchman in the course of another strike-fund financed, heavy session, hidden away in a corner of the snug in the Holly Bush.

Shuey took a last drag of his cigarette then flicked its stub away.

'So are yuz up fer it?' He slurred, breathing out smoke as he staggered down the front steps of the pub.

'Aye, let's dae it,' Kennedy snarled.

Shuey trotted after the bigger man as he strode away.

As they approached their target, Kennedy stopped to let Shuey catch up with him.

'So what're we gonna chuck?'

Shuey had everything prepared.

'I've got yuz an Exocet ready. Come over 'ere.'

From the foot of the privet hedge, bordering the garden of a terraced house, Shuey dragged out a ten-inch length of broken kerb-stone.

'Here y'are, that should dae the job.'

Kennedy looked across the road at the neat, little semi'. A bedroom window was lit-up upstairs, downstairs all the lights were off. There was no one else in the road, apart from the pair of them.

'Go on then,' Shuey whispered, hoarsely.

Kennedy looked at Shuey then bent to pick up the lump of concrete. He stood up, gave Shuey another, angrier look then swung away and started weaving off, towards the house opposite. As he stepped off the kerb, his foot slipped on the icy road and he twisted his ankle as he fell onto the freezing tarmac.

'Aargh!' He groaned.

Shuey darted through a gap in the hedge, into a front garden, and crouched down in the flowerbed behind the privet.

'Keep the fuckin' noise doon, ye fuckin' loon!' He muttered.

Kennedy clambered back up and tried his weight on his injured ankle. He let out an agonised grunt.

'Aargh,' he moaned. 'A've sprained me fuckin' ankle.'

'Stop making such a racket,' Shuey whispered. 'Go on. Hit the basta'!'

Kennedy limped across the road, over the verge and pavement and into the garden. Without hesitation, he lifted the kerb-stone back over his head like a footballer taking a throw-in, then heaved it and loosed it at the lounge window.

Sally was settled in her basket, next to the hearth of the coal-fired boiler in the lounge. Upstairs, Horace was already in bed as Pat came back in from the bathroom, in her dressing gown.

It sounded like a bomb going off; the whole house seemed to shake. Sally started a furious barking. Pat froze with a stricken look.

Horace was upright in an instant, sheets thrown back and legs swung out of the bed and into the slippers on the floor. He rushed through the bedroom door to the stairs. As he pounded down the stairs, Pat called after him.

'Horace! Be careful! You don't know how many of them out there.'

She ran to the edge of the bedroom window and peered out from behind the curtain.

It was a testament to the quality of their double-glazing that the heavy missile had starred the pane with cracks but not penetrated it. In his surprise, Kennedy had delayed for a couple of seconds, looking in bewilderment from the window down to the lump which had bounced back to land in front of him.

Looking down from the bedroom, Pat saw the big figure of the perpetrator hobbling away, illuminated by the lounge light that Horace had switched on, on entering the room.

From beyond the end of the road a police siren sounded. A flashing, blue light started playing on the houses opposite. Mesmerised by the light, Pat missed the skinny form scrabbling, monkey-like, on all fours, across the front garden of the house opposite before disappearing into the entry between the two sections of the terrace.

The two officers in the panda car saw a man freeze, in the middle of the road, like a huge rabbit in their headlights, before turning to attempt a getaway. Thirteen pints and a sprained ankle meant that he was never going to outstrip the pursuing Ford Escort and the policemen who jumped out when it had overtaken him.

Shuey saw none of this as he made good his own escape but heard the car's engine screaming during its brief, low-geared pursuit and the shouts of the officers as they arrested his henchman.

A policeman and policewoman, from amongst the officers who arrived in support, came and inspected the damage and took statements from Horace and Pat.

Horace secured the front-door behind them, as they left. Pat could see that he had been more shaken by the attack than she.

'I'll make us a pot of tea, love' she said.

She hadn't been able to bring herself to offer to make the two police officers a cup.

They went through to the kitchen together. Pat put the kettle on while Horace sat down at their kitchen table.

'So it was Malcolm Kennedy,' Pat said.

'Yeah,' Horace said. 'But you can bet yer bottom dollar that McFadden'll be be'ind it. But pinning anything on that slippery character'll be another matter.'

Pat shook her head.

'And to think that a son of ours could throw his lot in with creatures like that.'

Diane

Feeling her panic rising, Diane pressed the phone handset to her ear as the ring tone continued. She couldn't bear it if she were out again.

At last, her mum picked up the phone.

'Hello, Tamworth 64283.'

Relief swept over Diane. She released the breath she had been holding.

'Hi Mum.'

'Hello love, are you alright, dear?'

'Yeah, fine. Except... ...I've got a splitting headache. I just want to lie down and close my eyes.'

'Oh dear. Do you want me to have the girls for a bit this afternoon?'

'Could you?'

'Of course I can, love. I expect you're just a bit tired.'

'Yeah, I expect you're right.'

'Shall I come round to you?'

'No, no. It's alright. I'll bring them round to you. I don't want to be disturbed for a bit.'

Doreen was watching out for them as they approached. Waving to Diane through the front window, she bustled round to open the front door and help her daughter get the children out of the cold and into the house.

'Hello, and how are Grandma's favourite little girls?' She called out to them.

Diane lifted Sophie out of the push-chair and passed her to Doreen to cradle then unclipped Emily from her seat and held her hand to pull her up onto the step, into the front porch and on through to the warmth of her mother's front room.

'Oh, thanks Mum,' Diane said with a shiver. 'That's a great help.'

She darted back out to grab the baby-changing bag for Sophie and a bag with a few toys in for Emily.

'Here you are,' she said to her mother. 'Sophie should be alright for a bit; she's just had a change. I'll just fold up the pram and bring it in.'

'Don't worry about that. I'll bring it in and use it as a cot for her, if I need it. You get off and have a lie down. I expect your Mum's got over tired with you waking her all hours.' Doreen said as she beamed down on the little baby in her arms.

'Oh Mum, you haven't got some aspirin I can have, have you?

'Of course I have,' Doreen said. 'Here, you take this little lady for a second and I'll go and get them for you.'

Doreen rushed out of the room and up the stairs to the bathroom cabinet.

Diane sat down on the sofa, holding her baby, and held her hand out to Emily who stumbled the few steps to her mother. She hugged Emily as tight as you could with a toddler and felt her baby stir in her arms.

She breathed in the sweet perfume of Emily's hair and kissed the plump warmth of her soft, glowing cheek. The little toddler moaned, indignant at the crush of her mother's embrace. Hearing Doreen thumping down the stairs, Diane released her hold on Emily and whispered to her.

'Mummy loves you, darling.'

Diane kissed Sophie on her forehead and stood up eagerly, still holding her in her arms.

Doreen put the bottle of aspirin on the table. Diane passed the baby to her.

'Thanks Mum.'

She seized the bottle of pills in both hands, bent over and kissed Emily again, stood up with a smile, kissed Sophie in Doreen's arms and her mother's cheek.

'Thanks, Mum,' she said, giving her daughters another smile. 'Bye.'

'Oh, alright, love,' Doreen said, raising her eyebrows at Diane's eagerness. 'And don't worry about us, us girls'll be alright together 'til you're ready, won't we girls?'

'Ok, thanks, Mum,' Diane said. 'Bye!'

'Bye, Dear,' Doreen said. 'See you later. You get your feet up for a bit.'

Back home, Diane opened the back door, stepped in, closed the door then secured it with the key.

Keeping her coat on, she grabbed a plastic beaker from the draining board and the shopping bag with the small items she had just bought from the dreaded, VG store. She had promised herself she would never go there again nor risk stepping out in the village after that awful confrontation, the previous week, but that one last trip had been unavoidable.

She hurried over to the sideboard, opened the left-hand door and took out the bottle of Jack Daniels that Colin's dad had given him last Christmas. It was opened but scarcely any had been drunk. Diane grabbed it by the neck and hurried upstairs.

In the bedroom, she dropped the shopping bag on the bed, placed the Jack Daniels and beaker on the bedside table on her side of the bed where the bottle of aspirin from the bathroom cabinet was already waiting and kicked off her shoes. The fleecy rug felt comforting under her toes as she took the cap off the black-labelled bottle and sloshed a generous measure into the beaker. She sniffed at the mouth of the bottle and grimaced.

'Yuk!'

She picked up the beaker and managed to force down a couple of gulps without holding her nose. She put it down and replenished it. Reaching under her pillow, she retrieved a dark brown bottle. She unscrewed the cap and tipped out three aspirin.

Perfect as pearls in the palm of her thin hand, she closed her fingers over them and then shoved them into her mouth and onto her tongue. Another gulp of Jack Daniels washed away the sour after-taste.

She had never been good at swallowing pills and had decided that she could not try to get them down more than three at a time.

Diane took her mother's bottle of aspirin from her coat pocket and placed it next to the other on the dressing table, then pulled out a third, big bottle from the plastic, VG bag.

She looked at the stock-pile she had accumulated so easily, hoping it would prove sufficient to release her from the black despair that gripped her, then went to work on it with a will.

Death had been on Jim Greaves' mind that morning.

When he got up the pit he went over to the fitting shop and found Mad Jack holding court near the department's communal kettle. He waited for him to finish his latest anecdote before moving in on the half-dozen boiler-suited mechanical staff.

'Good afternoon, men.'

Jack swung round.

'Hey-up! Look 'oo's 'ere! The Undermanager's come to visit our 'umble abode.' He called out over the fitters' cheery replies. 'What can we do fer yer, Mr Greaves?'

Jim turned his back a little to the rest of the staff present.

'I wanted to ask you a favour,' he murmured.

'You don't want 'im to fix yer mower for yer, do you, Boss?' The nearest of the fitters called out.

It was becoming apparent that it was going to be difficult to be subtle about the conversation Jim had in mind, in the presence of the group of keen-eared fitters and engineers.

'Righto Mr Greaves,' Jack said, saving him. 'Let's adjourn to my office to discuss the matter in privacy, away from these cheeky kids.'

Jack led the way into the steel and glass-panelled compartment that formed the office. Jim followed him in and closed the door while Jack rounded the big, scruffy, old desk and flopped down into the office chair. Jack leant forward to listen, resting his elbows on the desk.

'It's a bit of a strange request,' Jim said, pulling a steel-framed chair from the wall and lowering himself onto its canvas seat.

'I was just wondering if you could show me how to start up one of the surface plant diggers and get started on the basic operations. I've driven tractors on the farm and I had a job where I did a bit of JCB driving when I was a youngster, so I shouldn't have too much trouble.'

Jim had remembered seeing Jack charge round with confidence in one of the massive diggers, when he had delivered the timber to the pickets from Barnsley, after their arrival on that first night of picketing, ten months earlier.

'When you were a youngster?' Jack said, with a sly smirk. 'Yer still are one. Any road, you need a bit more skill with them machines in the surface plant. They're five or six times bigger'n what you're talkin' about. What was it you wanted shiftin', any road?'

'Did you hear about it on the news, those two children, killed picking coal from the slag heap at that pit?' Jim said.

Jack frowned and nodded. With their fathers' concessionary coal stopped for being out on strike, two boys in Yorkshire, one ten years old, the other aged twelve, not so different to Jack's own sons in age, had been trying to extract something burnable from a spoil heap to provide some warmth for their homes. They had been buried alive under an avalanche of waste material when the lumps they had pulled from the hole they had burrowed into the side of the tip had disturbed the steep, mountainous heap.

'I was thinking of doing something, under cover of darkness, when the rest of the management's gone home,' Jim said.

Jack gave him a hard, sideways look.

'You want to dump a bit of housecoal where the strikers can get at it.'

'Yeah, something like that,' Jim said, sheepishly.

'It'd be the sack for yer, if you was found out,' Jack said.

'I suppose it would but it must be bloody miserable for the families of the strikers in this weather, with no coal and no money. And it might save someone from ending up killing 'emselves on our premises, so I could say I was doing it for safety reasons if I got caught.' Jim said. 'I was just thinking of dropping some near the break in the hedge, over on the south side, next to Furnace End Lane, away from where the police are keeping their eyes on the pickets.'

Jack snorted a laugh.

'What?' Jim said.

'I was just thinking, it sounds well; Saint Jimmy, Patron Saint of Striking Miners,' Jack chuckled.

Jim blushed.

'And their kids,' he said with a smile.

'Alright,' Jack said, easing his loose-limbed frame up from his chair. 'Leave it wi' me.'

Jim stood up, watching Jack as he headed for the door.

'So you'll show me how to start one up, then?' Jim said, with a frown.

Jack looked back and grinned.

'Nah, leave it to me. I know where yer mean, I'll drop some o'er a bit later on. And, if we get away with it, I'll try and drop some over next week.'

'Thanks Jack,' Jim said. 'You're a star.'

'Yeah, well it's better'n 'avin' you stickin' a ten-ton machine in the reser'.'

12:57

Doreen had been doing everything she could to help Diane cope with the draining demands that had hit her with Sophie's arrival a few weeks earlier, so she had been delighted to be able to give her daughter some respite and to have her granddaughters to herself for the afternoon.

Having been so worried about how low Diane's energy and spirits had slumped, Doreen had been relieved to see her daughter looking a bit more positive, despite her headache, when she had dropped the girls round.

After a while, as she busied herself in the kitchen with Sophie back in the pram and Emily in her high chair, singing out and beating on a plastic bowl with a spoon, Doreen started to feel a little, nagging discomfort. The more she thought about it, the more difficult she found it to understand why Diane seemed so motivated and cheerful when she was tired and suffering from a headache.

Diane was never devious or closed so Doreen kept dismissing the slight suspicion that her daughter might be up to something. But even if she had something planned, what could it be? Surely it wouldn't be anything to worry about.

But the feeling wouldn't go away. After an hour and a half, she was finding it impossible to suppress her intuition. Hating herself for not trusting her daughter and wishing she didn't feel it necessary to disturb her, particularly as she would have only just dropped off if she had lain down for a sleep, Doreen decided to phone and check.

'Wait there a sec', girls, Grandma'll be back in a minute,' she said as she darted away to the phone on the little table in the lounge.

She dialled Colin and Diane's number, the handset pressed to her ear. After a couple of seconds, the connection was made and she heard the ringing tone. She was going to feel such a fool if she was disturbing her daughter's nap but it was better to be safe than sorry.

The phone kept ringing.

'Come on, Diane, answer the thing,' Doreen muttered, hunched over the phone.

She started to feel that she might not be being foolish and, perhaps, there were some grounds for her fears. She replaced the handset, then picked it up and dialled the number again. Perhaps Diane was having a nice, hot bath. Doreen knew she had not misdialled but thought that she might still save herself some embarrassment.

But she didn't really believe that now.

'Oh God, come on Diane,' Doreen groaned, as she listened to the uninterrupted ring tone from the other end of the line.

'What are you up to?'

She smacked the phone down and ran through the kitchen to the backdoor. She scanned the kitchen; there was nothing hot on the cooker, nothing switched on.

'Just wait a second Emily darling, I'm just popping next door. Look after Sophie.' Doreen said, bending down to the toddler.

She ignored Emily's rising wail that started as she closed the door gently and rushed round to see whether Nita was in, next door, so that she could get her to come and sit with the girls for the few minutes it would take her to rush round to Colin and Diane's.

She stepped over the knee-high wall between the two front gardens and trotted the few steps to the front door, glancing into the front window, hoping to see her neighbour in her lounge. She rapped urgently with the flimsy knocker on the door. No answer.

Doreen started back to her own garden, keeping an ear turned back to Nita's door in case she responded.

Back over the little fence, she looked back towards Nita's front door; nothing.

She would have to take the girls with her. As she returned to her own backdoor, it was clear that Sophie had joined in with Emily's howling.

Doreen hoisted Emily from her high chair with a cheerful cry.

'Come on little lady, let's just go for a little walk.'

Doreen's manhandling of the pram out through the double French windows at the back of the house with both girls on board only served to agitate them even more. Doreen pushed the two doors closed loosely and pushed away to get round to the front of the house.

'Hush, hush now girls,' Doreen called out over the howling of the two tiny children.

'If anyone hears this racket, they'll think I'm doing you both in,' she muttered.

Doreen sped the pram down the pavement and into Diane's road with the frenetic gait of an Olympic walker.

She didn't waste time ringing the front door bell, she went straight to the side-door which she always used to enter her daughter's house. She twisted its handle and pushed. It was locked.

Doreen whipped out her own set of keys. Thank God she'd brought them. She found the right key and shoved it into the lock. The key on the inside of the lock that Diane had used to lock the door stopped Doreen's key short. Doreen gave her key frantic shakes and twists in the keyhole until she was rewarded by the musical chime of Diane's key bouncing off the kitchen floor. In an instant, she had the door unlocked and was through it and singing out to the pram.

'Don't worry, Grandma won't be long.'

'Diane!' Doreen shouted as she ran into the lounge and through to the hall to thunder up the stairs.

If Diane was still asleep that would wake her.

As soon as Doreen flung the door open and stepped into the bedroom she caught the whiff of whiskey. She looked over at Diane's body on the bed, still in her coat, then took in the Jack Daniels and aspirin bottles on top of the bedside table, a last few, loose, white pills scattered amongst them.

She managed to stop herself from screaming. Instead, she shouted again.

'Diane!' Wake up!'

She rushed round the bed and started slapping at her daughter's face.

'Come on, Diane. Wake up.'

Doreen pulled on one of Diane's limp outstretched arms to draw her upright. Diane gave a weak, incoherent murmur.

Doreen let her slump back and hammered back down the stairs and into the lounge. She grabbed the handset of the telephone and dialled 999 with the two little girls still wailing outside the backdoor, as an accompaniment.

'Hello. Which service do you require?'

'Ambulance! It's an emergency!'

The Ambulance

Wednesday, 9 January, 1985: 18:03

A power station's smoking chimney, clouds of vapour billowing from four massive cooling towers, vast stockpiles of coal; as the cameras panned over the scenes the television newsreader announced, with evident satisfaction, that the previous day had seen the Central Electricity Generating Board set a new all-time record, having met the nation's highest recorded demand for power with its "best ever day of production".

Paul jumped up, leapt over to the television and turned it off.

He couldn't listen to any more of that, it would take too much more of the fight out of him. He couldn't bear to think what it was going to do to the resolve of those who remained out on strike.

It seemed like a one-way street, now they'd passed the tipping point at which those working at Whitacre Heath outnumbered those still out.

The longer he waited, the more the hurriedly-scribbled, cryptic message he had found, on arriving back home, nearly four hours earlier, worried him:

Paul,
got to go out.
M and J sorted.
Love S
X

It was a relief to hear the scratch of her key in the door lock.

As he darted to meet her he was met by Martin, stepping in through the front door with a cheerful shout.

'Hiya, Dad!'

Susan was bent over the pushchair, extricating Jenny.

'Martin,' Susan called as she lifted the toddler. 'Straight upstairs and change out of your uniform.'

She looked pale.

'Is everything alright?' Paul said.

'Ooh, wait a minute. Let me get into the warm and I'll tell you.'

'Tell me what?' Paul said.

'Just let us get in, love.'

Paul stood aside to let her in with Jenny then stepped outside, flicked the catch on the pushchair, collapsed it and swung it into the hall.

He shut the front door and hurried in to Susan, in the lounge.

'What's wrong?'

Susan sat down on an armchair with Jenny on her lap. Paul sat down on the settee, side-on to Susan's chair. Jenny lunged towards him and beamed at him as Susan grabbed her to stop her from falling.

Susan looked into his eyes.

'It's Diane.'

Paul shook his head and said, 'What is? What's happened?'

'She's taken an overdose. Doreen called the ambulance.'

'What? This afternoon?' Paul said and Susan nodded.

'I saw it. We saw it come and go, on the pit gate. What's 'appened to her?'

'I don't know. She's in hopsital.'

'What?' Paul said. 'Is she alright?'

'It's too early to say.'

Tears started in Susan's eyes. Paul stood and went to sit on the arm of her chair and put his arm around her shoulders.

'That's terrible,' Paul said. 'What on earth was she doing, doin' something like that?'

Susan sobbed quietly for a while.

'They said there doesn't have to be a reason, as such. But she's got all sorts of things to bring her down. Doreen thinks she's had a bit of post-natal depression.'

'Where've you been? Have you been at the hospital?'

'No, Doreen called me to ask me to go and look after Emily and the new baby. She'd got no one else to turn to. I had to phone the pit, to get someone to take Colin straight to the hospital.'

Paul was silent.

'I took Jenny round to your mum's and got them to meet Martin from school and look after him 'til we got back.'

Paul was amazed. The barriers he'd insisted on between them and his father and Colin and his wife had all been dissolved for an afternoon, as easily as that.

'It's the first time I'd held the baby – Sophie.'

As she said it, she started to cry again.

'It's crazy,' she said, shaking her head.

Martin appeared at the door, halfway through changing, shorts replaced with jeans but still wearing his white shirt and little school tie, eyes wide and worried.

'It's alright, son,' Paul said. 'Mummy's worried about – Auntie Di. She's had to go into hospital. Go and finish getting changed while we finish talking about it.'

They both sat listening to Martin's steps as he climbed the stairs, Susan sobbing quietly.

'I keep thinking it's my fault,' Susan moaned quietly.

'No, you can't think that. It was them as stopped talkin' to us, we both know that.'

'I know.' Susan said. 'And I said that to her last week.'

'What? You saw her last week? You spoke to her?'

'Yes. I wouldn't have done – but I had to.'

'Why? What happened?'

'It was in Mundy's. I was in there and Diane came in. She'd obviously run out of nappies and come in for a little pack. I thought she'd be in and out without seeing me but those two big, fat, silly mares, Sheryl Skinner and Alice Parkin, had to go and start bullying her.'

'Well, what did you do?' Paul said, unable to believe that his wife would have joined in with them.

'I told them to pack it in and clear off, helped her pay up and got her out of the shop. She was about to talk to me and I said I couldn't because of that, because it was them that had stopped talking to us, when Colin decided to go back.'

Paul held her tighter as he sat over her on the arm of the chair as she cried.

After a minute Paul spoke.

'I'm not going back.'

Susan contained her sobs.

'What do you mean?'

'I can't talk to them and stay out. They should 'ave stayed out and they know it, my dad and Colin.' Paul said. 'You and I never brought any of this on and what you did for Diane, standing up for 'er, would have made 'er feel better, not worse.'

'I don't wish 'er any harm. It's up to you if you want to see them but I can't have anything to do with 'em, ever again. We've come too far. And it's them as burnt their bridges.

Open Shop

Wednesday, 13 February, 1985: 05:52

Jim had been forced to start his car, get the heater blowing on the windscreen and use a scraper to chisel the ice off before being able to set off for the pit, that morning.

The pickets' brazier was burning low; they'd be cold this morning.

Jim wound down his car window pulled up next to them.

'Morning, lads,' he murmured.

'Mornin' Boss,' a couple of them replied, morosely.

'Why don't ye just fuck off? We don't want to be associatin' wi' the likes o' you.'

The croaking voice came from the gloom at the back of the group of pickets. Jim knew who it belonged to before the skinny figure came lurching forward.

'We don't want you stoppin' here, tryin' to con' our members back to work,' the Lodge Secretary sneered.

'You're the only con-man around here, McFadden,' Jim growled.

'Huh!' McFadden snorted. 'A've told ye before, ye'll be getting' yersel' some trouble before this lot's all over.'

'Yeah, I've heard you've been getting others to do your dirty work for you,' Jim said before putting his car in gear and saying to the rest of them. 'See you tomorrow lads.'

'See you, Boss,' all the rest of the pickets called sadly, as Jim pulled away.

07:45

Jim had been uncharacteristically relaxed about getting underground that morning. He and Les Parker were just starting to ready themselves to stand up, leave the dirty offices and get underground when Bernie Priest, Ken Goodall's Senior Overman, arrived back from the canteen.

'Your old friend Mr Flavel has just asked me if you were still about, on the bank, Mr Greaves,' Bernie said. 'If you want to avoid him, I suggest we make tracks and get underground, sharpish.'

It had been no great surprise to anyone when Dick Flavel had returned to establish the bridgehead of a small UDM presence at Whitacre Heath, three weeks earlier, and then get his union's sparse membership to rubber-stamp his Secretary-ship. Now he had his own union office, the same size as that of his former NUM colleagues, McFadden and Deeming.

Despite having only twenty-six members out of a total workforce of over nine hundred, in the absence of NUM representation, the UDM was the only workforce union participating in the Colliery Consultative and the Safety Committees. Dick was using this to try to elicit more defections from the NUM.

Jim put the report books he had been reading and signing back into the shelves in the dirty communal room whilst Les placed their mugs in the big stone sink. They both stood waiting with Bernie for Ken Goodall. The older undermanager was in his captain's chair, finishing a cigarette as he adjusted the straps on one of his knee pads, prior to joining them for the walk over to the pit top. As Ken stood up and picked up his yardstick, Dick Flavel strolled into the communal room.

'Good morning, gentlemen,' he said, with a relaxed, confident smile.

'Good morning, Dick,' they all chorused.

'Who do I need to speak to about 22's, is it you, Mr Goodall or is it Mr Greaves?'

'Me,' Jim said.

'It's about the dust,' Dick said.

'What about it?' Jim said.

'The men say it's unacceptable.'

'It seems dustier because there's a dried out gob above the face, with us having worked out the top seam first. So that kicks out a bit more dust when it caves. But all the dust suppression is working on the machines and the conveyors and the airborne dust sampling's not shown up a problem. It doesn't look that bad to me and I'm down there every day.'

'That's a matter of opinion, I've 'eard it's thick down there.'

'Have you seen it yourself or is this all second-hand?' Jim said.

'I can't get down there, yet, meself but my members say it's diabolical down there.'

'Your members say that? How many have you got on 22's?'

'There's UDM men on there and a couple of 'em have been in to ask me to get some action taken.'

'With all the dust suppression working and the face well-ventilated, what action were you expecting me to take?'

'I don't know, that's your problem.'

'Well, your old friends in the NUM haven't said anything.'

'That's because their rep's are all on strike,' Dick said, falling for Jim's bait.

'Are they? I wondered why I hadn't heard from them recently!' Jim said. 'I tell you what, I'll have a look at it this morning and see what we can do.'

'I'd be obliged if you would and if you could let me know what you plan to do so I can tell us members.'

'Alright, I'll get back to you,' Jim said.

'It's time we were down the pit,' he said to the others before leading them out of the dirty offices to make their way to the shaft.

'He's not like a union rep',' Bernie said, as they stepped out of the building. "E's more like a bloody professional footballer's agent, just lookin' after the interests of a few players. Is this the future of mining, d'ye reckon?'

'I thought he was about to threaten to stop the wheels again,' Jim said. 'He could have called a mass walk out by half a dozen men.'

'You've never forgiven him for that statement, have you, Boss?' Les Parker said.

'Too bloody right I haven't.' Jim said. It's bloody ridiculous using that at all, let alone as your first resort, like he does.'

'I thought you'd be in favour of a bit of competition for the NUM,' Bernie said. 'Y'know, on the basis of a bit of divide and rule.'

'I think it's a terrible idea, the last thing we need is two unions. You can't beat having one strong union and managing it firmly, then you only have to reach an agreement on anything with one party. And, if they're strong, they'll make sure their members honour it.'

'There's a lot in what Mr Greaves says,' Ken Goodall said as they reached the airlock into the shaft. Jim waited on the handle of the outer door before releasing the air pressure until Ken had finished what he had to say.

'Scargill will look a right idiot if the only thing he achieves is ending what's effectively been a closed-shop for the NUM, with them as the only recognised union.'

'Well I've got no time for this new lot. The NUM's the miners' union.'
Jim said. 'They just need to boot that bloody Communist, Scargill, out.
It was a sensible union under Gormley. This UDM lot are a bunch of
turncoats – and professional politicians like Flavel. I wouldn't trust the
fuckers as far as I could throw 'em.'

The Attack

Thursday, 21 February, 1985: 22:47

Jim was coming to the end of another week of nights. The General Manager had placed him in charge of the pit again after being contacted by police who had been tapping McFadden's phone and overheard him contacting the NUM H.Q. in Barnsley, pleading for flying pickets, again, to bolster a last-ditch attempt at stopping the drift back to work by Whitacre Heath's remaining strikers

After three nights, with no trouble and no increase in the picketing on any shift and the nightshift safely onto the pit premises and ready for their penultimate shift of the week, Jim had decided he could leave the surface, get down the pit and do something useful.

As he fastened the last strap on his kneepads there was a knock on the door of the undermanagers' bathroom. Joey Bondar shoved the door open and leant round it.

'Sorry Boss, Roger's phoned over from Control, says you're to get over to 'im, sharpish,' his overman said. "E wouldn't say what it was about but it sounded like it was urgent.'

'Oh, right. Thanks Joey. I'll get straight over to him.'

Jim went as he was, leaving the door of his dirty locker wide open. He didn't collect his helmet but took his NCB donkey jacket, in case he was going to have to spend time outside, at the picket line.

Darting out of the door of the building, Jim jogged the thirty yards to the top end of the main office building, through the door and down the corridor. At the sound of his feet, Roger jumped up and stepped out of the Control Room door to meet him.

'Boss, Paul Wood, the fitter's bin round to the Control Room window. 'E wuz on the picket line. Says McFadden turned up pissed-up and acting erratic out there, late on, and turned nasty as the last o' the men were comin' in. 'E said to let yer know that someone came and told 'em on the picket line that, as McFadden stomped off, 'e was saying he was going to the Undermanager's house to brick the windows in 'n' torch it. They reckon 'e meant your place, Boss.'

'I'm on my way.'

'I was gonna call the police. Do you want me to do that?'

'Yeah, you better had. But I expect I'll get there before them.'

Jim ran to his locker in the undermanager's bathroom, grabbed his car-keys from the pocket of his sports jacket in his clean locker and ran back out to his car, in the central car park.

He roared down the pit access lane. Unusually, there was no one on the picket line.

As he drove the four miles of winding, country roads to his home with speed and care he recalled the threats McFadden had made on the night when the Barnsley men first came to Whitacre, a year earlier and on the picket line, in the previous week. He had considered refusing Roger's offer of calling the police, thinking that it might be better to deal with this himself, without witnesses.

Pulling up alongside the short drive of his house he saw an abandoned car further down the lane. He scrambled out of his Escort and started striding towards the vehicle. A man stepped out into his path from the gateway in the gap in his neighbour's hedge. Jim braced himself for the attack.

The figure eased towards him and spoke in a hoarse whisper.

'It's alright, Boss; it's me.'

'Paul,' Jim muttered. 'Roger gave me your message. What's going on?'

'Don't worry. We've got 'im. We got a mate's van, beat 'im to it, parked up and waited for 'im. Leave McFadden to us, we'll sort 'im now. I just didn't want anythin' like this crackin' off, y'know, without you knowin'.'

'Yeah, thanks,' Jim murmured. 'What the hell did he think he was doing?'

'There's no point tryin' to work it out,' Paul whispered. 'It's not you, it's 'im. 'E's a nutter.'

Jim shook his head.

'Anyway, thanks Paul.'

'No problem Boss. I'm just sorry 'e 'ad to try to drag you into all o' this.'

Jim was just thinking that it looked like Paul and his mates had managed to seize McFadden without disturbing any of the local residents when his own front door opened and Gabriella appeared and looked out.

'What are you doing here?' She called over to him, quietly.

'I'd better go, Boss.' Paul muttered, stepping back into the shadows. 'Say sorry to your missus, if we frightened her.'

'Okay Paul, see you. And thanks.' Jim murmured before turning to stride across the lawn to Gabriella, still framed in the doorway.

'Are you alright?' he said, as he got closer.

'Yes, of course I am. What's going on? What are you doing here, dressed like that?'

'Did you hear anything?'

'I only heard you out here, just then. What's happening?'

'Nothing to worry about, it was just that idiot McFadden, the NUM Lodge Secretary. He was drunk and came round here being stupid.'

'Who was that you were talking to?'

'Paul Wood, you know, one of my lads who's still out on strike. You've met him before.' Jim said. 'He and a few others came and grabbed McFadden to stop him making a fool of himself.'

Gabriella's smile was sceptical.

'He called me your missus.'

'Yeah, I know,' Jim said thoughtfully. 'I suppose we'd better do something about that. It's about time we got married.'

She snorted in amusement.

'Is that supposed to be a proposal, Jim Greaves?'

'Yeah, it is,' Jim said, with a smile.

'Okay then,' she said. 'But it's not very romantic, proposing to me dressed like that. You'll have to do it again when you're dressed properly and can get down on one knee to do it.'

'No chance.'

In Jim's mind there was no more fitting attire for a pitman proposing marriage than in his pit-black. He took her hands in his and, keeping her away from the grime and coal dust on his workwear clothes, kissed her on the mouth.

'You only propose once in a lifetime,' he said.

Gabriella looked him deep in the eyes and said, 'Are you staying for a coffee, since you're here?'

'No thanks, I hadn't better come in the house with this lot on, I'd get coal dust everywhere. And anyway, I'm supposed to be at the pit, in charge.'

'Okay,' she said, kissing him again on the lips. 'I'll see you in the morning, if you're back in time.'

'I'll make sure I am,' he said.

As he strode back to his car, he heard the builder's van that Paul and his mates had borrowed start up and roar down the lane towards him. As it came into the light from the house, Jim saw Paul in the passenger seat. He raised his hand and Paul acknowledged him with a curt nod.

Belatedly, a blue, flashing light appeared at the top end of the village.

London

Ian had stayed with his cousin in Doncaster for the first two weeks in January, after Jackie had thrown him out, then managed to find lodgings with a union official in Barnsley.

Keeping the strike alive and encouraging the union to press for the reinstatement of sacked miners gave him some last vestige of purpose. Some days he would go into 'Arthur's Castle', the NUM's National Headquarters building in Barnsley, but, more often, he continued to picket, away from his own pit.

On one of his visits to the office, he had tried phoning Paul Wood, down in Warwickshire, to make sure that he and his lodge were mobilising their remaining striking members for the important rally in London. He wasn't surprised when he got a number unobtainable tone, he had been expecting to find Paul in the same boat as the rest of them; unable to pay the bill, his phone disconnected.

All through February, the NUM, supported by NACODS, had tried, repeatedly, to re-open talks with the National Coal Board, but to no avail. The rally was an appeal to Members of Parliament and the NCB's directors to get them to provide help in finding an end to the strike. The union had laid on coaches to carry thousands of striking miners to the capital for the event.

Ian never saw Paul Wood that day but he was approached by another face from the past.

It happened as he was hanging back, on the edge of Parliament Square, away from where the tension was mounting to a point where hostilities seemed about to kick off between the police and the demonstrating miners. A gaunt, hunched figure approached him, his hair and beard matted with filth, his skin grey with ingrained grime.

'Hello, Ian, remember me?'

He had become averse to letting people down but gave a shake of his head, unable to identify the tramp standing before him.

'Er, sorry, pally. I'm afraid I don't recognise you, old lad.

The man's face fell.

'It's B-Brian, Brian K-K-Kettle. W-Whitacre 'Eath. L-last year.'

'Brian.' Ian exclaimed. 'I'm sorry, old pal.'

He thrust out his arm and grabbed Brian's cold, soiled hand and shook it warmly.

'I'm sorry, Brian, I didn't recognise yer. What're yer doin' here? I mean, like this?'

'I've been d-down 'ere fer ages. I left Whitacre.'

'What on earth made you do that?'

Brian shrugged and looked away.

'N-nothin' made any sense anymore. I just 'ad to leave.'

When Brian had seen the miners arriving with their "Coal Not Dole" placards and red and yellow NUM badges stuck on their coats he had followed them, unable to resist the urge to spend a little time near some fellow pitmen.

'I 'ang around Waterloo S-Station these days, on the other side o' the river.' Brian said.

Ian could see that meant sleeping rough.

Why don't you come back with me, we can get you 'ome somehow?' Ian said.

'N-no,' Brian said, suddenly wary. 'I c-can't go back again.'

'Of course yer can. This lot's got to be over soon. You ain't been sacked or owt have yer? You could go back down the pit with yer mates, then.'

'Nah,' Brian said, shaking his head. 'I lost 'em. I c-c-can't go back.'

'I'd better be g-goin',' he murmured.

He turned and staggered off down the pavement, revealing the dirt down the back of the heavy, old, tattered overcoat he was wearing.

'Brian, don't run off, old lad,' Ian called after him. 'I bet we could find some other Whitacre men who'd take you back.'

Brian increased his pace, scurrying away towards the bridge back over the river. Ian trotted the few paces needed to catch him and laid a hand on the tattered sleeve of his coat.

'Here, Brian, take this,' he said, handing him a pound note. 'It ain't much – I don't have much meself now. But it'll get you a pint an' a bite.'

Brian didn't bother to try to refuse the gift but Ian could see the sadness in the Warwickshire man's eyes at being forced to accept his charity.

Watching through squinting eyes as Brian turned left, to head back over Westminster Bridge, towards his stomping ground around Waterloo, Ian reflected on the cruel toll that the whole conflict had taken. Then he switched his gaze to another unfolding tragedy as the massed ranks of riot police started to drive the crowds of miners together with their riot shields, penning them in. As the miners pushed back and then retaliated the long sticks of the police started to fall with force.

Being on parole, Ian knew that he couldn't risk being taken and prosecuted again whilst he was still hoping, against all the odds, to be reinstated again to the world he loved. Despite that, it felt wrong not to be with them, amongst his mates, among his people, in that struggle. He needed to be able to feel it was still his fight but that was proving to be increasingly difficult. He forced himself to turn away.

Striding briskly up Whitehall, towards Trafalgar Square, he found that he couldn't close his ears to the howls and chanting of the miners' and the clatter of the plastic shields being used to pen them in.

Another dozen police, battle-bus, Transit vans hurtled past him, in the opposite direction, carrying even more riot police to join the fray; as if there weren't enough of them down there already.

Black and Blue

Sunday, 24 February, 1985: 23:47

N eil normally waited for Julie in the car but, tonight, he was determined to get some warmth back into their relationship; determined to make a fresh start.

He jogged up the stairs onto the platform as the train drew in and stood well-back, so he could scan in both directions, taking in the few passengers as they alighted along its full length.

He spotted her, his face lighting up as he strode towards her.

She didn't return his smile as she dragged herself towards him then stopped.

'Hi,' Neil murmured, taking the last few steps to her.

He leant down and kissed the frozen cheek she turned to his lips.

He put his arm around the cold shoulders of her mac' as they set off towards the stairway. The differences in their heights and gaits made it awkward to keep his arm around her on the way down the steps. He let it fall away and drop to his side.

'Was it a tough one?'

'You could say that,' she muttered.

They reached the foot of the steps and turned towards the exit, their footsteps echoing with those of the other passengers in the passageway.

She was no more forthcoming as they walked out into the dark car park.

'We're over here,' Neil said, with a light wave of his hand.

As he opened the passenger door of the car for her, she cast him a look.

'You don't have to do that for me. I'm not an invalid.'

Neil walked round the back of the car. She had always appreciated his gentlemanly manners, as she had used to call them. It felt so hard to do anything right these days; he knew if he hadn't held the door for her she would have criticised him for the omission.

He opened the driver's door, eased himself into the car and looked at her, beside him. In the dim, orange glow of the car park's lighting he could sense the misery that gripped her.

'What's the matter, Julie?'

She fumed but remained silent.

Eventually, she blurted it out.

'I'll tell you "what's the matter". I've just spent eight hours dealing with the results of what you and your mates do when you come across a group of miners.'

Neil had known that the Met' would be policing the demonstration in Westminster but, being on the early turn, he had been deployed elsewhere and off-duty by the time the miners had been due to march.

He knew what Julie was about to tell him and that, despite his lack of direct involvement, he would be guilty by association, in her eyes.

'I had to go down to Casualty. They couldn't deal with the influx of injured miners they were getting. Some of them came across Westminster Bridge in ambulances but most of them had to be helped over and even carried across by their mates.'

'They go there to petition Parliament and get corralled, out-numbered and beaten up by their own police. In Parliament Square, the home of democracy, of all places.'

'Julie, they're not blameless you know. I've told you before, some of them are out for having a go at the police.'

'So where were the injured police then? Julie said. 'I didn't see any.'

That wasn't actually true, she had seen three but they had been kept away from the miners and their injuries had been superficial. And, anyway, she had avoided them and made sure she was only involved in treating the pitmen.

'I saw grey-haired, old men, in their late fifties, old colliers like Uncle Ted. When they lifted their shirts to show me their backs you could see.' She said, a bitter bite to her voice. 'They'd been beaten black and blue.'

'They hadn't had any weapons but they'd been penned-in and thrashed.'

'One poor old chap had had his false teeth smashed with a blow from a truncheon to his mouth. His lips were mashed. They told me that hundreds of them have been arrested.'

She shook her head.

'What the hell do they think they're doing? It's like a South American state or a banana republic or something. You lot have been waiting for something like this and now you're happy; you've turned this country into a police-state.'

Neil knew that anything he said would only inflame Julie's fury and disgust. She knew that would be why he was remaining sad and quiet next to her.

'Haven't you got anything to say about it?' she snapped.

Neil shook his head.

'I wasn't there, Julie.'

'Oh, that's alright then. And what would you have done if you had been? I bet you wouldn't have objected – or told them to behave themselves, would you?'

Neil breathed in then started again.

'I was going to say, I wasn't there, but, people talk about us becoming militarised but the NUM are running this like a military campaign. They've created their own private army and they're attacking the state.'

'Do you really believe that nonsense?' Julie shrieked.

She took a deep breath and continued with a studied calmness.

'I've seen men who've been beaten with great, big lumps of wood – unarmed men! These are our people. And they say they were attacked. Who do you expect me to believe, them or the police, people like Carl Bloody Bessemer who go into coal mines and steal things? How many times have you let slip what a disgrace it's been, back out there? And still you try to defend them!'

'Julie, I don't like seeing them taking us on.'

'Taking you on? Don't you realise how ridiculous that sounds? How could they be "taking you on" when they're unarmed and you lot are armed to the teeth, trained to fight as a group and determined to teach them a lesson.'

'Look, Julie, I'm as sick of it as you are. But they're not blameless. We've had to protect the men who wanted to work from intimidation and violence, men who have a right to work if they want to, especially since they've never even had a vote on whether to strike or not.'

'And you think that makes it alright for you to do what I've seen today, do you?'

'No,' Neil murmured, shaking his head. 'Of course I don't.'

She let her head drop down.

He put his hand onto one of hers, where it rested, cold and motionless, in her lap. When she started to shake gently he knew that she had started to cry as the exhaustion and emotion from the day took their toll.

She snatched her hand away.

'Just take me home,' she blubbed.

Neil started the car and pulled away, gently.

The next day he needed to be gone early and Julie would be back late again. He could only hope that twenty-four hours would give time for her anger to abate a little.

Resignation

Jim was in the dirty offices at the end of the first shift of the week. He was just thinking about going to get a bath when Joey Bondar darted back from the pithead baths.

'Hey, Bosso,' he called out, making sure he caught the attention of everyone still in the communal room. 'Yer mate's done a runner.'

'What y'on about, yer silly bugger?' Les Parker said.

'McFadden. I just seen Jerry from Control, in the baths. He says McFadden 'as resigned from 'is position as el Supremo of the union an' buggered off back to Kent.'

'Good riddance, best place for him,' Joss Kemp said. 'That militant bunch are much more up his street.'

'Yeah, but a bit of a surprise that 'e's gone back down there,' Joey said. 'I'd 'eard 'e'd been up to no good down there and 'e'd had to scarper and that's 'ow 'e ended up 'ere, at Whitacre.'

'That to-do with him last week'll be behind it, no doubt,' Les said.

'Jerry says that he heard it on the Q.T. it was that and 'im drinking the strike funds and livin' off 'em like a playboy. After that business round at yourn, Boss, the lads who copped 'old of 'im uncoupled a few of his ribs and gave him an ultimatum: either 'e jacked or they'd shop 'im.' Joey said. 'It looks like the posse ran 'im out o' town.'

Jim nodded at Joey. He couldn't pretend to be surprised. Paul had said that he and his mates would sort the matter. And Paul Wood never messed about.

The End

Sunday, 3 March, 1985: 12:59

It was the end of Sunday dayshift on another of Jim's weekends in charge of the pit. As he entered the Control Room, Roger Dayton spun round on his chair.

"Ey-up Boss, 'ave you 'eard the news?'

As ever, Jim raised his guard, expecting to be on the receiving end of some silly, pit humour at his expense. It was only when he had walked round to lean on the other side of the Control Room console that he saw that Roger's excitement was unaffected.

It was news to Jim that hundreds of miners had travelled to London, that morning, to converge, just round the corner from Tottenham Court Road Tube station, in Great Russell Street. Once there, they had stuck it out for hours, outside Congress House, the TUC's modern but imposing headquarters building, through the bitter cold morning with its persistent bouts of drizzle; waiting, doggedly, to hear the verdict of the NUM's National Executive on the future of the strike.

Roger turned on the transistor radio on the top of the console, just in time to catch the momentous, opening line of the BBC News bulletin.

'*The Miners' Strike is over. Miners' leaders have voted to end the longest-running industrial dispute in Britain without any peace deal over pit closures.*'

'*The final vote by the National Union of Mineworkers' National Executive was close; 98 to 91, in favour of the return to work.*"

The decision to call an end to the strike had come after a tempestuous, three-hour debate, the South Wales delegates' motion calling for an organised, but immediate, return to work narrowly beating a more hard-line proposal by those from Yorkshire. The subsequent insistence by representatives from Kent that the strike should continue until miners sacked during the dispute were granted an amnesty allowing them to return to work had been dismissed.

The NUM president, Arthur Scargill, had told a press conference that he did not regard the outcome as a defeat. BBC News played a recording of his piping voice with its Yorkshire accent.

'The campaign against job losses will continue but miners will return to work on Tuesday.'

'We have decided to go back for a whole range of reasons. One of the reasons is that the trade union movement of Britain, with a few notable exceptions, has left this union isolated. Another reason is that we face not an employer but a government, aided and abetted by the judiciary, the police and you people in the media and at the end of this time our people are suffering tremendous hardship.'

'You can say that again,' Jim muttered, shaking his head.

As he and Roger listened to the news in the Control Room, Arthur Scargill was leaving Congress House, in London. Many yards apart, unaware of each other's presence, Paul Wood and Ian Flint were both members of the vast throng of miners that had braved the cold and wet for hours and now surged forward to cheer their leader as he emerged onto the steps of the building's main entrance.

As soon as their President announced the NUM Executive's decision, the crowd's mood turned. Ian started a chorus of booing that was taken up by those around him, drowning Paul's shouts of "Traitor!"

Eventually, the miners took up the repetitive chant that Paul started, in his fury.

'WE WON'T GO BACK TO WORK'
'WE WON'T GO BACK TO WORK'
'WE WON'T GO BACK TO WORK'

When the noise dropped enough, Scargill signalled to the miners to quieten so he could address them.

'Thank you for your support!' He shouted. 'Our strike has been a tremendous achievement. Men and women have fought a fight that has not been seen anywhere in the world. We've forced the withdrawal of the threat to close five pits by putting them into the new colliery review procedure, we've stopped the NCB from implementing its closure programme and, most of all, we mobilised the NUM.'

'The workers in this struggle have demonstrated to the working class that if they make a stand they can prevent attempts to butcher their industry.'

It really was the end. And, despite Arthur's bold words, that end was abject defeat.

As the cruel wind whipped up, the drizzle intensified, the raindrops failing to mask the bitter tears of disappointment on many a collier's face. But still they waited.

As the union executive's delegates came out and dispersed into the crowd in the pouring rain, outside TUC headquarters, the miners were still there to jeer, jostle and abuse them. Finding himself among them, Arnie Campion was desolated to be on the receiving end of the harsh brand they applied to him and his colleagues with their chanting:

' *SCAB! SCAB! SCAB! SCAB! SCAB!*'

Sunday: 21:55

That evening, after three pints at the Horse and Jockey, Jim had gone home with Gabriella. After dinner they were both reading novels with an L.P. on the stereo providing background music.

Realising it was nearly time for "News at Ten", Jim switched on the television. Like the BBC Radio News earlier, the first item was the big story of the day; the end of the Miners' Strike.

As the camera panned over the miners who had stood together, all that morning, waiting for the outcome of their leaders' deliberations, it passed over the sharp, intelligent face of a lean pitman. For a moment, Jim felt sure that he had just spotted Ian Flint, then dismissed the thought; the expression on the face he had glimpsed had been far too passive, too weary to be that of the keen-eyed, motivated, Barnsley man he had encountered first that night at Whitacre, a year earlier.

The ITN report quoted Michael Eaton, one of the National Coal Board's most senior directors and its chief spokesman. He said that the Board would not accept that normality had returned to the pits and no pay increase would be forthcoming until the miners called off their overtime ban. Furthermore, there would be no amnesty for miners who had been sacked during the dispute.

Jim expressed surprise to Gabriella at Mike Eaton's hard-line attitude. It seemed to be uncharacteristic from what he knew of the man who had graduated in mining engineering from Jim's university mining department many years before him.

Then the NCB man went on to talk in a way which seemed to Jim more like a proper mine manager, saying that he did not think that the dispute would leave a legacy of bitterness or guerrilla warfare and that he believed that normal relationships would return. Jim believed and hoped that would prove to be the case in Warwickshire.

Mr Eaton welcomed the end to the dispute and condemned Arthur Scargill's insistence that the fight would continue.

'Our men are fed up of being on strike and simply want to return to normal living,' He said.

'There is no victory. The coal industry has lost. It's the victim. We have lost markets, we have lost good labour relations over the course of this dispute."

The bulletin returned to the studio.

'The strike began in March, last year, over plans to cut production costs by closing up to twenty pits,' the newscaster reported. 'Major structural changes in the industry are still planned.'

'The Treasury has estimated the strike has cost the country one and a half billion pounds although the NUM claims that the costs have been far higher. The Treasury estimate includes the extra costs of running power stations on oil rather than coal, extra policing, as well as money lost by the steel industry and rail network.'

It was no secret that Michael Eaton had been chosen to act as the public face of the NCB because of how badly its Chairman, Ian MacGregor, played to the media. There had been a strange incident, months earlier when the latter had been filmed and photographed entering the NCB's headquarters, Hobart House in London. As an ill-judged joke, he had pretended to attempt to conceal his identity with a plastic shopping bag, raised to cover his face. The way it had been published on the front pages of national newspapers and shown on television had not revealed that the action had been in jest and had reinforced the impression that Mr MacGregor was remote, secretive and eccentric, if not slightly mad. Consequently, it surprised Jim that the NCB had let their Chairman be interviewed on the programme, saying that the priority was to get back to normal and safe working quickly.

'Every day many more miners have been returning to work, demonstrating to their leaders that they want the dispute brought to an end. That is also a clear signal to them to call off the overtime ban that the union introduced in November 1983.' Mr MacGregor said. 'We would then be able to get down to the crucial task of ensuring the future success of the industry.'

'We need to restore coal production to former levels, to regain coal markets we have lost and to plan ahead to ensure that Britain has the high volume, low cost coal industry which, alone, will guarantee a secure future.'

The news item concluded with the Prime Minister saying she was very relieved the strike was over.

'I want a prosperous coal industry,' Mrs Thatcher said. 'The miners would have been back earlier if the strike had not been kept going by intimidation and I am very glad now they can go back.'

Going Back

Dawn, Tuesday, 5 March, 1985: 06:12

He picked his way along the marshy footpath at the edge of the moor, gazing across to where the colliery's headgear reared up from the misty gloom of the valley's floor.

Absorbed in his thoughts, his foot landed, heavily, in a peaty puddle. The black water splashed over his foot, drenching his sock through his worn-out trainer.

'Fuck!'

The deep, rumbling murmur of exchanges carried up to him in the still air, from along the main street of the pit village, where the dayshift men had gathered in a disorganised mass.

If they'd known he was up there they'd have said he was crazy. But it wasn't the first crazy thing he'd done in the last few, crazy months.

He'd had to be there, to draw a line under it all.

Women had turned out in several small groups, along the pavement, ready to cheer their men back in.

The two standard-bearers, at the front of the body of men, raised the poles at either end of Railsmoor's NUM Lodge banner, its crimson cloth bordered with gold tassels. Ian couldn't see the scroll embroidered under the heavily-romanticised view of their colliery's headgear in its middle but he knew its words well enough.

"Strength Through Unity".

That just about summed it up.

In the end, it had been those on other side that had had the unity and so it had been they who had prevailed. Except for the enclave of hardliners in Kent, the men down south, in the Midlands and Nott's coalfields, had never really been with them, never been as militant. And, apart from the rail unions, they had been on their own.

Two hundred and eighty dayshift men shuffled round to face down the road, in the direction of their brief march to the pit yard; a pitifully short march back, after such a long, hard fight.

Ian could have done better at school but he'd always wanted a real man's job; all he'd ever wanted to be was a miner, it was his life, how he saw himself. And he knew he'd been good at it; one of the best.

Everyone at the union office in Barnsley was still saying that they'd carry on pressing for the reinstatement of sacked miners, like him. And the Kent miners were still out, demanding an amnesty for them.

The Kent men might be up for the fight but Ian got the impression that the Union's leadership believed there was no realistic prospect of obtaining that concession. Maggie had spoken.

He'd needed to be there to see them going back without him, to accept the reality; it was over. They'd lost their fight, he was never going back underground, his mining days were done.

The people in the union office were all fine with him but he'd decided to leave immediately. He had no business there anymore, now that he was out of the pit. He felt too much like a hanger-on.

He was starting to shape a plan for the future. He'd arranged to go over to stay with his cousin in Doncaster. He'd try to get some work while he went back to school there, try for some 'O'-levels and then, perhaps, some 'A'-levels.

Further down the road, he might try to go on and get onto a poly' course as a mature student. He'd always enjoyed reading fiction, proper literature, so he thought that, one day, he might even be able to try for a degree in English.

He'd started wondering whether, with that qualification, he might be able to get into journalism, on a local level, on the basis that if you can't beat 'em, join 'em. Or perhaps it might open up a way into teaching or lecturing.

Jackie had made child-minding arrangements for their son so she could start back as a teacher to support Matthew and herself.

Things were starting to thaw between them. Perhaps, in the future, they might even be able to be friends again. But Jackie was a proud, strong-minded lass and Ian reckoned he'd pissed her off too much to have any hope of her ever taking him back again.

The crowd below took more form as the men readied themselves. They'd be going back without him and men like Davy and Sammy, seized at Orgreave nine months earlier and sacked as a consequence of the convictions they'd received, based on the fabricated statements of members of the South Yorkshire Constabulary.

Ian had encountered Sammy only once since that excruciating, New Year incident with his wife. He'd bumped into him in Barnsley, one day in Feb'. Sammy hadn't seized hold of him and pummelled him, as Ian would have expected. Instead, he'd just stopped, stepped out of Ian's path, eyed him intently and acknowledged him with a sad, "Ian". There was nothing Ian could say to make it better.

"Sammy." He had replied.

He remained gutted by the disappointment he'd seen in the eyes of the man who'd been his first supervisor on his face-training, the man who'd been his pit-dad, ever since his very first shift down the pit.

"BOOM! BOOM! BOOM! BOOM!"

Like an artillery salvo; the four smart raps on the bass drum rang out through the valley. As the Railsmoor Colliery Brass Band stepped off and took up the tune of 'The Red Flag', the column of miners followed with their Lodge banner leading the way.

Ian caught the thin cheers of women. He turned away, shaking his head.

What the fuck had the Lodge officials been thinking? How could they do this; march back as if they'd just won some great victory when they'd actually suffered a massive, inglorious, total defeat?

As he pressed on, determined to catch the first bus to 'Donnie', the anthem's words ran through his head:

It suits today the weak and base,
Whose minds are fixed on self and place
To cringe before the rich man's frown,
And haul the sacred emblem down.
Then raise the scarlet standard high.
Within its shade we live and die,
Though cowards flinch and traitors sneer,
We'll keep the red flag flying here.

He hurried back along the muddy path, listening to the band's rendition. Unable to prevent the words of the alternative, derogatory version of the workers' anthem from forcing their way into his head, he sang its last couple of lines, to himself:

'The Labour Movement's just a farce,
-you can stick the Red Flag up yer arse.'

Paul's Strike

I t would be the end of the following week before Paul drew his first pay-packet, so it had been another modest meal for them, that evening.
'I expect it was strange, going back, after all this time,' Susan said, when she came back down from putting Martin to bed.

Paul nodded slowly with grim emphasis but made no comment.

Whitacre Heath's remaining two hundred and seventy-six strikers had gone back, unceremoniously, as directed by their union, at the start of the week.

Somehow, it had been established throughout the workforce, officials and management that nothing would be said and everyone appeared to be trying to return to normal.

Paul was convinced that he would never be able to face dealing with the men who had gone back before the end. Feeling this most strongly towards those who had been closest to him, he remained determined to maintain the barriers between himself and Colin and his father for what he saw as their betrayal of their class and their families – and of their relationships with him.

For their parts, Horace and Colin both knew Paul subscribed to the view expressed months earlier, by the NUM's Vice-President, the forthright Scotsman, Mick McGahey:

"We won't always be on strike – but they'll always be scabs."

Paul had given it a lot of thought over the last few weeks of their final slide to utter defeat. The portrayal of the NUM leadership as the barrier to negotiations in the last months of the strike had infuriated him when it seemed that it must be obvious to anyone with even half a brain that it had been the Tory Government and its NCB Chairman who had been intransigent and provocative.

Susan sat next to him on the settee, staying silent.

Eventually, it all burst out of him.

'It's obvious now that the strategy of the Tories and MacGregor – and the whole establishment – was to put pressure on us with a load of propaganda from all the media to force us to hold a national ballot in the hope that we'd lose it. It must have frustrated the hell out of 'em when we didn't fall into that trap.'

'If another country suddenly attacked Britain on a load of fronts then the Tories and the Healeys and the Kinnocks would laugh at anyone who said they should hold a ballot to see if the people wanted to fight or not. They'd say that anyone saying that was a traitor who just wanted to help the enemy by stopping the country from defendin' itself.'

'And they'd be right.'

'The Executive were right to vote against holding a ballot, it was their duty to defend the industry and give the NUM the fighting lead it needed. It was Scargill's job to mobilise the miners to resist MacGregor's attack – not demobilise those who'd decided to act. If he'd called a ballot it would have sent the wrong message. Lettin' us decide, region by region, whether to strike or not got more than eighty per cent of us out in a matter o' days.

'None of 'em told MacGregor and his Tory bosses that there should be a ballot to see 'ow many miners wanted their pits closing.'

'And what those nutters are doing is turning off a source of energy that could supply the country's needs for another three 'undred years. When North Sea gas runs out, this country'll be ruined. They'll never be able to turn the coal industry back on.'

'They accused Scargill and the rest o' the leadership of scaremongering and militancy. They said they were undemocratic for not holdin' a national strike ballot and incompetent for takin' us out on strike in the spring when the weather was warming up and coal stocks were high. But the class struggle is always going to be like any other fight; you don't necessarily get to choose the time and place of yer battles.'

'If we 'adn't struck when we did then the time and the reason for striking would've passed and by the time the "right moment" came to strike the pit closures would've been done and dusted and we'd 'ave effectively accepted them.'

Paul gritted his teeth.

'It would have been a poor Soviet general who said the USSR shouldn't defend itself against the Nazis in 1941, just 'cause the timing wasn't right. They'd have shot any general who'd said that. And rightly so.'

He paused and thought.

'We knew the attack was comin' and we'd done our best to set us-selves up for it by reducing coal stocks with the overtime ban.

It's obvious now; it was only 'cause the ban was startin' to bite that the Tories got the NCB to announce loads of closures – more and more each day – to force us out on strike through the spring and summer. It was the Tories who chose to make us an enemy. What were we supposed to do, surrender meekly or accept the challenge and fight?'

'Every fight has the possibility of defeat. Who'd say that's a reason for giving in and lettin' the enemy do whatever it wants? No one expected Thatcher to do that with the Argies – and that war wasn't a foregone conclusion.'

'How could we have been right to betray, not just our kids and the future but everythin' our grandfathers had fought for, by waving a white flag?'

'We were defeated 'cause the Tories and their lackeys threw the whole state machine at us: bribing the scabs and providing police protection to get 'em back to work, politicians of every colour and all the press pilin' abuse on us, courts ignorin' the Law and attackin' the union and seizin' its assets and the police turning into a paramilitary organisation – and lovin' it!'

'Another thing that did for us was the divisions, not just in the NUM with the Nottinghamshire miners but the betrayal by the TUC and Labour leadership who deserted us. It was like they'd learnt nowt from 1926.'

'The best they ever did, including that NUPE lot you went and spoke to, was pay lip-service. The TUC's Liaison Committee was supposed to monitor and support the NUM but it turned its coat and actually blamed us for carrying on with the strike, while that traitor Kinnock accused us of violence at exactly the same time as the police were using cavalry charges against peaceful, unarmed, picketing miners and tearing lumps out of 'em. You saw it for yourself up at Orgreave.'

'And then there was the press and media. Instead of tellin' people what it was all about and what each side was fightin' for, they reduced it all to a personality clash between two people; "the evil Scargill" and "Maggie Thatcher, the Falklands heroine and saviour of the nation".'

Paul shook his head.

'And what did we get out of it? According to the union, eleven

401

thousand pitmen arrested, over seven thousand injured, with over two an' a half thousand of 'em getting broken ribs, broken arms and torn shoulder muscles from being dragged, 'andcuffed, across the ground, hundreds jailed on falsified evidence and over seven hundred sacked. Poor lads, like Brian, decent men – the best – completely broken by it.'

'Despite that, we were still right to fight. It was the right thing to do for our country as well as our class – to try to save a vital, strategic industry and source of employment. 'Cause it's not just the pits that'll go, it's all the support industries that provide equipment and other stuff to the mines, they'll all be gone – and all the export trade they do around the world.'

'The most heart-breakin' thing is that as well as betraying us and losin' the fight for us, the scabs allowed the enemy to split us up. Our strong pit communities – all broken up and scarred forever.'

It had been a long speech, by a man of few words, to an audience of one who needed no convincing of his arguments. It took them nowhere but she'd listened, hoping that allowing her husband to release it all and share his interpretation of what it all meant might provide him with a little solace.

"And, God knows, we need some of that," she thought as she laid a hand on his and leant over to brush his cheek with her lips.

Optimism

Friday, 8 March, 1985: 18:25

Hearing Gabriella's car pull up outside, Jim jumped up from the sofa, rushed to open the front door and meet her with a kiss.

'So, the Kent miners have gone back, then,' she said as she stepped into the house. 'I heard it in the car. That's everyone isn't it?'

She dropped her briefcase and handbag at the foot of the stairs with the wry smile that women reserve for wayward children who have, at last, desisted from some mischief.

Jim had moaned to her about the Kent Coalfield; its men so militant that, even when everyone else had conceded defeat, they had had to have another few days of futile dispute. He was convinced that it was no coincidence that, generally, the most productive and efficient mines were those in the least militant areas like Nottinghamshire and the South Midlands and the least efficient were in the most militant areas like Kent and Scotland.

'I bet it's been good to get back to normal, this week, hasn't it?' Gabriella said as she took off her coat.

'Yeah,' he said, his tone failing to convince.

He walked down the hall to the kitchen, picked up the kettle, topped it up at the tap and plugged it back in, thinking.

They had not gone back to life as it had been. He knew they never would.

He resented Scargill and the Tory politicians who had, between them, turned his industry into a battleground. He resented them for causing the many thousands of pitmen and their families to endure many months of unnecessary hardship. Their courage and determination had been dwarfed by the monumental futility of it all. What right did he, or any other pitman, have to demand that their pits be worked until the coal was completely exhausted, however costly it was to mine?

403

Throughout it all, Jim had kept reminding himself how pitiful it had been to live in the Britain of the 1970's; a Britain in which management and the democratically-elected Government appeared impotent and the unions held the power.

He had remembered what it had been like when Britain had descended from being a modern, post-war country to being "the sick man of Europe"; industry on a three-day week, scheduled power cuts every other night of the week, no heating in your house, studying for 'A-Levels' by candle-light with no real hope of anything better.

His dad had told him, at the time, the only sensible course of action for a young man would be to emigrate and get away from it. But Jim had stayed, determined to play his part in restoring his country to prosperity. And if the conflict meant that Britain would never be in thrall to the unions and fall so low again then perhaps that made it all worthwhile.

By the end of the first month of the strike, Whitacre Heath had got thirty per cent of its men back, working. Thereafter, there had been a fairly steady rate of drift back until the ratio had been reversed, with just thirty per cent holding out, on strike, at the end. Those die-hards had only come back at the start of the week.

There'd been no noticeable animosity amongst the men of Whitacre Heath. With their usual resilience and humour, the pitmen had gone back to pulling together in their face-teams but, beneath it all, there were many bonds between friends and family members that had been broken, never to be restored.

The most important thing was that, against the odds, they had kept Whitacre Heath and all its faces open. And, with all the workforce back, the roadway drivages had started advancing again, for the first time in a year, heading out to open up new reserves for the future.

Yet again, the NCB was raising the targeted output required for each pit to stay open. But when Margaret Thatcher and Ian MacGregor said that they were committed to creating a smaller, highly-efficient, economic coal industry Jim saw no reason to disbelieve them. If they'd been intent on the industry's destruction, they were both so forthright that they would have said as much and the massive programme of planned investment in new pits seemed to confirm this.

Meanwhile, it felt like the pit and the whole industry were suffering from a massive hangover.

When the weather improved, as the year progressed, they might shake off the residual gloom. In the meantime, it seemed justified to Jim to hope that a coal industry in which the emphasis was to be on high performance and efficiency might hold an exciting future for a young manager who was competent, committed and caring – and passionate about production.

The Note

Watford, Tuesday, 12 March, 1985: 15:25.

Neil had been on earlies.

He parked his car in their space, pulled the key out of the ignition and paused.

There was something wrong.

He looked up at the windows of their lounge and bedroom on the second floor of their small, apartment block. The curtains were open but he couldn't see into the rooms from down in the car, so it was nothing he had glimpsed up there that had set his intuition jangling.

He tried to shake it off. Grabbing his bag from the passenger seat, he got out of the car and strolled over to the entrance to the apartments.

Inside the building, he broke from the door at a run then raced up and round the four half-flights in the echoing stairwell, his keys still in his hand.

He unlocked their front door, shoved it open and saw the space on the wall where one of Julie's pictures had hung, the lamp gone from the top of the coffee table. Various other objects were missing but, as well as the few items that he had brought to their relationship, she had left everything they had bought together.

He found the note on the corner of their simple, light wood sideboard. Every note he had ever received from her had always been on her signature, light mauve stationery, this was plain, white.

He unfolded it. The writing was her usual, generous, flowing hand, curvaceous and feminine, so redolent of her figure and character. But her frustrations sprang out at him; the first sentence a hammer blow.

Dear Neil,
The last year has made me realise that we cannot go on together.

I know that you chose your job for all the right reasons but it feels like I and everyone I know back home are on one side and you're on the other.

I know you feel you have to be loyal to the Police, even when you've been unhappy about things you've seen and things you've had to do. But you seem to be becoming more and more like them as time goes on. However hard you try you won't change them – but they'll continue to change you.

I can see now that I'll never be happy with you in the life you've chosen . It's the career you love and I'll only hold you back and make you unhappy if I stay, so I have decided it's best if I leave.

I am sorry to be letting you down and hope that you'll accept that what I am doing is best for both of us.

With love,

Julie

Taking the note with him, he rushed out of the flat, locked the door, hurtled down the stairs, three at a time, and strode out of the building and over to his car. He unlocked it, snatched the door open, threw himself into the driver's seat, started up, reversed, turned and accelerated away.

The note lay on the passenger seat as he raced along. He didn't need it to remind himself of what she'd written; the words were fixed, indelibly, in his memory. But it was the symbol of his mission.

By the time he was on the M1 and heading south, his plan was fully formed, with all its necessary contingencies.

Julie wouldn't let the team and patients down at Saint Thomas', so she'd be planning to stay in London. She'd either be in the nurses' accommodation or staying with one of her friends, most probably Geraldine, her best friend from when they'd trained together.

He'd try Geri's place and then work through the list of other potential hosts that he'd compiled in his head. If he drew a blank at those, he'd be waiting for her at the hospital's entrance, at the end of afternoon-shift, in the hope that she'd gone on to work after moving out.

His first stop was going to be the police station, where he'd get things moving by writing and submitting his letter of resignation.

The more he thought about it, the more right it felt. Suddenly, it felt like he'd been deceiving himself for months in order to keep on believing in his vocation. Now, resolved to leave, it felt like he was on the road to liberation.

He even had the seed of a potential plan for his future career. He'd leave the Met' and go into banking, not back onto the branch management ladder but into the City. He'd heard about investment banking and, whilst he knew nothing about it, it sounded like a lucrative occupation. He had developed a taste for life in London and reckoned that, with the increased salary that he could hope to be bringing in, in a few years, he and Julie would be able to buy a place in the Capital.

He'd had his spell of trying to make a contribution to society, now he'd focus on building a comfortable life and raising a family with Julie.

Julie was right, he had loved being a policeman but she'd underestimated his love for her. He'd prove he was the same person she'd fallen in love with.

He had no intention of letting her go. But, as he sped on down the motorway, he kept returning to his biggest source of doubt until it dominated his thoughts.

Looking back over the last few days, Neil had the inescapable suspicion that there'd been someone else involved in Julie's decision to leave him.

Who Are You?

Two days earlier, Julie had been holding another man's hand.

Sunday evening, 10 March, 1985

He had been brought into Casualty early on Friday morning and quickly, thereafter, up to Intensive Care with high-level concussion, a fractured skull and several broken ribs, having taken what the young doctor had described as a hell of a beating. It wasn't clear what internal injuries he had, apart from a punctured lung. The days and nights after the attack spent lying, hidden, in a bin area behind an office building near Vauxhall Bridge, had resulted in him contracting a severe case of pneumonia, a condition complicated by his damaged lung.

He would have been a big, strong man once but hardship and deprivation had robbed him of all of his muscle and bulk.

Julie was always caring with all of her patients but felt particularly drawn to this man. She sat by his bed, rubbing his palm with her fingers, trying to inject some fight into him. With no family, friends or other visitors to take an interest in him, he was in danger of dying alone. Whenever she could make a little time, she would come and sit with him, trying to get him to talk to her in his rare, brief moments of hazy consciousness.

She knew that her colleagues were wondering at the way she was behaving but she didn't care; she was determined that he would know that there was someone there for him.

When his breathing became less laboured, Julie hoped that this might mean that he would benefit from some proper, restorative sleep. She lifted his arm and laid it by his side, under the sheet and blankets, drawing them over him to keep him warm.

'You just sleep and I'll come back and see you later.'

She remained watchful of him, as far as she could, for the remainder of her busy shift. When it was time to finish she came back, drew a chair round and took his hand out from under the sheet and tried to rub some love and life into it. She looked at him with sad eyes.

'You're not fighting for me are you?' She murmured.

'You've got to try.'

After she had been to the locker room and changed, she went back up to the Unit. Ignoring the dubious looks she received from the sister and nurse on the next shift as she passed the open door of the ward office, she made her way to his bed.

She hadn't known why she had needed to come back to him until she was standing next to his bed.

'Goodnight,' she whispered.

Laying a hand on his shoulder, she bent over and kissed his forehead.

She stood up quickly and snatched a breath.

'Sleep tight.'

She turned and walked out quickly, calling a quiet "Goodnight" into her two colleagues at the desk in the office as she headed out.

'Yeah.' 'G'night,' came the sceptical replies.

Monday, 11 March, 1985: 13:08

Julie had arrived early, compelled by the need to go up and check on him before going to change into her uniform for the start of her shift. She walked in through the swinging, clear-plastic double-doors and up the short length of corridor to where it opened out into the Unit.

His bed was occupied but not by him.

She darted back the few steps to look round into the open door of the office at the sister, sitting at the desk.

'Hi. Where's the man who was in number five gone?'

'Who?' The sister said. 'Oh, you mean the pneumonia case?'

'Yes, where is he?'

'He's gone, I'm afraid. Passed away. Just before four, this morning.'

The sister saw Julie's face go deep red as the tears started in her eyes.

'Sorry Julie. They said he drifted off peacefully enough.'

'Oh, okay.' Julie croaked. 'Sorry. Thanks.'

Julie walked out of the hospital's main entrance, turned left and headed for the embankment. She walked over to lean on the stone wall above the Thames and gazed over at the Houses of Parliament on the other side of the river.

He'd had too few lucid moments for her to be able to get much out of him but she knew that he had been one of the men that the elected representatives in the building opposite had turned into an enemy.

"I wuz a pitmon," was the way his accent had sounded to her Yorkshire ear.

She had tried to get him to reveal the rest of his name. She had told him that she could get word to his family or friends. She had said to him that he must have friends. He did have, he had said, until he had thrown everything away.'

He had murmured to Julie that he came from somewhere she had never heard of; a pit village in Warwickshire. She knew that Neil's force had had several spells in that coalfield but knew little of what had happened there apart from his accounts of Carl Bessemer's stupid lamp-stealing and car-crashing escapades.

She couldn't imagine how a strong pitman could end up in a decline in London that had led to such a tragic, anonymous end. His family, friends and workmates would never see him again, never know what had become of him.

She knew that it was unfair to blame Neil for his death but she just couldn't help it. Julie knew at that moment; it was the end for them.

She dashed away the tears from her cheeks with the back of her hand and glanced at her watch. She pushed herself away from the wall and made a brisk start to the hospital; she couldn't be late for her shift, there were still patients who needed her care. She trotted a few paces then stopped and shook her head, still wondering at how his life could have come to end like that.

When he had first arrived in London, he had managed to get some casual labouring work on a building site at the Elephant and Castle. Living in a squat he had spent all of his pay on drink. One of the other occupants had introduced him to the escape to be found in marijuana and, from that, he had graduated, quickly, to heroin.

Soon, he was unable to earn the money he needed to satisfy his need for release and had turned to crime, until his rapidly escalating addiction had made him useless to the hapless gang he had fallen in with.

When he had become too dirty and unpredictable, the rest of the people in the house had decided to deprive him of his right to squat, turned on him and flung him out.

Five hours after Brian had met up with Ian Flint at the miners' demonstration, he had been sitting, alone, on the cold, damp pavement under the grimy arch of the bridge over the Thames. He should have been more careful. As a new member of the homeless in that area, he had not been accepted by the strange under-society that existed on those backstreets and alleyways. He had dropped his guard and made the mistake of taking out the pound note that the Barnsley man had given him, not to gloat over it but to draw some warmth from that gift from a fellow collier, a friend, failing to realise that his action would be spotted and the money viewed with avarice by the nearest of the group of four, drunken, homeless men and the hag with them.

Taking no chances with him, they had waited until he had fallen into a doze before one of them had crept up with a short piece of scaffolding and struck him a cracking blow to the side of his head. What the subsequent attack by the whole group had lacked in organisation and skill had been compensated for by its duration and desperation.

Eventually the first assailant had dropped his weapon. The scaffold pole was still ringing loudly on the pavement as he had reached into Brian's ragged greatcoat to draw out the pound note and hold it up, like a prize. Two of the others had lunged for it. The resultant brawl had lasted only a few seconds; the five, bickering vagrants had frozen on hearing a police siren, dispersed and scampered away in both directions along the riverside and off into the shadows of its back-alleys.

Ian Flint would never know that his act of comradeship on that February afternoon in London had sealed the fate of his former comrade; that as he had pressed that pound note on Brian, he had been handing him his death warrant.

Hope and Harmony

'And I would just like to remember some words of St. Francis of Assisi which I think are really just particularly apt at the moment:
"Where there is discord, may we bring harmony. Where there is error, may we bring truth. Where there is doubt, may we bring faith. And where there is despair, may we bring hope".'

(The Right Honourable Margaret Thatcher, immediately before entering 10 Downing Street, after her first Election victory on 4 May, 1979)

PART THREE

Inrush

The Manrider

Tuesday, 5 March, 1985: 06:44

The driver was already braking as the coal-carrying Loco Road opened out into the big, square-work junction. The train drew to a halt, opposite the huge, wooden doors. The men bound for the North End of the pit struggled out of the tight, low openings in the sides of the carriages with their snap bags. Jim Greaves was among them, determined to get 15's Face off to a prompt start and gain the shift an extra couple of minutes of coal-turning.

The first man to the walkway air door in the massive, timber barrier that separated the intake air of the Loco Road from the return air of the North Main Supply Road leant his shoulder against it. Two others joined him, thrusting on it to force open the way through to their part of the mine. The stream of men wasted no time in passing through, stepping high over the wooden cross-member at floor-level, into the pressure chamber and pressing on towards the second, identical barrier.

The great structures included the massive, hydraulically-rammed doors that allowed access for supplies from the Loco Road. The distance between the barriers was sufficient to accommodate a run of six vehicles so the men crowded in.

Through the second door, they spilled into the hot, stale, return air with its distinctive smells of coal, machinery and rough and ready humanity. Past the rope haulage engine house, offset to the left-hand side, the manriding train of open carriages waited on the steep rise of the manriding station, its train guard ready to go.

'Come on, let's have you on this fucking train,' Joey Bondar snarled at the crowd trudging up the slope on the manrider station's boarded floor that provided a safe, even walkway over the track's sleepers. 'Let's get this show on the road.'

One of the men shouted a greeting to his overman, "Ey-up Shit-Face.'

'Get on, Shit-'Awk,' Joey called back.

Jim was still stomping up the incline, having deviated into the entrance of the manrider's engine house to shout "good morning" in to the engine driver.

Joey had abstained from taking his usual seat with the guard at the front of the train, reserving the second two-man seat instead, so he could seat his Undermanager beside him.

As Jim walked up, alongside the train, he saw Kevin Clark, an electrician sitting in the train, complete his struggle to secure two brand new, protective gloves over the steel toe-caps of his pit-boots. Once in place, the shiny, red rubber looked convincingly like a chicken's claws and Jim couldn't help chuckling with the men around him as Kevin waggled his feet, twitched his elbows like wings and started clucking like a hen.

Having captured Jim's attention and a wider, ready audience, Kevin engaged him in a loud dialogue.

"Ey-up Mr Greaves. Yer know 'ow you was against us gettin' water-money on 15's?'

'Yes,' Jim called back, cautiously, as he carried on walking along the side of the train.

'Well, it's made it a very sought after location, as a workplace.'

'That's good, then,' Jim conceded, reluctantly, as he clambered up into the carriage to squeeze in beside Joey.

'Yeah, a lad on 56's asked me who he needed to see if he wanted to get a job on 15's.'

Jim waited, expecting that he would be the supposed target for a host of applicants.

'I told 'im, anyone who wants a job on 15's in all that water needs to see a fucking psychiatrist.'

'Rap him off!' Joey barked at the guard over the resultant hilarity.

The train guard used the metal blade of his T-shaped ringer to complete the circuit between the two, bare, signal wires, ringing the bell in the engine house three times to indicate "manriding" to the motor driver. He looked back down the train to make sure that everyone was on board and there was no one left in an unsafe position in the confined spaces alongside the departing train, switched on the big, battery-powered headlight that he had housed in the steel box on the front of the leading carriage and then gave another two clear rings, signalling they were to be wound inbye.

From behind them, in the engine house, the distant sound of the rope-haulage motor picking up reached the men's ears, as the return rope in the wheels hanging from the arch legs to the side started running towards the engine house. When the slack throughout the rope's five-mile length had been drawn in, the manriding train lurched forward, then surged in a number of cycles as the sudden acceleration threw more slack and the gradient slowed the heavily-laden train.

Then they were off, to the accompanying clatter and rattle of the numerous wheels for the return rope and the steel rollers between the rails that protected the main rope that drew the train.

The carriages breasted the incline and dropped into a steep dip. Jim had never travelled underground on a more undulating road. It climbed and dropped repeatedly, like a big dipper, throughout its entire length, through a series of hollows known as "swillies".

On the way in, Joey bent his Undermanager's ear about proposed work for the weekend and the overtime budget they could expend before moving on to entertain him with some pit gossip.

As the train sped along, their helmets bumped together repeatedly as they ducked their heads down under the low arches, the plug of tobacco Joey was chewing failing to mask the smell of stale beer on his breath from the session he had obviously enjoyed in the pub, the night before.

After ten minutes the train slowed for a stopping point. The train guard signalled the command to stop by holding his ringer against the bare signal wires for a continuous signal of one ring. The train pulled up and swayed backwards and forwards on its rope as the three-men of 14's Development's heading team alighted with their three supply lads, electrician and fitter and several button-lads for the conveyor belts in that vicinity, exchanging cheerful banter with the men remaining on the train to travel further inbye. The headers and their mates side-stepped their ways up the tight space between the carriages and the deformed arches that supported the roof and roadside to get to the cut-through from the return airway to the intake side's belt-road to access the face district they were driving.

When the guard was sure they were all clear he rang the train off again.

'We should make sure that 14's men ride at the front of the train, so they don't have to struggle up the length of the train,' Jim said to Joey. 'It would save time and get your face team in a minute or two earlier.'

'Fair point,' Joey said, nodding. 'I'll tell 'em an' get the guard to start doin' that, Boss.'

Two miles away, the engineman slowed the train as they approached the lights in the road that terminated the main road at a 'T'-junction. Taking the left-hand branch led to 15's face, a thousand yards further in.

The train stopped and fifty men piled out of the train. Seventy yards in, the face's three supply lads peeled off at the turnout, the siding used for shunting supply vehicles, in order to arrange them in the sequence in which they would be needed to be moved up towards the face.

As Jim edged his way up the tight gap between the manriding train and the steel arches supporting the roof and side of the road, Joey remembered to collar the train guard and pass on the instructions his Undermanager had given him.

The North End

Jim and Joey marched up the recently-driven maingate, to get to the face.

15's was the first face at Whitacre Heath to work out towards the old Tame Valley Colliery. Its installation had only just been completed when Mr Everitt had told Jim that he wanted him to take over the management of that part of the mine and it was manned, mainly, by teams of men and officials transferred from 85's in the South End of the pit, when that face had finished.

The district had a fresh feel about it. The short, uncluttered gate road still stood tall, its arches yet to bend under the strain of the caving behind the advancing face. On the five-foot high face-line, the chocks, side pans of the conveyor and the shearers all still bore a coat of fresh, matt white paint, evidence that, although the face equipment had been used before, it had all benefitted from a full overhaul at the NCB's workshops at Swadlincote, just across the county border, in neighbouring South Derbyshire.

The first half dozen cuts had been taken. The faceline of conveyor and roof supports had moved out under the exposed strata, bringing the working face to life.

When they arrived at the maingate pack, Jim spoke to the two packers at that end of the face and told them to build the tightest possible pack to the roof, this close to the face start-line, urging them to take care in achieving as complete a seal as possible.

When the face installation had been completed, with all the chocks in position on the face, pairs of facemen had uncoupled the square work's face-side legs from the steel roof beams and then reversed them back up on the panzer to be removed at the tailgate end of the face. Joey told Jim how he had deployed men to uncouple the brackets securing the other legs, behind the roof supports, for the first thirty metres up the face-line. This was to prevent the steelwork from maintaining a triangular air passage that would increase the risk of spontaneous combustion, in the high risk area of the face start-line, when the face-side ends of the steel beams dropped off the back of the advancing chocks.

The development, installation and launch of new face districts always cheered pitmen, with the assurance these undertakings gave of continuing pit-life. But there was something to blight this optimism, on 15's; half-way through the installation of the face, water had started to seep up through the floor, making the rest of the installation work more difficult. As soon as the face had started advancing, the water problem had worsened.

Walking under the canopies of the roof supports, a slurry of coal fines was apparent in the space between the chock's bases. Passing in front of chocks, ready to be pulled in, where the conveyor had been snaked over, as he stepped over the extended stroke of each ram, Jim found that his boots were paddling in several inches of inky water.

'Where's all this water coming from?' He said.

There was too much of it and it lacked the milky appearance for leaks in the hydraulic system powering the chocks to be the source.

'It was always going to be a wet district,' Joey said. 'We're 'eadin' towards some old workin's that have probably got water in 'em.'

Jim stooped and looked, to his left, at the shiny, damp surface of the coalface.

15:07

The importance of the legal obligations of colliery owners and general managers to prevent inrushes was reflected by their being set out at the very front of the Mines & Quarries Act. Earlier incidents and flooding disasters had revealed the need to ensure that collieries created and maintained up to date plans and sections of their workings. Every NCB mine fulfilled this particular, statutory requirement by having its own, in-house, surveying department.

That afternoon, back on the bank, having bathed and changed, Jim left the undermanagers' bathroom and went straight over to the main office building to call in on Derrick Hill, the Colliery Surveyor. Being under forty, Derrick had achieved his position of seniority at a relatively young age.

'I was on 15's Face this morning and I visited 13's Proving Head,' Jim said. 'My overman mentioned that the face is working towards some disused workings. I was wondering if I could see a plan showing where those workings are in relation to 15's.'

'Wilf,' Derrick called to his oldest subordinate. 'Can you get the plan of the North End out for Mr Greaves to have a look at, please?'

'Certainly, Derrick. Good afternoon, Mr Greaves.'

"Afternoon, Wilf,' Jim said.

Wilf Albrighton was heavily built, dressed in shirt sleeves, his trousers supported by a dull brown pair of braces as well as a broad, heavy, leather belt. He padded over to join them at the Surveying Department's large, high, central, drawing table. Just like the great meeting table in the General Manager's office, it had an impressive, eight-foot long version of the Colliery Plan displayed under its clear, Perspex-covered surface. Drawers built into the table were repositories for other drawings and plans. Wilf drew one of them open and leafed through the corners of a few sheets to find and extract the one they needed.

Members of the surveying team had prepared this latest version of the plan of Whitacre Heath's North End, by hand, on a massive sheet of stiff, high-quality, white paper. Drafted in pencil, its lines had been completed in black ink, with pens of various gauges, then a variety of expensive, coloured pencils had been used to highlight particular features, painstakingly, with perfectly even shadings. It was a technical work of art and craftsmanship.

When it was laid out on the table, held open by leather-covered, lead blocks placed on its corners and edges, a stark picture was revealed.

Wilf talked them through the plan, pointing out particular points to note with his white, plastic, surveying scale-rule.

The old Tame Valley Colliery lay to the north west of Whitacre Heath's North End. The planned advance for 15's Face was in a north-westerly direction, straight towards the old Tame Valley workings. The grid of old roads that were part of the abandoned mine reached out, grasping like eager, fingers for 13's Proving Head, a short, dead-end road about forty metres in length with a line of advance parallel to, and a few hundred yards to the right of 15's Face.

According to the plan, 15's' planned take would extract the seam parallel to the spur of abandoned workings for three-quarters of a mile. The intention was to finish 15's Face there, leaving a fifty-metre, protective pillar of coal between it and the old, disused workings.

'How confident can we be in the old Tame Valley plans you used to make this plan?' Jim said.

'Well, they were legally obliged to keep accurate and up-to-date plans then,' Wilf said.

Jim couldn't help thinking that Wilf looked old enough to remember the old mine still working.

'But if there was ever a time when they'd become inaccurate it was as the pit approached closure.'

'Was that because they'd chance not getting caught out by the Mines Inspectorate and try to save some surveying fees?' Jim said.

'Chance it, yes, and it would have saved the fees but the big saving for them would have been in Royalties,' Derrick said. 'In those days, before the industry was Nationalised, they had to pay the owner of the land undermined by their workings, so in the mine's last months, or even years, they'd reduce their overheads and boost their profits by robbing coal and not declaring it. So they'd end up having plans that didn't show the full extent of the workings.'

'So what confidence can we have in what we're looking at on this plan?' Jim said.

'Well our measurements are up to date, as of last week,' Derrick said. 'But I wouldn't put much faith in the old pit's plans, would you Wilf?'

Wilf shook his head, frowning.

'I wouldn't want to have to stake my life on 'em.'

'Do we know where the old shafts were?' Jim said.

'They were here, less than three quarters of a mile from 13's Proving Head,' Wilf said, sticking his finger on the plan. 'And, on paper, the distance between the closest old workings and 13's Head is fifty-five metres.'

Jim paused for thought. He always tried to avoid being overly-dramatic but it all looked horribly close and unassuring.

'It makes you wonder what we're doing up there,' he said. 'Do we know what's above the old workings now? Would they have capped the shafts to seal them?'

'I can tell you why we're up there, it's to extend the life of the pit,' Derrick said. 'I helped to develop the proposal to attempt to prove additional reserves in the North End that went to the Colliery Consultative Committee. They accepted it and then the General Manager authorised it with support from Steve Wheatcroft, the Area Chief Mining Engineer.'

'If proven, this will add about six million tonnes to our reserves, an additional eight or nine years onto the pit's remaining life.'

'But to answer your other questions, they didn't bother sealing shafts in those days and, even if they had, the caps would have been removed by what happened afterwards.'

'Why? What was that?' Jim said.

'Whately Quarry,' Derrick said. 'A massive extraction of prime gravels.'

'Is it still working?'

'No, it closed in the early sixties,' Wilf said. 'It's a bloody great lake, covering over a hundred acres now. And the old shafts are right in the middle of it. Take a look at it on your way home. You could call in for a drink, the Warwickshire Water Skiing Club has got a nice clubhouse there. So, you're right, I suppose it does make you wonder what we're doing up there.'

From the copy of the proposal that he took from Wilf, Jim found that Ken Goodall, the Statutory Undermanager, had been responsible for the preparations, so he went to Ken's clean office to find out more.

Ken confirmed that he had arranged for a pumping survey to be undertaken to quantify the total pumping capacity from the various parts of the mine and up the shaft, the make of water from all parts and, therefore, the mine's residual pumping capacity. On the basis of that survey, the Colliery General Manager had authorised the installation of an upgrade to the North End's pumping system to deliver a significant increase in its capacity.

A second, wider diameter pipeline had been slung under the existing pipe-range hanging from the arch legs, all the way from 13's Proving Head and back to the pit bottom sump. They had installed more powerful pumps with duplicates to provide a safety-net in the form of redundant capacity. A massive underground tank was to be built as close as possible to the disused workings to act as a sump, in the hope of accommodating any sudden, unexpected increase in the make of water.

After the pumping capacity had been increased, Mr Everitt had authorised the proving operations to commence. However, Jim found, to his disgust, that the building of the big sump had not been postponed but simply neglected.

13's Proving Head had been opened out from an existing junction on the return airway, the road used to access, and transport supplies up to, 15's Tailgate, the face's supply road. When the heading was only nine and a half metres in, Ken had arranged, as planned, for the Area Drilling Team to drill a proving hole in the direction of the old workings. He gave Jim a copy of the report on that first borehole.

The hole had been drilled, eighteen months earlier, at twenty degrees to the left of the heading's north-westerly direction of advance. A standpipe with a cut-off valve had been installed into a carefully-drilled hole and secured with resin. This was so that the water could be shut off, if the old workings proved to be waterlogged. The standpipe had been pressure-tested to 500 pounds per square inch because, if there was water in the old workings with a continuous head through the old shafts to the surface, the water pressure could be as high as 400 pounds per square inch.

The Drilling Team had drilled the inch and a quarter borehole through solid coal for eighty-five metres. Then they had hit water. They had measured a flow rate of ten gallons a minute, about a bucket-full every thirty seconds.

They had closed the cut-off valve on the standpipe so they could measure the water pressure. Twenty-four hours later they had found that the static head beyond the standpipe had risen to 270 pounds per square inch.

Ominously, the report noted that water had started to seep through the face of the heading itself. To Jim, this sounded like a worrying indication that the surrounding seam was not competent and that this seepage might be due to pressure being released through it.

This appeared to be confirmed by what had happened next; after another forty-eight hours, the pressure beyond the standpipe was measured and found to have dropped to 120 pounds per square inch. This had been accompanied by a significant increase in the make of water in the heading, with water rising out of the floor and running down the slight upward slope of the heading.

The report concluded by indicating that the borehole had been sealed to stop the seepage of water and make it safe, a complete seal having been achieved by pumping Shallow Oilwell cement into the hole, under high pressure.

17:45

Jim took a detour on his way home and found the gateway into the Warwickshire Water Skiing Club. He drove into the carpark, parked up, got out of his car and stood, alone, gazing at the great expanse of water and marsh land

The next day, he deployed men to dig a huge sump-hole in the roadway at the entrance of the heading in the hope of accommodating any sudden flood of water.

Planning

Wednesday, 13 March, 1985: 08:52

There were only five in the meeting but that was sufficient for what they were going to cover.

Jim had invited Derrick Hill and Wilf Albrighton for their knowledge of the plans and geology and two big, useful-looking pitmen with a particular technical bias; the Area Drilling Team consisting of its leader Todd Crisp and its only other member Chris Baxter.

Todd and Chris worked out of the Area Headquarters at Coleorton Hall, within the small department of the Area's Chief Mining Engineer, making them more privy to Stephen Wheatcroft's perspectives and intentions.

Jim had asked Ken Goodall for the use of his clean office. Wilf had brought the plans and section diagram, showing the geology of the North End. With the coffees Jim had made with the kettle in the Control Room, across the corridor, they were ready to start.

'Right gentlemen, I'll take notes of the meeting and get them typed up,' Jim said.

'That's all very formal and professional,' Todd said.

'You have to be. This is a serious subject.' Jim said with a cheery grin. 'Anyway Todd, what are our instructions from your leader.'

'Mr Wheatcroft has agreed with Mr Everitt that we're to drill the next proving borehole, a 3.175 centimetre diameter hole, that's an inch and a quarter in old money, five degrees off the heading's centre-line and rising at 3¼ degrees, which I believe is the slope of the Seven Foot seam in that area, is that right Wilf?'

'That's right, it rises at that gradient 'til it steepens and outcrops at the surface, two miles away, at the north western boundary of the Warwickshire Coalfield,' Wilf said, pulling the geological section diagram out of his pile of documents and placing it on top.

'That's it, Wilf,' Todd said. 'A picture's worth a thousand words.'

'Let's have a look at where you're going to be drilling,' Jim said, standing to find 13's Head with his finger on the plan.

'After the first borehole, they drove 13's Head for another thirty metres,' Jim said. 'According to your plans, how wide is the barrier, now, between the old workings and 13's Head?'

'Fifty-five metres,' Derrick said, placing his ruler to confirm the measurement between the narrow heading and the grid of old workings. 'Fifty-five and a half, to be precise. But as I've told you before, there's nothing precise about our knowledge of the old workings.'

'And the purpose of the borehole is to prove whether the old workings contain water or not,' Jim said.

'We know they contain water,' Todd said. 'We're trying to find out how dangerous it is and what pressure it's at – whether there's a head of water back to the surface. But, mainly, it's to make sure we've got a safe barrier of coal between the old workings and 15's Face as it works towards them.'

'Todd's right,' Derrick said. 'We need to make sure that we leave enough of a pillar of coal between the old workings and 15's Face, when it finishes, to hold back whatever head of water we find when we drill.'

'Let's have a look at the practicalities of the job,' Jim said.

'We'll be drilling the hole seven feet up in the face of the heading, using a drilling machine mounted on a horizontal frame,' Todd said. 'Small rams, powered by a hydraulic power-pack, provide the force on the drill. We'll work on the wooden, ripping platform, across the end of the road, to do the drilling and the frame will provide the stability and rigidity we need to keep the direction and gradient of the borehole accurate.'

'And, learning from last time,' Chris said, 'we're going to use a special drill-rod with a non-return valve immediately behind the drill-bit to prevent any uncontrolled flood of water coming back at us through the bit and the drill-string.'

'What are the timescales?' Jim said.

'We can get all the kit delivered by the end of this week,' Todd said. 'If you can get it transported to site by next Wednesday, we can have it set up by next Friday and drill the hole on the following Tuesday.'

'Right, so that's a target date for drilling of – the twenty-sixth of March.

But I still find those words ominous,' Jim said pointing to the label on the old workings on the plan:

"Tame Valley Colliery,
waterlogged Seven Feet waste"

The Borehole

Tuesday, 19 March, 1985: 13:03

Walking out from 15's Face, Jim arrived at the junction into 13's Head.

An elementary, open-sided bridge enabled the route from the face to pass over the conveyor belt that served the heading. The conveyor disappeared, down the road, through ventilation barriers, to deliver onto the main, coal-carrying, belt conveyor.

Jim was pleased to see that the full inventory of equipment of drilling equipment had been unloaded, tidily, at the entrance of the heading. The Drilling Team, assisted by 13's lack-lustre heading team of Crowley, Hedges and Coton, would be able to get it all into position and installed, ready for drilling to fulfil the Chief Mining Engineer's expectations, in a week's time.

Tuesday, 26 March, 1985: 06:08

As arranged, Todd and Chris reported to Jim in the dirty offices, before the start of dayshift on the twenty-sixth of March.

'So, today's the day then, men.' Jim said.

'Will you be joining us, Mr Greaves?' Todd said

'I'll try to join you for the drilling. If you give the Control Room a call when you're about three-quarters of an hour away from starting, I'll make sure they get word to me so I can get round to you.'

'It won't take us long to get ready to drill so you can expect a call about ten.'

Jim had confirmed to Ken Goodall and Mr Everitt, in their respective, statutory capacities of Undermanager and Colliery General Manager, that the drilling team would be attempting to locate the workings with the borehole that morning.

As usual, as he made his way through 15's face, Jim took a turn on a shovel, spooning the slurry of coal fines onto the panzer from the gap between the panzer and chocks ready to be pulled in after the recent snake of the armoured face conveyor; the wet conditions a permanent reminder of the waterlogged workings they were mining towards.

At nine-fifty, the Control Room phoned 15's panzer driver and told him to call Mr Greaves on the face tannoy and let him know that the drilling team were ready for him in 13's Heading. Jim arrived there forty minutes later.

13's Heading was a dead-end of about forty metres. The Dosco MKIIA heading machine that had been used to drive the short length of road was parked up in the heading. As big as a tank, it dominated the road where it sat, six metres back from the face of the heading.

With its Caterpillar tracks, the paddle conveyor that ran round its front apron to feed its central, chain mini-conveyor and, mounted on a turntable, a movable boom that, with its cutting head, was over ten feet long, it was reminiscent of television's Thunderbirds' "Mole" tunnelling machine.

Jim walked through the narrow gap between the Dosco and the side of the twelve-foot wide road to find Todd and Chris with Terry Harrison, the deputy in charge of the heading, and the three infamous headers, Crowley, Hedges and Coton, standing idly by.

Before they started drilling, Chris talked Jim through the technique and the equipment in more detail.

'We're going to drill the borehole through this standpipe,' Chris said, slapping a three-inch diameter, steel pipe installed in the middle of the face with a pressure gauge protruding from its side and a gate-valve bolted near to its end.

'It's six feet long and we secured it in the seam with resin and pressure tested it through the pilot hole to 600 pounds per square inch for twenty minutes to make sure we've got a hefty factor of safety, in case we hit a body of water under high pressure. The gate-valve's there so we can shut off any flow when we withdraw the drill-rods.'

One of Derrick Hill's assistant surveyors had been in with a theodolite and level to line up the drilling machine accurately with the standpipe and with the bearing and gradient planned for the hole.

The drill-string would consist of an inch and a quarter diameter bit and five-foot drill rods that would be added and screwed together as the hole advanced, with the rod containing the non-return valve installed first and attached to the drill-bit, as Chris had described in their planning meeting.

The drilling machine was like the normal bull's head type of electric borer that Jim was used to using underground. However, instead of being hand-held, this one was mounted on a highly-engineered, oblong, horizontal framework of strong angle-iron. This was installed on the wooden platform of scaffolding planks, with a hydraulic Dowty prop attached, by clamps, at each of its four corners. The four Dowty props had large steel extensions, known as top-hats, to enable them to be pumped up to the full height of the road.

An attachment would pump water through the drill rods and out of two small holes in the drill-bit to flush the debris from the drilling away from the end of borehole in order to optimise the straightness and accuracy of the drilling.

Jim joined Chris and Todd on the platform as they manned the drilling rig.

Being a dead-end, the heading had no ventilation circuit for the air to follow. The air was delivered to the face of the heading by a large electric, auxiliary fan which drew air from the intake and blew it through air troughing, the flexible plastic tubing about thirty inches in diameter, which had been hung over the conveyor road out of the heading. This played a cooling blast onto them as they stood, hunched on the platform, under the roof of the heading.

With the first, non-returning rod and drill-bit coupled, Chris switched on the borer and turned on the flushing water. The hydraulic ram on each side of the frame provided a measured, steady and equal pressure to the borer.

It took about five minutes to drill the length of each section of drill-rod. Each time they had drilled to the full extent of a drill-rod, Todd secured the drill-string temporarily with a clamp on the drill-rig and unscrewed it from the boring machine, which they moved back on its firm base using the hydraulic rams. Then they would screw another five-foot drill-rod onto the back of the string, ease the borer forward onto it, reconnect and drill again.

They continued to operate this cycle until, with the eighth drill-rod attached, Chris switched the borer on and the drilling started again.

[The Official Report into the incident would record: *"at a length of 36 feet the drill hit a void and a very large quantity of water was delivered under great pressure via the hole. Mr Todd Crisp, the leader of the drilling team, reported the rush of water as "immediate and frightening" and described it as being accompanied by "an alarming whistling noise and a stink".*]

'What the fuck is that?' Gary Hedges shouted over the howling noise. His two mates, Crowley and Coton, were already well on the way to the junction, forty metres down the heading, with Terry Harrison, the heading's deputy, hobbling hurriedly after them.

'It smells like somethin's died,' he cried.

Terry and the other two headers disappeared round the corner at the junction.

'It's probably just hydrogen sulphide from the old workings,' Jim shouted over the screaming of the evil-smelling water jetting out of the standpipe around the drill-rod.

Todd and Chris were hastily trying to withdraw the drill string.

'We need to get this closed off,' Todd blurted out. 'And quick!'

They drew the drill-string back with the boring machine, a rod length at a time, thrusting each one towards Jim as soon as they had it uncoupled. Jim handed them down to Gary, who stowed them back in their steel, storage box.

With the drill-string recovered and clear of the hole, Todd shouted above the noise.

'We need to measure the unrestricted flow rate through the borehole.'

Jim jumped down from the platform to do this with Gary. They stood ready with the delivery end of a four-inch diameter hose, attached to the valve on the standpipe, and thrust it into a forty-five-gallon drum as soon as Todd shouted that the valve was fully open. The drum stood about four feet high and thirty inches wide. Todd used a stopwatch to measure the time.

It filled in an instant. As it started to overflow, Todd stabbed the watch's stop button with his thumb and sang out.

'Five seconds.'

Jim did the mental arithmetic as Todd scribbled away in his book.

'About five hundred and forty gallons a minute,' Jim called up to Todd.

After a couple of seconds of scribbling, Todd shouted.

'Yeah, that's what I make it.'

'That's almost ten gallons a second. You'd fill a bucket in just over half a second from an inch and a quarter diameter hole.'

'That sounds about right,' Todd called back. 'You wouldn't want to get your hand in the way of it.'

433

Jim clambered back up onto the platform. With the drill-string recovered and clear of the gate-valve, Chris and Jim turned the big handle on the valve, staunching the flow of water and enabling them to measure the head of water with the pressure gauge on the standpipe.

With the water flow stopped by the standpipe's gate-valve, the noise was reduced again and Todd read off the pressure on the gauge's dial.

'That's a static head of four hundred and ten pounds per square inch – just over twenty-eight bar, nearly thirty times more than atmospheric pressure!'

He looked grimly at Jim.

'So, what do we know?' Jim said. 'From the stink and the amount of pressure, we appear to have hit the old workings and four hundred pounds of static pressure means that there must be column of water backing up to the surface through the old shafts.'

'I'm sure you're right,' Todd said, 'but we'll take a water sample and get it analysed to verify where it's coming from. The old workings are forty-five metres closer to us than anticipated, at the point where we broke through. Which proves that the old plans are wildly wrong. Which, I suppose, is no surprise.'

'Yeah, but we don't know what level of inaccuracy there is elsewhere,' Jim said.

Everyone there knew the Undermanager would be thinking of the old workings in the vicinity of 15's Face.

Senior Management

Back on the surface, Mr Hope, the Operations Manager, had told Jim of the senior management response to the outcome of the drilling. This had enabled Jim to work with Wilf, in the Survey Office, on some quick but valid analysis.

Knowing that the morning's drilling would be discussed in the daily management team meeting, Jim had invited Derrick, the Colliery Surveyor, to attend with Wilf. Mr Everitt, Mr Hope, Bill Metcalfe, Ken Goodall, Angus Dewar, the Colliery Mechanical Engineer and Phil Grant, the Colliery Electrical Engineer were settled around the massive meeting table in the General Manager's office.

'I phoned Steve Wheatcroft, earlier, to let him know the results of the drilling,' Mr Everitt said. 'He instructed me to attempt to drain the old workings through the borehole for three months. 'E wants us to see if it's possible to reduce the level of water in them.'

'Gaffer, I think we should present the figures we prepared earlier to Mr Wheatcroft and try to dissuade him from doing that.'

'You'll be lucky,' Mr Everitt grunted.

'The old Tame Valley pit isn't the only one that threatens us,' Jim said. 'Tame Valley was linked up, underground, with Two Gates Colliery in the early 1900's, to improve the ventilation of both mines. And the gravel quarry took out the tops of both of Tame Valley's shafts, so neither of them are capped now, if they ever were. That means that the old workings are connected straight up, into the worked-out quarry, which is a great, man-made lake. Wilf, could you take us through the stat's we worked out?'

Wilf distributed the purple-stained, duplicated sheets of figures, still damp from the sharp-smelling, Banda printing chemicals.

'Mr Greaves and I thought it was important to give you some idea of what we're looking at, in terms of the total amount of water likely to be present in the two mines and the quarry and the make of water due to natural precipitation which will determine the rate at which it's replenished,' Wilf said.

'That means rain, Phil,' Mr Everrit said to his Electrical Engineer. 'Right, so what are we looking at?'

'Our figures are rough estimates and we've erred on the seriously conservative,' Wilf said. 'But if we assume that the two pits, Tame Valley and Two Gates, both had a peak of production for forty years of their lives and that, in that time, they each produced an average of at least four hundred thousand tonnes per annum, that would have created underground voids of over thirty million cubic metres.'

'If we assume that subsidence over the decades has caused ninety per cent closure, that would leave voids of three million cubic metres. Filled with water, that would equate to about seven hundred million gallons of water.'

Angus Dewar, the Mechanical Engineer, whistled and said, 'So that's what would flood in if we breached the pillar between us and those old workings?'

'Yes, because Whitacre is deeper and lower than both of the abandoned mines, all of that,' Wilf said. 'And then there's the lake in the old gravel quarry. That's over eight hundred metres long with an average width of about four hundred metres. Taking a very modest estimate of four metres as being the average depth of the lake, that gives us over one and quarter million cubic metres of water. That's an additional three hundred million gallons.'

'With seven hundred million gallons in the old workings and about three hundred million in the lake, we're looking at something like a billion gallons of water,' Jim said.

'Bloody amazing.' Mr Everitt said, 'I wouldn't have thought there was that much water in the whole of Warwickshire.'

'I'd say that's the very least we could expect to be facing,' Derrick said. 'As Wilf said, he and Mr Greaves have been very conservative in all of their estimates. I reckon it could easily be twice as much as that.'

'With our pumping ranges from the North End of the pit, we can shift about three hundred thousand gallons per week,' Jim said. 'So, the best that we can hope to pump in three months would be about three and a half million gallons. That's about a third of a percent of what's there. We wouldn't even make a dent in it.'

'And then there's the rainfall,' Wilf said. 'If we ignore the run-off of water draining into the lake from the surrounding land, which would be considerable, and just take the typical annual rainfall over the lake's surface area, this would replenish the lake with sixteen million gallons – about five times what you'd have pumped out.'

'Gaffer, the thing I'm worried about is the competence of the borehole,' Jim said. 'We know that with the first hole, drilled when the heading had only advanced nine metres and the barrier was much thicker, the surrounding strata deteriorated. We've reduced the thickness of that pillar radically now and it seems to me that we're expecting a lot from a narrow pillar of about thirty-five foot of coal with four hundred pounds per square inch on the other side of it.'

'We know that pillar of coal's fissured, which isn't surprising, because the old mine used shot-firing to extract the coal. There's no way of knowing how much erosion there's going to be from three and a half million gallons of water rushing through an inch and a quarter hole at massive pressure for three months but it's bound to enlarge the borehole considerably.'

'I'd worry that the water would start to create a path for itself through breaks in the seam. If it does that and bypasses the hole and stop-valve, we might never be able to shut it off.'

'Or we could end up with an inrush.'

No one dismissed Jim's concerns, the inrush at Lofthouse Colliery was still fresh in everyone's minds. The parallels were inescapable; an NCB mine in Yorkshire had advanced too close to a disused nineteenth century mineshaft, the fissured, protective pillar of coal had broken down and millions of gallons of thin slurry had swamped the pit. It was only six years since the 'Mines Precautions Against Inrushes Regulations' had become Law as a consequence of that disaster.

'I'll let you into a secret,' Mr Everitt said. 'When I spoke to Wheatcroft, I tried to tell 'im that we'd be wasting our time trying to drain those workings. Actually, my rough estimate of the amount of water involved wasn't a million miles from yours. I'd be wasting my time trying to revisit it with 'im. We'll just have to give it a try and, in three months' time when the pressure hasn't dropped at all, I'll be better placed to persuade him to let us seal up the borehole.'

Jim was reluctant to speak up against the General Manager but he took a deep breath.

'Sorry Gaffer, but that may not be safe,' he said. 'We know that the water started finding a route round the first borehole within forty-eight hours. We don't even know where the workings' closest point to us is. We know that the plans were forty-five metres out, where we drilled into them, but we don't know if the two old collieries took even more out to avoid royalties at other parts of their workings.'

'At least let me line the hole with steel pipe first, to prevent the hole enlarging and breaking down the pillar.'

'Mr Greaves, I think that's an excellent idea, because that's exactly what I suggested to Mr Wheatcroft,' Mr Everitt said, causing Jim to relax for a moment.

'But 'e refused and told me not to fuck about and just get on with it and start draining those workings.'

Jim sat and fumed. Harry Everitt held the statutory position of General Manager, not Wheatcroft. If Jim had been in the Gaffer's position he would have acted on his conscience and disregarded his senior's instruction or, better still, not asked the question in the first place and just lined the hole anyway; he didn't believe in asking questions to which he didn't want the answer. Besides, he could not believe that a General Manager would end up being sacked for safeguarding his men and his pit.

It sounded to Jim as though the Area Chief Mining Engineer was insane, determined to kill them all, but managed to hold himself back from actually saying it.

Planning the Wall

Wednesday, 27 March, 1985: 07:03

T he following morning, as soon as he had finished reviewing the nightshift's performance and discussing arrangements for the dayshift's production, Jim joined his senior overman at the table in his dirty office.

Les listened intently as Jim revealed Mr Wheatcroft's amazing expectations, the water volumes in the disused workings and the lake above them and his concerns about the competence of the protective pillar of coal.

Given the deterioration of the standpipe seal and the surrounding strata with the previous borehole, Jim had decided to reinforce the face of the heading, across its whole cross-section, by building a metre-thick, concrete wall against it. Willie Stokoe, the Nightshift Senior Overman, was sitting at a table in the communal area, just outside Jim's office, signing a bundle of deputies' statutory report books.

'I want the wall keyed-in to the strata all the way round the roof, sides and floor for additional stability,' Jim said.

Willie looked up with a sneer on his face and shouted through to them in his Geordie falsetto.

'You'll be needin' to put a round o' shots in that floor. It's as hard as bell-metal, mun.'

'I don't want any shots fired in there,' Jim called back. 'I don't want anyone doing anything that could break up the seam any more than it already is. I'm worried enough about whether that pillar's going to hold up. They'll just have to use pinch bars and picks.'

Willie smirked and snorted. Shaking his head, he returned to his cursory review of the nightshift's reports.

Gerald Chipman called in on the dirty offices, a little later, to discuss the day's supplies requirements. Jim collared him and placed his orders for the materials and equipment for the wall, including thirty tons of bagged, Readymix concrete, an electrically-powered concrete mixer and heavy RSJ girders, corrugated steel sheets and lagging boards for the retaining framework. He requested a number of ten-foot long, two-and-a-half-inch diameter threaded steel pipes with threaded collars to couple them up, so he could have them installed in the wall to enable them to seal the strata around and past the wall, in the future.

Jim phoned Todd Crisp and informed him of his intentions.

A metre wall would not have the strength to stop an inrush in itself, given the huge pressures involved, but it should help to hold the strata together at the face of the heading. It also meant that the drilling rig could be left undisturbed, keeping the drill aligned with the hole's gradient and orientation.

'That sounds prudent,' Todd said. 'I'll get Chris to organise all the necessary bits to extend the pipework so we'll still have a working stop-valve and pressure gauge.'

'Thanks Todd,' Jim said. 'He'd better get it all to me today, though. I want to move fast and get started on the wall. We need everything on site ready for tomorrow dayshift.'

'Okay. I'll get Chris to come over with it and get down there to install it all for you tomorrow.'

Jim went down the pit, travelled in to the North End and straight out to 13's Head.

Terry Harrison, the deputy for the heading, came in early to do the pre-shift inspection for 15's Face and to supervise the three militant headers, Dan Crowley and his mates, Gary Hedges and Dave Coton who had started to work an early shift. It suited the headers to get up the pit at an earlier time, away from the throng of men that hit the pit bottom to form long queues waiting for the cages at the end of every shift and it suited Jim not to have Crowley spreading sedition amongst the face teams on their journeys in and out of the pit.

The big, old deputy moved, tortoise-like, to join Joey Bondar who was waiting for Jim at the entrance to the heading, looking worried.

'I was just callin' Control to get them to get you up 'ere,' Terry wheezed. 'I called Joey to come and have a look first, as overman.'

'Why, what's wrong?' Jim said.

'Water's started flowin' out the face of the 'eadin'.' Joey said. 'It's comin' out, right where the seam meets the stone floor.'

'Come on, let's take a look,' Jim said, leading the way.

Joey walked with him and Terry waddled along after them.

'How much is there?' Jim said.

'It looks like about a bucket-full a minute,' Joey said.

'Good morning gentlemen.' Jim said as he reached the three headers.

'Here he is, come to write us a wet-note to get us up the pit early today, working in all this water,' Dan Crowley said to his mates.

'No chance,' Jim said. 'I need you down here getting these footings in for the wall.'

'Why won't you let us fire some shots?' Crowley whined. 'It's like bell-metal.'

Jim noticed Crowley's use of the same pit expression as Willie Stokoe had used to describe the floor.

'We've got a thin pillar of coal holding back millions of gallons of water at high pressure.' Jim said. 'We've got to have those footings dug out but we can't do anything that could disturb the pillar.'

'Now keep quiet for a minute so I can have a look at this water,' he snapped.

'Aha,' Crowley crowed, 'there you are, he's admitted there's water in here!'

Jim shot a scowl at the header then returned to his inspection.

'Give it a rest will you, lads,' Joey said. 'This is serious.'

'This isn't looking good,' Jim murmured to his overman. 'The face of the heading looks like it needs holding together. We'd better put some strata bolts in it.'

Jim pointed out the positions for three pairs of holes to be drilled into the seam. The headers started drilling the holes. They made such hard work of it that Joey could see that his boss's patience had expired.

'Fuck me, it's like watchin' a bunch of schoolgirls fuckin' about,' Joey said. "Ere, let me do it.'

The overman grabbed the bull's-head boring machine off them and he and Jim worked it together, to get the holes drilled more rapidly. Each strata bolt was six feet long, with a steel core surrounded by a wooden sleeve. Jim used the boring machine, with an attachment, to spin them and secure them with three eighteen-inch long ampules of resin in each hole, leaving their threaded ends protruding from the holes.

The headers sawed three nine-foot split bars to length and used the bull's-head borer to drill a hole in each of their ends then screwed the nuts down to hold plates against the timbers to force the flat surface of each timber tightly against the face of the heading in an attempt at shoring it up and holding it together. This took three times as long as it would if done by motivated, competent facemen or headers.

Jim showed them how to place bags of cement to channel the water seepage so they could carry on picking out the footings.

12:32

Back up on the surface, as he organised the next shift's manpower with Mike Taylor, the afternoon-shift's senior overman, Jim told him to deploy men to continue digging out the footings for the wall in 13's Head, obtaining Mike's assurance that he would make sure they minimised the vibrations as they worked.

When the afternoon-shift officials had left the dirty offices, Jim phoned Todd Crisp.

'Todd, there's been an increase in the make of water in the heading. There's water actually running out of the face at floor level, about two or three gallons a minute.' Jim said. 'I could do with you coming to Whitacre with Chris Baxter tomorrow, so you can extend the standpipe fittings ready for us to get on with building the reinforcing wall.'

'I wish I could,' Todd said, 'I've already tried to change it, but Mr Wheatcroft's insisting I go to Birch Coppice and crack on with some exploratory boring we're overdue to do for them. But I've invited the NCB's National Bore-Master, Maurice Huntley, who's, effectively, my professional superior, and he'll be with you and Chris tomorrow.'

19:44

That evening, Jim received his usual daily end of afternoon-shift report on the phone from Mike Taylor.

'How have you got on with the footings in 13's Head?' Jim said.

'They've managed to get down a bit more so we've got about six, perhaps seven, inches of a trench dug into the hard floor.'

'That's less than half the depth I'm looking for. I'm going to have to get the nightshift to send some men in there to carry on picking it out.'

22:25

Having received no call from Willie Stokoe, Jim phoned the nightshift's senior overman and instructed him to send men to the work in 13's Heading.

He got the worrying impression that Stokoe was intending to disobey his orders and use explosives in the heading so he reiterated them in the clearest terms.

'Willie, I am giving you a direct instruction. I want you to make sure that no one fires any shots in that heading. Is that clear?'

'If you say so, Meestair Greaves,' Stokoe whined in his high-pitched, Geordie accent.

'I do,' Jim snapped.

The Inrush

Thursday, 28 March, 1985: 06:09

The dirty offices were packed with black-faced, nightshift officials, their tired eyes squinting, after their shift underground, against the glare of the room's lighting, as they briefed their pristine-skinned, dayshift colleagues.

Jim weaved his way through the throng of deputies and overmen, in his pit-black, and found Chris Baxter waiting, changed and ready to go underground, wearing NCB wellingtons instead of pit boots. They needed to get straight down the pit, extend the pipework on the standpipe in the face of the heading and start building the retaining wall. Jim just hoped they could get it all done in time.

He and Chris had just finished their mugs of tea and were about to head for the pithead when Joey Bondar sidled up behind Jim, laid a hand on his boss's shoulder and leant his head towards him.

Jim felt Joey's hot breath on his cheek as overman murmured in his ear.

'Boss. I thought you'd want to know. I just 'eard as Stokoe ignored what you told 'im. 'E sent the early turn 'eaders in with a can o' powder and told the pre-shift deputy to draw det's.'

'What?' Jim shouted.

Some of the officials paused and turned to look at him.

'If he has, I'll sack the bastard,' Jim snarled.

He grabbed the phone and dialled the number for 13's Head. It rang out for over a minute with no reply but Jim held on.

"Allo,' Terry Harrison's slow voice answered. '13's 'Ead.`

'Terry, it's Mr Greaves here. Have you fired any shots in the head this morning?'

'Yes, Boss.'

Jim closed his eyes and said nothing.

'Mr Stokoe, er Willie, told us to bring in powder and det's and fire the floor.'

Hearing no reply, Terry continued.

'He said the floor was like bell-metal and the nightshift couldn't make any impression on it. It'll be alright, I only fired a small round.'

'How many sticks?' Jim said through clenched teeth, his eyes still closed.

'Only four sticks in four holes.'

Jim suppressed a wince then shook his head.

'You were there yesterday when I said I wanted no shots fired in that head.'

'Yeah, but you can only obey yer last order, can't yer.'

Jim drew a deep breath.

'Terry, stay by that phone and keep your eye on the face of that heading while I travel in to you.'

'How many?' Joey said, when Jim had replaced the phone receiver.

'Four holes, four sticks.'

'Oh.'

'Where's Stokoe?'

'Lyin' low, over in Control.'

'Wait here. I'm going to sort him out, then we'll get down there.'

Jim made no attempt to check his anger as he strode through the lamp cabin and out across the pit yard car park to the main building. He stomped down the corridor and gave the Control Room door a shove. It flew open and banged against the table by the wall.

Jerry Malin, the dayshift Control Room attendant, seated at the console, spun round on his chair. The mechanical and electrical shift-charge engineers stood, leaning on the other side of the console.

'Where's Stokoe?' Jim growled.

''E made sharp when 'e seen you come stormin' out the lamp cabin door, Boss,' Jerry said.

'Oh, he did, did he?'

'Yeah. 'E said, "that young man looks like he's over-excited – time I was gone". Then 'e made sharp and fucked off.'

Jim pulled the door closed behind him and trotted down the corridor and out of the main entrance at the bottom of the building. There was no sign of Willie.

Jim couldn't waste time playing hide and seek with a grown man, he needed to get down the pit and assess the effect of that explosion in 13's Heading. He stomped back over to the dirty offices.

A couple of minutes later the door of the ladies' lavatories in the main office building eased open a couple of inches. With his escape route clear, Stokoe skulked out and left the building by its main entrance.

Back in the dirty offices, Chris Baxter introduced Jim to the National Bore-Master, a sorrowful-looking man called Maurice Huntley.

Jim picked up his flame safety lamp and hooked it onto his belt. He, Chris and Joey each carried a component needed to extend the standpipe in the face of the heading as they made for the pit-top, taking the Bore-Master with them.

Having alighted from the cage in the pit bottom, they rushed round to catch the last manriding train, to ride inbye along the Loco Road.

Disembarking from the train at the stopping point, they went through the small doors in the big ventilation barrier that separated the main intake from the return airway of the North End of the pit.

As they stomped up the steep to board the front carriage of the manriding train on the endless rope haulage, Joey laid his hand on his Undermanager's shoulder in an attempt at lightening to lighten his boss's grim mood.

'Fuck me, Bosso,' he muttered. 'That National Bore-Master's well named, ain't he?'

'How's that?' Jim said.

'Well, they must call 'im that 'cause he bores for England. 'E does look a tedious old twat!'

'Yeah,' Jim said with a light smile. 'I suppose it is a bit of an unfortunate title.'

At the front of the train, they found Mad Jack ensconced with a gang of mechanical staff that Jim noticed included his senior fitter, Paul Wood.

Jim's group greeted them and placed their pieces of pipe-work and other components in the train. As they clambered in, Jack called out with a grin.

'Come on Mr Greaves! The mining-side 'oldin' us up again.'

Fifteen minutes later, as they clambered off the train at the inbye end, Jim linked up with his old Canley colleague.

'What brings you up here today, Jack?'

'Tribology. We've measured an increase in ferrous debris in the oil sample on a gear-end on 15's maingate machine.'

'Already?' Jim said. 'You want to leave that alone, Jack. It's only just been reconditioned.'

'I know. That's what I'm worried about.' Jack said. 'Where are you bound for?'

'13's Proving Head. I need to get a wall up to secure the face of that heading. It's looking decidedly dodgy. And some idiot's had shots fired in there this morning.'

Jack grimaced.

When they got to the junction into the head, Jim, Chris and Joey laid down the items they were carrying. Chris and Maurice stayed at the entrance of the heading to check the inventory of supplies which had been delivered for them.

'You lads carry on into 15's,' Jack said to his little army of fitters. 'I'm gonna take a quick look in this 'eadin' with Mr Greaves.'

Jack never travelled without a companion from his own team so he thought again and called out to Paul.

'Tell you what Woody, you come wi' us un' all'.'

Jim inspected the floor as he strode up the short, incline. It was noticeably wetter than the day before.

Beyond the great bulk of the Dosco heading machine, Jim could see four cap-lamps, with little sign of movement, up by the face of the heading. As he led the way through the tight gap between the Dosco and the arch supports of the road, Jim was not surprised to find that the lights belonged to Terry Harrison and the three headers, Crowley, Hedges and Coton, who were all standing around, doing nothing.

'Those lads did a good job on nights with those shots,' Crowley called out as soon as Jim was in hearing distance. Jim refused to be drawn.

He stood and let his eyes roam over the face of the heading. He could discern no difference in its condition; the seam looked terrible, it was worryingly riven with breaks but he could not say, with any confidence, that it was noticeably worse than on the previous day.

Jim clambered up onto the boards of the drill-rig platform. Four feet off the ground, the cool blast of air from the air troughing hit him. His head was close to the roof, just under the steel bows of the heading's steel arch supports.

When the road had been driven, the cutting head on the heading machine's boom had failed to form a good arch profile because the weak, fissured coal seam had fallen away from the roof, on both sides, leaving a broad expanse of smooth, flat shale above the arches, like a well-finished, black-glossed ceiling.

Although the indolent headers had packed the shoulders of the arches with packing bags filled with stone, to support the wide, open surface, a lot of flat roof was visible in the three to four feet between the last arch and the face and in the metre-wide spacing between the second and third arches, where the corrugated sheets and packing had been removed to key-in the wall that Jim had ordered to be built.

Colliery officials were trained to use all of their senses, particularly their hearing, to keep the pit, their men and themselves safe. Jim had no idea what he expected to find up there but he needed to carry out a full inspection.

He listened hard.

He frowned. He was sure that he had just heard something, something weird and faint, but the rush of air being discharged from the troughing made it impossible to be sure.

He shook his head. If it wasn't his imagination, it sounded like nothing he had ever heard before, certainly nothing like anything you would expect to hear down a mine. And why had the noises only started after he had got up there, close to their source, up in the roof?

Then he heard it again, more clearly this time.

The hairs on the back of his neck sprang up.

Something hateful and malevolent, up in the roof, had sensed his presence and was indicating its indignation with a faint but rising chorus of mean, eerie whines and unearthly squeals. Whatever was making those noises didn't belong in this world, let alone down a mine; it sounded utterly alien, diabolical, even.

There was evil above them.

He leapt down from the high platform, bending his knees to absorb the impact as he landed on the floor of the heading, staggered a couple of steps then stood upright.

Crowley sniggered.

'Ha! You've got some money ain'tchyer?'

The three headers cackled at the old, familiar collier's taunt that attributed any supposedly undue caution to a person's determination to spend whatever money they owned in this life.

'You can laugh if you like,' Jim said, gazing up at the roof, 'but I just heard something fucking strange up there. And I didn't like the sound of it,'

'It was like nothing I've ever heard before,' he murmured.

The headers quietened themselves, still smirking but looking uncertain now. Crowley roused himself.

'Dah!' He jeered. 'You're lettin' yer imagination run away wi' yer.'

His two mates chuckled.

A sudden crackle followed by a fizzing hiss, from the roof above, silenced them.

Their eyes widened and shot upwards, their mouths falling open.

Particles and small lumps dropped out of fine fissures that were opening up in the two areas of flat, bare shale above them, running away from the face, scoring zig-zag patterns, back down the roof of the heading. In a few seconds, the previously smooth roof was scored by a mass of criss-crossed cracks.

As the rest of them ducked, Jim stood tall, having sensed something like this was about to happen.

'Whoa!' Crowley called out.

Joey, Jack, Paul and the three headers all stared up in awful fascination at the rapidly deteriorating mess above them. Terry Harrison, the big, fat deputy, was already off; waddling away at speed.

'What the fuck's that?' Crowley shouted.

Stinking water started to spurt out of some, then all, of the cracks in the roof and cascade down onto them.

'WHOA!' Crowley shouted again, louder this time.

'There you go,' Jim called out, in a satisfied tone. 'You're not fucking laughing now, are you?'

With a muted but loud, deep-noted thud, a ton of coal jolted away and rolled out of the face of the heading in one great, irregular block. Jim and the headers jumped back, barely avoiding being trapped beneath it.

The great lump left a gaping hole in the face of the heading. With another, sharper thump, a lump of coal, a foot wide, shot out of its mouth, about four feet off the floor. A stream started to flood out of the jagged maw, like a waterfall.

This broke the spell; the three headers darted off to make their escape.

It was almost comical to see them reach the constrained gap between the Dosco Roadheader and the side of the road, struggling and elbowing each other to be the first away and out of the heading. They caught and overran Terry, the deputy, who carried on waddling after them with uncharacteristic haste.

There was a swishing noise as water started to spray in jets from the coal around the borehole's standpipe in the middle of the face of the heading. Joey and Paul started to walk briskly to the walking route past the Dosco, casting terrified glances back over their shoulders. Jack stayed, motionless, watching Jim as he stood, observing the rapid disintegration of the face. As the water level rose over their mid-calves, Jack's self-control expired.

'Come on, Mr Greaves,' he shouted over the increasing roar, grabbing hold of his young colleague's arm and pulling him away, down the heading. 'You can't stay here!'

Jack shoved Jim before him, past the Dosco and down the forty metres of the heading, his fingers biting deep into Jim's arm all the way until they were out of the heading and standing on the bridge over the heading's conveyor belt.

Out in the main road there was no sign of the headers who had obviously kept running and were gone.

Jim looked at the big emergency sump hole he had ordered to be excavated under the junction. It had been filled in seconds, affording no useful protection whatsoever. The belt road down to the Main Intake belt road looked like a river.

'Boss, I'm goin' as well,' Terry said. 'It's three miles to the pit bottom but if we get goin' we might just make it.'

'You go, if you want to, Terry,' Jim said. 'But don't you dare take that train! Go out on the belts. I need the train here so I can try to evacuate the men off 15's Face and 14's development.'

Then he swung back on Mad Jack.

'Well done, Jack' he snapped. 'You've pulled me out of there and now I'm going to have to go back in. I have to know what's happening in there. I'm going to have to try to do something to hold it back.'

'Well,' Jack said, looking unnecessarily abashed. 'If you're goin' back in there, I'm comin' with yer.'

'I'll wait 'ere for yer, Boss,' Joey said. 'Just call me, if you want me to join you in there.'

The overman sat down in the dry road, still wide-eyed, on a crude bench formed by a plank resting on two oil drums, next to Colin Beighton, the electrician for the area.

Paul Wood stood at the centre of the entrance to the heading, watching as Jim and Jack jumped down into the water and started to wade back up to the source of the flood.

The depth of water running down the heading had increased in the couple of minutes they had been away from its face. Jim strode up against the flow with Jack following close behind him.

Paul started paddling up, pensively, behind them. He bent down and looked under the Dosco heading machine, between its caterpillar tracks.

"Scuse me, Boss,' he called out. 'Take a look at this.'

Jim rushed back to him and bent down to look. In the stream of water flowing under the machine there was the incongruous, ornamental sight of a fountain of water spraying up out of the floor, playing onto the underside of the conveyor slung from the back of the Dosco heading machine.

The deluge flowing down the road was rising rapidly. It filled the gap under the heading machine and swirled round its big, caterpillar tracks.

08:23

Jim gazed at the face of the heading, appalled; it was in a terrible state with the occasional piece of coal still dropping from it.

He splashed over and picked up the phone receiver from its white box on the last arch leg on the right hand side, nearest to the face of the heading. He dialled the number for the Control Room.

Jerry Malin answered.

'Hello Boss,' he said, before even hearing Jim's voice.

'Jerry, get a message out to all the faces and everyone down the pit to evacuate the mine immediately. The pillar between 13's Head and the waterlogged old workings is breaking down. Get everybody out of the mine as fast as you can. I don't think there's much time left.'

'Can you tell me what's happening down there?'

'No, just get the message to evacuate out as quick as possible! There isn't much time. Get someone else to find Mr Goodall and get him to phone me in 13's Head and I'll give him a report.'

'Righto, Boss.'

Jim started to stagger back down to the junction, followed by the Jack and Paul. The bleep of the phone rang out. Jim went back and answered it as Jack and Paul carried on wading back out.

'Is that Mr Greaves?' Harry Everitt's voice barked out of the phone's handset.

'Yes, Gaffer.'

'What's happening down there?'

'Despite orders I gave Stokoe last night, he had a round of shots fired down here by the early shift, this morning. The roof in the heading's opened up with cracks all over it, about a cubic metre of coal rolled out of the face, a few minutes ago, and we've got a stream running out of the middle of the face and water pouring out of the cracks in the roof. I doubt whether the protective pillar's going to hold for much longer. We need to get all the men out of the pit as quick as possible.'

'How fast is the flow of water?' Mr Everitt asked.

'I'd say several thousand gallons a minute. But we've got no way of measuring it.'

'Don't talk so daft, in a road that size, that would be up to your knees.'

'Yes,' Jim said. 'That's where it is, Gaffer.'

The phone went silent for a few seconds.

'Jim, I need to ask you. Will you see if you can get whatever volunteers you need and stay down there to do what you can to try to hold that water back long enough for us to get the dayshift to the surface before it comes in?'

'Yes, Gaffer. I'll do that for you.'

Jim put the phone down and hurried back to where Jack and Paul were waiting. They looked impatient, desperate to be out of the heading. He followed them down to the junction, surprised to find himself resolved to the commitment he had just given, reconciled to losing his life that morning.

At the junction he gathered the men around him; Chris Baxter, Maurice, the National Bore-Master, Colin and his former friend, Paul, Joey and Mad Jack.

'I've been on to Control and told them to evacuate the mine. The Gaffer's asked me to stay down here and do what I can to hold the waters back, to try to give the men time to escape. He's told me to get anyone who's daft enough to agree to stay and help me do it.'

They looked horrified at what they knew he was about to suggest.

'I'm not going to put any pressure on any of you,' Jim said. 'But I reckon we've got the right skills in this group. And I could do with your help, if any of you are prepared to volunteer to go back into that heading with me.'

They stared back at him with wide eyes, as if he were crazy. He had no idea what each of them would say; he had given his word to the Gaffer expecting to have to stay and face it alone, anyway.

Eventually, Joey broke their stunned silence.

'You're always gerrin' me into trouble, Boss,' he grumbled. 'This looks like another mess you've got me into. I'll stay with yer.'

'Yeah,' Chris said. 'Count me in.'

452

Paul and Colin both spoke quietly, avoiding each other's eyes.

'Yep,' Paul said.

'And me,' Colin muttered.

Maurice nodded looking reluctant and even sadder.

Mad Jack shook his head then walked over to the white box of the phone on the junction's upright stanchion and phoned the fitting shop.

'Hello Bill. Jack 'ere. Tell 'em I won't be out for the Mech' Eng' Meetin' this a'ternoon.'

He put the phone back on its holder and sauntered back over to where Jim waited, smiling at him.

'You knew *I* wouldn't leave yer. They don't call me Mad Jack fer nowt.'

Jim looked round at the group, amazed that they had all agreed to stay behind as casually as he had given his commitment to the Gaffer, grateful, beyond words; as grateful, in that moment, at the prospect of not being alone at the end as for any of their expertise.

He looked round them all, nodded then whipped the pocket book out of his shirt pocket.

'You'd better give me your tally numbers.'

He wrote them in his book as they called them out then went over to the phone, called the Control Room and gave Jerry the names and numbers of his volunteers.

They were still there, at the junction, in the mouth of the heading, when the men from 15's Face District started to pour out, on their way out of the mine.

Some of the first men were jovial, even when they saw the shocking sight of the swirling river running out of 13's Head, under the basic, open bridge and down the belt road.

Warbler Worrell bustled past without a word, casting a fearful look at the waters flooding out of the heading.

Des Proctor, the face chargehand broke off to shake hands with the Whitacre Heath men he knew, including Jim.

'Cheerio lads,' he said, with a silly grin. 'It was nice knowin' yer.'

'Fuck off, you morbid bastard,' Joey said as he shook his hand.

The last faceman out of the district was the oldest man on the face.

Horace Wood limped into the area lit by the electric lights that hung from the junction with the heading. Pulling up, he looked down and saw Mr Greaves standing back in the heading, up to his knees in the water. Before he could speak he recognised the figure, standing a couple of steps into the heading, facing his Undermanager with his back to the junction.

'Paul?' Horace said.

Paul remained motionless. Jim Greaves saw his fitter's grim mouth set in a tighter line.

'Paul,' Horace called again, a little louder. 'Aren't yer comin'?'

Horace didn't move, waiting for him to reply.

'Sorry Horace, 'e's stayin' 'ere. With the Boss.' Colin said. 'And the rest of us.'

Paul remained immobile. Horace took a step towards his son then stopped.

'Take care, Paul,' Horace said. 'Take care lads. Boss.'

Paul glanced sideways and saw the tilt of the head and the raised eyebrows under the peak of his young Undermanager's helmet. Paul knew what that expression meant; it was the same questioning reproach that his mother had always had for him whenever he was about to get something wrong. He knew exactly what his boss's look was intended to convey; was this how he was going to leave it, this last opportunity to close the rift between himself and his father?

Now, instead of the respect he had seen in his boss's eyes when he had volunteered to stay, he saw only sadness and disappointment. And those eyes continued boring into him.

Paul spun round and clambered up from the flood and out of the heading then strode off, outbye, towards the manriding train station.

Stan Townsend, the Face Deputy, had waited, a patient member of the audience to the little drama that had just been played out. He stepped forward.

'I've been all the way round the district, Boss,' he said. 'It's all clear, there's no one left in there.'

Stan looked unhappy. Jim titled his head, frowned and stared up more intently at the deputy. It was apparent that Stan was struggling to bring himself to say something more.

'Boss, I'm afraid there's some water running out of the roof on the face, near the tailgate end. About a couple o' gallons a minute, p'rhaps a bit more. Not good news. Sorry Boss.'

'Okay. Thanks, Stan.' Jim said. 'You'd better make tracks, you don't want to miss that train, when it goes.'

Stan leant down and shook his Undermanager's hand then did the same with the other Whitacre Heath members of Jim's group before stomping off without another word.

Now that they were left behind, despite the fellowship of their group, they all felt gripped by the sudden, lonely melancholy of being doomed and left to their fate.

At that moment, Jim realised that he had broken with the age-old tradition; the accepted practice in perilous situations of seeking volunteers without families. He was the only member of the team who wasn't married, the only one who would die childless down there.

He thought about it. The arrangements that he and Gabriella had made would all be wasted. There'd be no wedding for them, he'd miss out on that date by three months.

With evident reluctance, Joey stepped back down into the torrent, flowing out of the head and down the belt road, to join his Undermanager.

'This is going to gather in them swillies, further outbye,' he said. "'Ow long d'ye reckon we've got before it seals us in, Boss?'

'I don't know,' Jim said. 'I'll get the surveyors to work out how long it'll be before that happens.'

Paul returned without a word and hung his father's slim, maroon, leather, snap bag in the side of the road before sliding back down, into the heading.

'Right, let's start fighting back, men,' Jim said. 'The first thing I want to do is to try to reinforce the strata a bit, try to buy ourselves a bit of time.'

He turned away and headed off. The rest of the team followed him, wading up against the flow of water.

At the face of the heading, Jim picked up a steel drill-rod and climbed up onto the platform. Joey strode over to the left side, where an electric bull's-head and cable hung on an 'S'-shaped metal hook, from a strut between two arch legs. He grabbed the boring machine, heaved it up and chucked it to land, with a heavy clatter, on the platform's wooden boards, then clambered up to join his boss.

Jim attached the drill-rod to the borer and lifted it up, thrusting the drill-bit, on its end, against the coal. Joey joined him in taking hold of the borer.

Jim squeezed the start lever on the borer's right-hand handle to start drilling the hole into the coal for the first of the strata bolts they would use to secure more split bars against the face of the heading to try to slow the rate at which it was breaking down.

Every couple of seconds, the drill-bit would snatch and stall as it hit a fissure in the broken seam and they would have to wrestle the bull's-head as it bucked and fought against them like a ferocious beast trying to spin itself in response to the resistance.

Eighteen inches in, water started pouring from the drill-hole. They had tapped into a flow of water, amazingly close to the surface they were working, with their very first attempt.

'You want to get that borer out from under that water, Boss, or you'll get yourselves electrocuted,' Colin called up to them.

Jim pulled the borer off the drill-rod and slid the rod back out and rested it on the platform.

Joey and Jim hunched over the hole, gazing at it, disconsolately, as it ran like a tap.

'We'll never get any resin to 'old up there, it'll just get washed straight out before it sets,' Joey said. 'I reckon we're wastin' us time with this, Boss.'

Jim looked grim and thought for a few seconds.

'Right then!' He shouted. 'Chris, are you ready with your expandable plug?'

'Yeah. I'm ready,' Chris called back. 'Let's give it a try.'

'We need to move fast now, lads.' Jim shouted. 'We've got to try to slow the flow of water through the pillar, to slow the erosion.'

While Chris started to set up the drill-rig equipment, assisted by Paul who had jumped up onto the platform with him, Jim clambered down and he and Maurice showed the rest of the group what they were going to try.

The expandable plug was a short length of drill-rod with two powerful, sprung-steel fingers on its end. These were designed to stay pressed, firmly, against the sides of the borehole. At any point, if the drilling team started twisting the string of drill rods, clockwise, in the hole, the sprung-steel strips were intended to stop the end of the plug and its rubber sheath from turning. With the innermost end of the rubber sheath held firm, turning the drill rods would wind a nut and washer along a thread, behind the sheath, pushing the sheath's outermost end along the plug's steel core and causing the sheath to balloon out and fill the hole. The team liked the look of this; it gave them all their first real glimmer of hope.

'But we may have a couple of problems,' Maurice said. 'First, if the drill-rig has moved even a millimetre, we won't be able to line up the drill-string with the borehole.'

'And, second, if the seam is fissured all the way through to the old workings and this stream is bypassing the borehole completely, from the other side of the pillar, then sealing the hole will make no difference.'

456

'So what happens then?' Joey said.

'Then we're doomed,' Maurice moaned.

Jim thought, again, about the working of the old mine. It had all been mined by shot-firing faces with the explosions from numerous, massive rounds of shots. As a trained shot-firer, himself, he knew the order of distance that their shock waves would have affected. It would be far greater than the thickness of the pillar.

He shook his head, rousing himself.

'Well, thank you for that, Maurice. That's certainly cheered every-fucker up.'

Thursday: 10:19

Jim phoned the Control Room and spoke to Jerry Malin.

'How's the evacuation going?'

'All out, Boss. Just. Except for you and the lads down there. Three hundred and four men out, there's just the seven of you left down there.'

"Grief!" Jim thought, *"That was fast."*

No one had hung about.

'Are you and your team gonna start headin' out now?' Jerry said.

'No, we can't move away from the heading, things are changing too fast. We need to try to get it back under control before we dare leave it behind us.' Jim said. 'Is the onsetter up the pit as well?'

Jim didn't want the onsetter in the pit bottom; when the pillar yielded, it would only take a few seconds for the water to flood all the way to the pit bottom nearly three miles away and fill both of the half-mile shafts to the surface.

'Yes, Boss. Like I say, there's only you and your team down the pit now.'

'Good.'

'That's it,' Jim thought. *'That's most important thing done.'*

Thursday: 10:34

Chris Baxter had checked the stuffing-box which enabled the drill-string to be moved in the hole whilst minimising the water leakage. This was needed to maintain a workable environment around the borehole rather than to try to stop the deluge pouring out of the face at them.

Jack and Jim hauled themselves up to join Paul and Chris on the platform. They shoved the expandable plug through the cut-off valve and stuffing-box and into the proving borehole and then connected the first

457

drill-rod and started pushing it up the hole in the seam with the hydraulic rams. They attached a second rod and applied pressure again.

After less than nine feet the drill-string hit something solid and immovable. It was jammed, fast, in the hole.

Whitacre Heath Colliery, General Manager's Office. Thursday: 10:38

'All this water he's talking about,' Mr Hope said, frowning and shaking his head. 'It can't be coming from just an inch and a quarter borehole.'

Mr Everitt ignored him and bent over the massive meeting table to pore over the plans and seam sections that Derrick Hill had brought in from the Survey Office. He spun the plan round, showing the positions of the old workings in relation to the North End of the pit with the actual point that the borehole had broken through into them.

'So what we're saying is, we expected to find the old workings fifty-five metres away but the old pits had actually mined forty-five metres closer to us than we thought. That's an error of over eighty percent!' Mr Everitt said. 'The important thing is that what Mr Greaves has said all along is right, we haven't a clue where the rest o' those old workings are. The old plan shows that they were irregular shapes. They could be closer to the heading than ten metres at some points. And it sounds very much to me like that pillar is breaking down.'

He shook his head, 'And we know how much water we're looking at.'

'Yes, Mr Everitt, a billion gallons, perhaps even two,' Derrick said. 'At over four hundred pounds per square inch.'

Harry Everitt's face took on a haunted look.

'You know, the inrush at Lofthouse happened twelve years ago, this month,' he said. 'They kept it quiet but I 'eard at the time that the pressure of the water bursting in on the pitmen would have destroyed their bodies, stripped the flesh from their bones, atomised them and blasted 'em to the other end of the pit in seconds.'

'Fuck me, we could be looking at another Lofthouse.'

'Not exactly Gaffer,' Bill Metcalfe, the Services Manager, said. 'We've evacuated the mine.'

'No we fuckin' 'aven't, there's a young undermanager and 'alf a dozen men down there with him,' Harry snapped. 'It's exactly the same number as was down the pit at Lofthouse, by the time it broke in. The old, black magic number; seven.'

Derrick decided to change the subject.

'Mr Greaves asked us to calculate how long they've got, if the pillar of coal holds, before the water collecting in the lowest swilley cuts them off.'

'And?' Mr Everitt said.

'It's likely to be about ten o'clock this evening.'

The General Manager grunted.

'Well let 'im forget about that. There's no point in tellin' 'im and depressing 'em with that kind of miserable bollocks.'

Paul's and Susan's house. Thursday: 11:09

Susan heard the knock at the back door. Her heels clacked rapidly as she crossed the kitchen floor to open the back door. It was a surprise to look down the single step onto the side-path and see her father-in-law standing there, his discomfort apparent.

'Hello, Horace, love,' Susan said, struggling to read his expression. 'Sorry, come on in.'

'Look who's here,' she called to the toddler sitting on the lounge floor. 'It's your grandad.'

'Hello, me little duckie,' Horace said, walking over to Jenny and dropping down to the characteristic crouch of a true pitman with impressive suppleness for such an old man.

He touched the toddler gently on her cheek and let her grasp hold of his finger.

'Do you want a cup of tea or a coffee?'

'No, it's alright, Susan, love.'

He looked up at her with sad eyes.

Her heart went cold; Horace was there to tell her that Paul had been killed.

Horace saw the anxiety in her eyes.

'It's alright, me duck, there's nothing to worry about,' he said, standing up with a quiet grunt. 'I just came to let you know that Paul's stayed behind, down the pit.'

Susan breathed out in relief but looked confused.

'He shouldn't be back yet, anyway, he's on days,' Susan said, looking increasingly quizzical. 'I wouldn't expect him 'til after three, at the earliest.'

'No, I know,' Horace said unhappily.

He opened his mouth to carry on.

'But you're on days as well this week, aren't you?' Susan interrupted him.

'Yes, that's what I were goin' to say. The dayshift's all come up the pit early except Paul and a few others.'

'They're not trapped down there or anything, are they?'

'No, no,' Horace said. 'He was fine when I left 'im. It's just that a few of them had to stop down there. There was a danger of an inrush of water so they got everyone to come out up the pit. They've stayed down there to stop it.'

'Stop an inrush? Where from? What time will he be coming out?'

'I don't know what time he'll be coming up. I expect they'll just stay until they've done it.'

'What have they got to do?'

'I'm not sure. He didn't have time to tell me.'

'So you've seen him? And he spoke to you?'

''Yes, he spoke to me,' Horace said with a sad, little smile.

'He was on 'is way to the face I'm on now, at the North End o' the pit. Him an' 'is boss, Mad Jack, got diverted into this road where the problem is.'

'He's down there with Mad Jack!' Susan said, alarmed by the implications of that revelation.

'They'll be alright, there not doin' nowt daft or owt, down there,' Horace said. 'They've got Mr Greaves, our young Undermanager with 'em. He'll be in charge, 'e'll keep 'em safe. And Colin's stayed down there with them, as their electrician.'

'Colin's stayed down?' Who else is with them?'

They've got Joey Bondar, the overman, down there and Chris Baxter. Chris's a good man – 'e were a Whitacre man before he went to work in the Area drillin' team; a big, useful, professional lad.'

Susan stood quietly, her mind racing.

'Why've they got the young undermanager managing it?' She said. 'Shouldn't they have someone more experienced in charge?'

"E's on it 'cause 'e was the one as was down there, when it started to break in. But I'd say he's the best we've got, anyway, on somethin' like this. He knows what he's doin', underground.' Horace said. 'An' 'e won't panic.'

The vagueness of Horace's explanations was leaving Susan with too many speculations.

'Tell me the truth, Horace. Should I be worried?'

'I don't know, there's not much I can tell you, me duck. It's a new one on me, an inrush. I've been involved in fires and falls but never a flood.'

Susan looked horrified at Horace's last word.

'Where's the water coming from?' She said.

'It's comin' from the old Tame Valley Pit, an old disused colliery. It closed back in the 1920's. It was so old even I don't remember it.'

'Another pit?' Susan said. 'What are they doing, close to an old pit?'

'It's where the coal is, duck. We have to go after the coal that's left.'

'How much water is it?'

'I've no idea,' Horace was able to answer truthfully.

He didn't have any information but he knew that the old pit was under the massive lake in the old quarry and feared they could be facing a colossal volume of water or mud under pressure. And like most pitmen, he knew all about that infamous Lofthouse disaster.

Susan caught a glimpse of enough of the worry in his face.

'It's a lot, isn't it?' She said.

'I honestly don't know. But I reckon it could be.' He said, nodding reluctantly.

'Look do you want to come over to our 'ouse to wait for 'im?'

'No, thanks Horace, I'd better wait here. I need to be here for him when he gets back.'

Susan walked over to the sink and looked out of the window.

'How will I know what's going on or when I can expect him back?' Susan said. 'Is there someone I can call?'

'I don't expect they'll tell anyone anythin' 'til it's done. They never do.' Horace said. 'They never want the papers or radio and telly getting wind that something's up 'cause they always get it wrong an' over-dramatise it.'

Susan turned back to him, shaking her head in disbelief at this.

'You could phone Control, I mean the Control Room, but I'd leave it a while,' Horace said. 'Paul wouldn't want any of us fussin'. ''E'd want us to leave the men in the Control Room and 'im and the rest o' them underground to get on with the job.'

'I bet they'll give you a ring if – anythin' changes. Or if it goes on for too long. Or when they know what time they'll be out, like. But in the meantime we just need to wait and be patient.'

Susan could see that she had no choice.

'Are you sure you don't want to come round with Jenny to us, 'til he gets back?' Horace said.

'No, it's kind of you to offer, Horace. But he'll be expecting us to be here for him when he gets back.'

'Alright, me duck. But you can drop round anytime for a cup of tea or a bite to eat if you get fed up wi' waiting here on your own.'

'Who's going to tell Diane that Colin's stayed down there?' Susan said.

'I was going to go and tell her meself,' Horace said. 'When I'd let you know, like.'

Susan shook her head and thought of what she had said to Pat and others after her appearance at the union conference. It was all supposed to have changed their lot forever; they would have their own parts to play, not just acting as supportive women, worrying and waiting for their men to come back safely. Whatever dangerous action was being played out, somewhere a few miles away and half a mile underground, had dragged their lives right back to where they'd been before the strike.

Susan thought about offering to break the news to Diane herself.

'It's probably best if you go and see her, Horace,' she said. 'Tell her not to worry, Paul and Colin can fix anything between them and they've got a good team with them. And tell her I'll call round for a cup of tea in a bit. But I think it's best if I give you a chance to talk her through it first.'

'Alright, me duck,' Horace said, hobbling over to the door. He stopped and turned. 'Let us know if you hear anything and I'll do the same.'

Horace walked up the path, along the side of the house across the front garden, and let himself out of the garden gate.

As he set off down the pavement, he remained convinced that he had done right thing, by not telling Susan that Paul had wanted his dad to tell her that he loved her. That would have really set the alarm bells ringing.

Paul hadn't actually said those words but Horace knew what he had meant, holding Horace's handshake and looking into his dad's eyes, a couple of hours earlier.

'Tell Susan,' Paul had muttered. '…Yer know.'

'I know, lad. I'll tell 'er.'

Horace could only hope that he would never have to pass that message on, that Paul would still be coming back to tell her himself.

13's Head. Thursday: 11:34

They all heard it and looked up; another deep, dull thump, some way behind them, up in the roof, above the corrugated sheets, back down the slight incline of the heading. Something had obviously just given. Water started running out from the edges of the corrugated tins over an arch, about twenty yards back. No one commented but they all know what it meant; the water was bearing down on them, rushing through the strata all around them and breaking it down.

There was an unpleasant, crackling noise. The water flow from the edges of the tins increased to a bizarre, picturesque cascade, a delicate curtain of water falling all the way across the heading, closing it in behind them.

'It's like the whole ground's on the move,' Joey shouted down from the platform.

'We just have to hope it holds together a bit longer for us,' Jim shouted back.

Maurice called to them from the driver's seat at the back of the heading machine, that he had climbed into to escape from the water.

'The problem is – you've restricted that hole with that plug and you're diverting the flow and all that pressure into the fissures. If you seal the borehole you could just blow the coal pillar away.'

'Well, we'd better just do fuck all then hadn't we,' Jim snarled, not even bothering to look in the Bore-Master's direction.

The expandable plug had continued to defy their efforts; it was impossible to ease it or even force it past where it had become jammed.

'Let's try to pull it out,' Jim shouted. 'We need to try to open up the hole with the drill.'

With difficulty they managed to withdraw the drill-string and strip out the two drill rods they had used.

When they had recovered the expandable plug from the borehole they all stared in horrified fascination at the state of the dense, rubber sheath. It looked like it had been mauled by a shark.

'How the fuck has it got in that state?' Joey said. 'It looks like somethin's been chewin' it.'

There was no reply. No one knew what was going on in front of them. What they were confronting was new; unimaginable and terrifying.

Paul looked at Jim. They could see it in each other's eyes, they had both realised the same thing; they were all starting to act as though the monstrous threat, thrusting towards them, was some kind of enormous, terrible, animate being.

The drill-string thrust in as they re-bored the hole. It felt like they were doing something positive and lifted the spirits of everyone in the team, except Maurice.

'This is alright as long as you're in perfect alignment with the hole,' he called up to them from the driver's seat at the back of the Dosco. 'Otherwise you'll just be drilling another hole and you'll never seal the original.'

'Good point, Maurice,' Jim shouted down from the platform, with heavy irony. 'Thank you for that.'

'Let's leave it at that,' Jim said, with six rods up the hole. 'We've drilled about thirty-three feet, that's about a yard from the end of the borehole. Let's try and seal it again.'

They withdrew the drill-string, removing the rods, one after another. When they had recovered them all, they put another expandable plug onto the rig and started easing it into the hole, a rod at a time. With the plug pushed thirteen feet into the hole, the water from the standpipe and the face of the head slowed and virtually stopped.

'That's amazing,' Joey shouted. 'Do we try to seal it with the expanding plug-thing now?'

They discussed it and agreed that, with the curtain of water still cascading down behind them, they needed to try to find a point closer to the old workings to achieve a seal. As soon as they pushed the plug forward the water started flooding out again.

'I think there's even more water coming now,' Maurice called out from the back-end of the Dosco.

13's Head, Thursday: 12:39

'This is now more of an art than a science. We've got to rely on our intuition on this.' Jim said to Chris, after an hour spent trying to find a point where it felt like the expandable plug's sprung strips might bind.

After more, gentle manoeuvring of the plug in the hole they found that the water slowed again when the plug was nineteen feet into the hole. They tried screwing the string to expand the plug. The flow from the standpipe increased to more than it had been at any time.

'How the fuck can that 'appen?' Joey moaned.

The main offices. Thursday: 12:45

Veronica, on Whitacre Heath's reception and switchboard, had started receiving calls from the local media, just over an hour and three quarters after the dayshift had left the team attempting to hold back the flood and evacuated the mine.

Initially, when told of their calls, Mr Everitt had instructed his secretary, Angie, to "tell 'em bollocks", which she had translated into a statement of "no comment".

Ninety minutes later, at a quarter to one, the Colliery General Manager was forced to relent. A reporter from the Coventry Evening Telegraph was threatening to run with what he had got from whichever collier had been his informant. He had succeeded in making it apparent that this was likely to prove far more sensational than anything Harry Everitt would want to see published.

Like most colliery general managers, Harry preferred to keep a low and positive profile in the local press. But he had the added imperative of avoiding too much attention being drawn from the Coal Board's senior management to the incident developing at his pit; he didn't want the national level directors having any reason to question Whitacre Heath's viability and considering it as a target for inclusion on their closures hit-list.

'You'd better let 'em in Mr Hope,' Harry told his Number 2. 'You'd better talk to 'em. You'll be more diplomatic wi' the fuckers than I would. And it'll be good experience for you, for when you get a pit of your own.'

Tim Hope smiled, thinking that the experience would be superfluous for him, when he became a general manager, if he did the same as Harry Everitt and delegated such tasks to his deputy.

Mr Hope faced the press in the conference room. A reporter and photographer from the Coventry Evening Telegraph and the same from the Tamworth Herald were joined by an interviewer and recording engineer from Radio Mercia.

'We understand that you've got billions of gallons about to burst in from another old mine,' the reporter from Coventry's daily paper said. 'And you've had to sacrifice a team of men to hold it long enough for the rest of the men to escape.'

'We've got a problem we're sorting out,' Mr Hope said. 'But I think someone has been trying to impress you by being overly-dramatic.'

'What about when it's likely to blow?' The Telegraph man said. 'The bloke we're speaking to seems to be confident that the water will break in by two o'clock this afternoon.'

'I think you need to be careful,' Tim Hope said. 'No one knows that it is going to break in and we're determined to stop it, so there's no way anyone could be specific about timings like that.'

'It sounds to me like you may have had someone over-stating things to you in order to get his name in the press. The rest of the colliery's management and I have known of the presence of the workings all along and we took the precaution of proving their position with a borehole. The reason we've played safe and withdrawn the workforce was a precaution because monitoring arrangements we put in place indicated that the make of water was increasing.'

'But isn't it true that the men left behind underground have been sacrificed to save the rest of the men?' The Tamworth Herald reporter said.

'No, we've kept a small team on site with the right mix of skills to make the pit secure,' Mr Hope said. 'I can assure you we don't put our men's lives in danger knowingly.'

'And who are they, these men,' Radio Mercia's interviewer said. 'Can you tell us who they are and where they're from?'

'No, that wouldn't be fair,' Tim Hope said. 'Not without their permission.'

'Is this the end for Whitacre Heath Colliery, Mr Hope?' The Telegraph reporter said.

Mr Hope took a slow breath.

'We have no reason to believe that the water problem will result in the closure of the mine. Gentlemen, I need to ask you to be responsible about this; please refrain from overstating the risks, particularly for the sake of the families of the men who've volunteered to seal this ingress of water.'

Paul's and Susan's house. Thursday: 13:56
Susan and Jenny had returned from seeing Diane a few minutes earlier. Susan had stripped Jenny out of her insulated suit, lowered her into her high chair and given her a drink in her plastic beaker before turning on the local radio.

The DJ gave a bit of light chatter before playing a Golden Oldie, "Here Comes the Sun" by the Beatles. Normally, Susan would have been singing along with that sort of song to amuse Jenny.

As the song was faded, the DJ said, *And now it's time for the Two O'clock News.'*

Straight after Radio Mercia's News jingle, the newscaster's opening announcement was about the incident.

'One of North Warwickshire's coal mines has been evacuated due to the threat of billions of gallons of water breaking in from the Whateley Water Park.'

'Whitacre Heath Colliery, nine miles south of Tamworth, employs nine hundred and fifty men. The colliery's management evacuated all of the mine's dayshift as soon as they arrived at their coalfaces, underground, this morning. A senior manager admitted that the management had been aware of the risk since an underground borehole hit the

waterlogged old workings of a disused pit under Whateley Water Park some weeks ago but they had continued to work the mine.'

'Our reporter, Jim Latimer, spoke to overman, Vic Worrell, a manager at Whitacre Heath.'

A recording of Worrell's excited voice burst in.

'I received a call from the Control Room, as soon as we got to the face, tellin' us I 'ad to evacuate all the men as quick as possible because of the danger of an inrush. I came past the team that volunteered to stay to buy us time to get out on me way out. They're brave men. I wouldn't have wanted to stay down there to face that lot when it bursts in…'

The radio station had cut the interview, just before Worrell had said, '… no one in their right mind would.'

'Stay tuned,' the newsreader said, *'as we keep you updated on developments with this dramatic story.'*

The news moved on to report on a school orchestra in the area that had gone through to the final stage of a national music competition. Susan left the radio on to listen out for any more information from the pit, oblivious to the remaining news items.

13's Head. Thursday, 14:16

Jim and Joey crouched on the platform.

Paul called up to them, 'Shall we stop for a bite of snap?'

'Most of us ain't gorr any,' Joey said.

'I've got mine and me dad's,' Paul said. 'We can share that.'

'And there's my snap as well,' Colin said.

The strong flow of cool air from the air-ducting had been playing on their drenched clothes all the time they had been fighting the inrush, chilling them.

Jim said, 'Let's go down to the junction and get out of the draught from the fan and warm up for a few minutes.'

They were all subjected to another drenching as they ducked and jumped through the curtain of water falling from the roof, further down the heading.

Under the overhead lighting installed at the junction, they all turned off their cap-lamps, to conserve their batteries' power. Jim and Jack sat on top of two, large, oil drums and the rest sat on two benches at the side of the road, with their backs against the tunnel lining of corrugated sheets.

Colin offered around tinned-salmon sandwiches and Paul handed out a mixture of cheese and ham sandwiches to the team and a Club biscuit to Mad Jack.

Paul offered Jim a sandwich.

'No, don't bother about me,' Jim said. 'Save it for the workers.'

Jack broke his biscuit and tossed half to Jim.

They sat quietly, taking a fifteen-minute break in their soaking clothes. Paul took his dad's thermos flask of tea from the maroon snap bag. He opened it and filled its two plastic cups then passed them around so that everyone could take a slurp.

When they started again, they were refreshed and cheered a little by a bite of food and the brief respite from the cold.

But Jim felt that there was no logic to what they were doing. They would have the plug in one position that looked promising, try to find a better point, deeper into the hole, fail and go back to the previous, more promising point only to find that even that now looked hopeless.

Throughout it all, the colossal weight of the immense body of water bore down on the slender, fractured pillar of coal to threaten them, like a time-bomb with a fuse of unknowable length.

Whitacre Heath Colliery Control Room. Thursday: 14:37
Tim Hope walked into the control room.

'We're famous.'

'I know, I 'eard it were gonna be on. We switched the radio on and caught it,' Peter Linton, the afternoon-shift Control Room attendant, said. 'I bet that cheered the Gaffer up.'

'Not half. It was just what we could have done without. The Gaffer says he's going to flay that idiot Worrell alive when he gets hold of him.'

'He wants to sack 'im, burn 'is clothes then send 'im on nights regular,' Peter said.

Tim Hope walked round to lean on the other side of the big console, with his back to the window. One of the surface phones on the console rang. Peter picked up the handset and answered.

"Ullo, Control.'

As he listened to the person talking over the phone his eyes raised to meet those of Mr Hope. Peter raised his eyebrows and gave a downward grimace with his mouth.

'What is it?' Tim Hope said, quietly.

'Hold on, me duck,' Peter said. He put his hand over the phone's mouthpiece. 'It's Veronica, on the switchboard, she says Susan Wood, Paul Wood's missus, is on the phone, asking to speak to Control.'

Veronica put Susan's call through.

'Hello, Susan love,' Peter said. 'It's Peter here, Peter Linton. I expect you want to know about Paul.'

He paused while Susan spoke, then said, "Old on a minute, me duck.'

He covered the phone's mouthpiece again.

'She says she doesn't want to bother us,' Peter said. 'But she just wanted to sort out 'ow we let 'em know if somethin' goes wrong.'

'Hand it over, Peter, let me talk to her,' Mr Hope said, holding out his hand for the phone receiver. He took it from Peter and put it to his ear.

'Hello Mrs Wood, it's Tim Hope here, Operations Manager at Whitacre Heath.'

'Hello Mr Hope,' Susan said. 'Sorry to trouble you.'

'You want to know about Paul?'

'I know Paul wouldn't want me bothering you but his dad, Horace, said that I should contact the pit's Control Room if I needed to know anything.'

'Yes, that's fine. Do you want to know how they're doing?'

'Well, yes, it would be nice to know something. I know the press and radio can be a bit dramatic but I heard a report on Radio Mercia a bit ago and, to be honest, it sounded rather worrying.'

'But the main reason I phoned was because if something did go wrong, down there, I just wanted to ask if you could let me know and leave it to me to tell Colin Beighton's wife, Diane. It's just, she's a friend of mine and she's – she's not been well and she's not terribly strong so I'd just like to make sure I was with her or I could let her know, if something did go wrong down there.'

'That all makes sense and I'll make sure that if anything were to go wrong that we'd do exactly as you've said. But I really don't believe it'll come to that. We've got a good team, the best possible team, down there. They knew that they could come away and leave it but it was their decision to stay down there and I know they believe they can sort it out. Let me tell you what's really going on down there.'

'The old workings are some way in front of one of the faces and we proved their position with a borehole and this showed that they contained water. We've been trying to drain the old workings but we started to suspect that there was a problem with the hole and the strata, early on in

469

the dayshift, so what Paul and the team are doing is trying to seal off that borehole.'

Susan felt no more assured; she could tell that the Operations Manager's account was as moderate and understated as he could reasonably make it, without actually being untruthful. But she did feel a little better for being able to talk to the men on the surface who were in direct contact with Paul and the team. Suddenly, her husband didn't seem quite so remote and lost to her.

'I'm not going to pretend that what they're doing is without danger,' he said. 'But I've got every faith in Paul and his manager and the rest of the team down there. And what they're doing is vital for the future of Whitacre Heath.'

'So, I don't suppose you have any idea what time they'll be able to come out?' Susan said.

'No, I'm afraid I don't. They're still trying to find the best way of sealing it off fully but I will make sure you know as soon as anything changes and I'll make sure that my colleague Bill Metcalfe, our Services Manager, calls you at the end of afternoon-shift anyway, because he'll be in charge of Emergency Control until then and then Ken Goodall, our Undermanager will be taking over on nights.'

'Alright,' Susan said. 'Thank you Mr Hope. I'll wait for Mr Metcalfe to call.'

'Alright, Mrs Wood, thank you. You can expect the call sometime between nine and half past this evening.'

'Oh, okay. Goodbye Mr Hope.

'Goodbye Mrs Wood. And don't worry, I'll make sure you're kept informed.'

Tim handed the receiver back to Peter.

Both men looked uncomfortable.

'You'd better get them to get me the numbers for everyone's contact – next of kin. I'd better give all of them a call and give them the same message,' Mr Hope said. 'That radio news broadcast is likely to have set some hares running. It'll worry everyone when they hear about it.'

Susan put the phone down and sat down on the settee. Her sharp gaze drifted out of focus.

The best thing she had heard was that the colliery management seemed to believe that it was likely that they would still be in contact with Paul and his colleagues in seven hours' time. But she did find it ominous that none

470

of the rest of the management were going down the pit to relieve or help Paul and the team that had been left behind with him, underground.

13's Head. Thursday: 17:10
Jim had been receiving brief phone calls from his management colleagues, in the Control Room, every half an hour, demanding updates. Mad Jack had answered the last one.

'Blue Elephant Tandoori, Atherstone, can I 'elp yer?'

The calls had stopped over an hour ago. Jim assumed they had got bored with hearing his confusing accounts of expandable plug movements or perhaps his refusal to accept instructions from a surface-based committee on what point along the hole to try to expand the plug had offended them.

They were on their own.

Some of the coal held up by the split bar secured with strata bolts across the top of the face dropped away as though it had just got bored with hanging there. The face was obviously less secure, as a consequence. Jim reflected that, whereas he had been intending to reinforce the suspect pillar with five metres of concrete, they'd just lost another half a metre; not much, on the face of it, but five percent of the pillar's total width.

It looked like they didn't have much time left.

'Bugger!' Joey uttered, on behalf of all of the tired, wet team.

Thursday, 17:47
The drill-string, with its expandable plug, was thirty-one feet into the hole, the furthest they had managed to get it and five feet from where the hole ended in the waterlogged workings, the source of the stinking, cold water that continued to rush at them.

'Give it a turn,' Jim said.

They rotated the drill-string slowly. There was no resistance, whatsoever. They kept turning; nothing.

'It looks like the hole's been eroded by the water that passed through it. Not surprising, given the volume that's come through it, even in the short time involved, and the pressure that's forcing it through.' Jim said. 'Let's start pulling it back and try to find some resistance to turn against.'

When they had drawn it out and removed four of the five-foot drill rods the water flow slowed considerably.

'Let's try here,' Chris Baxter shouted.

Jim joined him and Mad Jack on the platform as they started to turn the drill-string to try to get the plug's springs to bind against the wall of the borehole.

'Come on, bite yer bastard.' Jack growled.

They kept turning the rods.

'It's no use, there's nothing there,' Chris said.

They all knew they had to face it; all their attempts at securing and expanding the plug throughout the length of the borehole had failed.

'Damn,' Jack said, as he started to climb off the platform. 'I've just remembered something.'

He went to the white box of the phone at the side of the heading and phoned the fitting shop.

'Hello Bill, it's Jack, 'ere. I need yer to do me a favour. I was supposed to be taking our Gert' out tonight, to celebrate our anniversary. I won't be able to do it now, so could yer get someone to sort out a nice bunch of flowers for her at the earliest opportunity. Put something suitably sloppy on the card – I can't come up wi' it now, I've got a bunch of jokers listenin'.'

'You'll have to take her out tomorrow.' Jim called over to him, fooling no one.

He turned to Chris.

'I think it's time we tried to ram your rubber bung up the hole.'

'You only get one chance with that one, you know,' Maurice told him

'That's true,' Chris said, 'But I think Mr Greaves is right. Look how fast things are deteriorating. We might as well try it now. It's gonna be our last chance, anyway.'

Jim knew that the rest of the team would be thinking, like him, that if they'd cut and run at any time up to ninety minutes ago, the time it would have taken to get to the pit bottom on foot, they could have made it out and up the pit safely. But now the strata appeared to be deteriorating too quickly. To leave it now and hope for the best would be madness. All the way they would be expecting the massive flood to hit them and eliminate them at any moment. And anyway, it had always been Jim's hope that they could save not just the shift of men who had evacuated the mine successfully but the pit, itself.

Jim phoned the Control Room, reported the disappointing position and informed them of his decision to cease trying the expandable plug in favour of trying the bigger diameter rubber bung. Mr Everitt walked into the Control Room just after his two deputy managers had held a quick discussion on Jim's proposal.

'We want you to continue to persevere with the winged expandable-plug,' Bill Metcalfe told Jim over the phone.

As Jim was starting to explain, angrily, to Bill why he would not be obeying that instruction, Mr Everitt shouted at his Services Manager.

'Will you, leave that poor fucker alone! He's the one who's down there. 'Im an' 'is men are the only ones who know what it's like. 'E knows better than any of us what e's doing down there. Tell him I say it's up to 'im what 'e does next.'

'Hold on Jim,' Bill interrupted Jim's raised voice on the other end of the line. 'The Gaffer says you need to decide what you do next. Just keep us informed, will you?'

'Right, thanks, will do.'

Jim put hung the phone receiver up in its box.

'Hallelujah,' he muttered.

Thursday, 18:38

'Right, let's give it our best shot.' Jim said to Chris and Jack, having pulled himself back up onto the platform to join them.

Chris and Jim screwed the large-diameter, rubber bung firmly onto the first drill-rod and placed it into the standpipe. The first metre of travel was easy. The torrent gushing from the standpipe reduced but the increased resistance in the borehole caused the water to find other paths, increasing the amounts spraying from around the standpipe and out of the face of the heading.

The two hydraulic rams on the drill-rig, pushing the drill-rod and bung into the hole encountered increasing resistance. The bung wasn't moving, it was stuck, fast, in the borehole. The amount of water bypassing the bung increased again, flushing through the strata and spraying fiercely out of the fissures in the face and the gaps in the resin securing the standpipe within the seam and blowing lumps of coal, of various sizes, off the face.

'Let's try to rom it 'ome with a bit more pressure,' Jack shouted.

Paul turned up the valve on the hydraulics operating the rig then Chris operated the rams again. The bung and rod didn't move, but the four Dowty props providing the anchorage for the drill-rig between floor and roof bent alarmingly as the rams pushed against them.

'Whoa! Hold it!' Jim shouted, 'We'll knock the rig down or put it out of line with the borehole. Pull the bung back out before we blow the standpipe out or break the strata up anymore.'

The Shepherd

I am the good shepherd: the good shepherd layeth down his life for the sheep.

13's Head. Thursday: 19:09

Jim went to the phone and reported the failure to the Control Room. He hung the phone back up in its box and closed his eyes. That was it; they were out of options.

In desperation he tried to pray, silently, for salvation or inspiration. He couldn't do it, he couldn't clear his mind with the rest of the team around him, watching him and relying on him.

'Hold on a minute,' he said. 'I'll be back in a bit.'

He hurried away, down the side of the Dosco, and jumped through the water pouring from the roof, behind them. They saw him reach the junction and turn right, not heading for the way out of the mine but going further inbye, towards the face.

'Where's he gone?' Maurice said.

''E's probably goin' to earth,' Joey said, using a euphemism for defecating underground.

With the end approaching, Jim had to do it.

He had turned right and headed inbye so they wouldn't suspect that he was making a run for it and heading for the shaft. Perhaps that was what they should do. But they were out of time, they'd never make it and he couldn't bring himself to leave that awful mess, uncontrolled and breaking down behind them.

He started to run, stooping under the twisted, steel arches, putting some distance between himself and the heading.

It felt wonderful to be away from that hell-hole and the noise of rushing water for the first time in eleven hours, to be running away from that terror.

He rounded the curve in the roadway then stopped running, stopped pretending. He was going nowhere. It was all over; they'd had it.

A quick look over his shoulder confirmed that he would have the moment alone he needed.

'How did I end up here?' He whispered, shaking his head as he scrambled down onto his knees on the dirt floor.

He thought about his home, friends and family. Then he thought about Gabriella. He would never see her again

"Stop that nonsense," he thought, he hadn't come here for that.

He should have known that this was where his ideals would get him; that his destiny would be to die young, making a sacrifice. But it still felt so premature. He had thought he would have time to do so much more.

The worst of it was that he had always imagined that if he were to be killed down the pit that it would happen suddenly, without warning, not with all this time to taunt himself and contemplate it all; never having another night in the pub with his mates, never seeing the sunshine, his home, his family or Gabriella again.

Him and Gabriella: he had pretended that he wasn't that important, nor she nor they as a couple. He had disciplined himself to treat their life together as secondary to the pit and now he had lost it all, both that life and the pit.

At least there had been a point to it, they had had to save the whole shift of men and it had been right to try to save their pit. Someone had to do it and, realistically, it was never going to be anyone else.

He closed his eyes.

"I am the good shepherd: the good shepherd layeth down his life for the sheep."

His silent effort in the heading had felt inadequate. This prayer needed saying out loud. He would only have this one, last chance. He had to make it count.

Head bowed, hands clamped together, he took a deep, earthy breath and rattled it out:

'Help me find a way to beat this. Let us live.'

He pictured Gabriella, smiling in her surface world.

'Just let me live to see my wedding day.'

Transported by the prayer, when he opened his eyes it was a bitter, little surprise to find himself still underground, still in deep trouble.

But his prayer had calmed him.

He got to his feet, knocked the dust off his wet knees, they mustn't know what he'd been doing, and started the walk back to his team in the heading, reconciled to oblivion now.

He stopped short.

He felt the reproach, he'd nearly forgotten her again.

It always seemed easiest to neglect her when he needed her most. Taking hold of his guardian angel's pendant, hanging from his neck, feeling the engraved image against his palm, he closed his eyes and murmured in a gabble.

'Saint Barbara, Patron Saint of miners, be with us tonight. Bring my men to safety. Help us to save our mine. Help us to save ourselves.'

He opened his eyes and strode forward.

They were all watching for his return, except Maurice who had climbed out of the water hours ago and taken up residence in the dry seat in the driver's position on the Dosco heading machine, facing towards the face of the heading.

The boring expert sat, slumped forward, his head in his hands.

As Jim strode back through the curtain of water behind them, the rest of them all saw the difference. Whatever he'd been doing, he appeared transformed; full of energy and confidence.

As he drew level with Maurice in the driver's seat of the Dosco the Bore-Master moaned.

'It's finished.'

'Is it fuck,' Jim snarled, casting him a glance. 'If that was true we'd 'ave lost the pit.'

Maurice nodded.

'In my professional opinion it looks very much like we have,' he wailed.

'Well, in my professional opinion, that's a load of bollocks,' Jim snapped. 'But if that's what you believe, you'd better fuck off right now. I'll phone them on the bank, tell 'em to expect you and to wind you up the shaft, if you can get to the pit bottom and call the banksman on the phone in time.'

'Go out on the route we came in on. The swillies will have water accumulated in 'em so you'll have to wade or swim up to a couple of hundred yards in places, if they're not already closed off.'

Maurice did not need to be told twice. He clambered down from the Dosco, dropped into the water with a splash, then stumbled off down the road, with the flow of water, exhausted.

Jim did not insult the rest of his team by asking if any of them wanted to join the National Bore-Master in his escape attempt.

'Right,' Jim shouted. 'Can anyone drive a Dosco Mark Two A?'

'I'll have a go,' Colin said.

Jim explained the inspiration that had come to him, as he was walking back to the heading. They would place a large piece of wood across the two Dowty props that acted as the back legs of the drill-rig, then manoeuvre the Roadheader to position its cutting head, on the end of its boom, against the timber, to stop the props from bending under the back-pressure from the drill-rig's rams.

'Chris and Jack,' Jim said. 'Can you get a sleeper ready to span the two legs?'

Jim knew that the big heading machine had not been used or moved for months.

'Has the Dosco got power to it?' He said.

'Yeah, it should 'ave,' Colin said.

Colin climbed up into the driver's seat and tried to switch on the machine but the circuit breaker tripped out as soon as the spotlight on its boom flickered.

'It'll probably be too wet,' Jack called down from the platform.

'Nah, I'll gerrit goin',' Colin said, jumping down from the Dosco then trotting downstream and out of the heading to go to the electrical panels that powered it.

A quarter of an hour later, just as Jim was about to go to see what Colin was doing, the spotlight on the Dosco came back on and, a couple of minutes later, Colin leapt back through the water falling behind them.

'Right, let's try again,' he muttered, clambering back up into the driver's seat then powering up the machine.

'You're going to have to be bloody careful,' Jim called to him. 'If you clout the Dowty trees you'll put the drill-rig out of line – and then we'll be completely knackered.'

Despite the awesome number of unlabelled control levers arrayed in two ranks in front of him, Colin showed no lack of confidence in being able to achieve this.

He nudged the levers that controlled forward motion. The great, tank-like Dosco lurched forward, its cutting head striking one of the Dowty props a glancing blow, then sat back on it tracks.

'Whoa!' Jack shouted, having narrowly escaped being crushed by the powerful machine. 'That's fucked it.'

'Let's hope it ain't,' Jim shouted. 'Colin, you're going to have to drive it a fuck of a lot steadier than that!'

'Sorry Boss,' Colin said, looking totally unperturbed. 'I've got the 'ang of it now.'

'Fuck me,' Joey said. 'I fuckin' 'ope so.'

Colin adjusted the position of the machine's great boom then started easing forward as Chris, Jack and Jim held a hefty baulk of wood gingerly in position. He drove the last six inches too sharply and closed with the timber with a thump that reverberated throughout the rig.

'Bloody hell Col', it'll be a miracle if we're still lined up with that borehole,' Paul said.

'You need to watch 'im,' Jack muttered. "E's a fuckin' finger-trapper.'

Thursday, 21:07

With the platform reinforced by the mass of the heading machine, they were ready to try to place the bung again.

The phone rang. Jim dropped off the platform and went over and answered it. It was Bill Metcalfe, the senior manager on duty in Incident Control on the surface.

'The National Bore-Master is with me, he's made it out to the surface,' Bill said. 'He says you haven't got a hope. He says it's imperative that you get out of there – and fast.'

"We can't leave it. It's deteriorating too quickly now.'

'He's the expert and he says you need to get out of there. Now!'

'Tell him "bollocks", we're staying here to sort this out.'

21:15

Susan was sitting on the edge of the settee when the phone rang. She jumped up, leapt over to it and snatched it up.

'Hello, Tamworth 63928.'

'Hello, Mrs Wood?'

'Yes.'

'Good evening Mrs Wood. It's Bill Metcalfe here, Services Manager at Whitacre Heath. I'm just phoning to give you an update on how Paul and the team are doing.'

'Yes?'

'One of the team has just come out of the mine because the Undermanager decided he didn't need him. I've just spoken to them underground to see if they want to follow him out but they are determined to stay and sort it out. They're just going to make another attempt at sealing it.'

Susan was silent for a few seconds, confused and worried by one of the team having come away from the others.

'I'm sorry Mr Metcalfe, but can you tell me if it's getting better or worse down there. Are they managing to stop the water coming? Or is it getting worse?'

'I'm afraid the best I can say, Mrs Wood, is that it's no better and no worse at the moment. If anything changes, I'll make sure you're informed.'

'Oh, okay,' Susan murmured. 'Thank you.'

'We'll phone to give you another update, hopefully at about eight o'clock tomorrow morning.'

When Me Metcalfe had rung off she put the phone down.

"At least he's still alive."

And while he was still alive, there was still hope. She picked up the phone and called Diane and then Pat and Horace.

13's Head. Thursday: 21:52

This time, when the rubber bung encountered the increased resistance, a metre up the hole, and the water started to come through the seam at them, the heading machine's great bulk stopped the Dowty props from bending and they were able to thrust it on, past the point of resistance. They kept advancing the borer on its rams, to force the bung forward. Each time, they would secure the previous drill-rod, draw the boring machine back and screw in the next rod.

With every inch that the bung advance, the amount of water hissing out of the face, roof and sides of the heading increased alarmingly. It was as though they were baiting a wild beast. They cast repeated, nervous glances at Jim, fearful that the upsurge in the water forcing its way through the strata would simply blow what remained of their fragile barrier of coal away.

Suddenly, the water flows steadied then started to reduce.

'The only thing that worries me is that, while we have the borehole open we're relieving some of the pressure,' Chris said. 'If we stop that hole off, couldn't we be putting all that four hundred psi of pressure back onto the pillar?'

Jim had been thinking the same thing himself.

'It shouldn't make that much difference, we were only relieving the pressure with an inch and a quarter diameter hole,' he said, with more confidence than he actually felt.

After an hour they had managed to push the bung fourteen feet into the borehole. At that point, it gave the biggest reduction to the flow from the standpipe and the heavily fissured face and roof.

'That's encouraging,' Jim said. 'But let's try to place it a little closer to the old workings.'

'Are you sure Boss?' Joey said. 'We don't want to go and bollocks it up again.'

'I know, but I want to stop it with as much of the pillar on the right side of the bung as possible.'

Disappointingly, as soon they started to drive the bung gently forward the make of water increased again and water started spraying from the cracks in the roof and face.

Jim felt like kicking himself and suspected that the rest of the team felt like joining him.

They kept pressing on with no apparent reduction in flow. They were starting to despair; the bung was less than five feet from the end of the hole. They drove it on a few inches and the waters coming through the strata and the standpipe suddenly dropped to at least as low as they had been before. The water pouring from the roof behind them slowed to a trickle and then just dripped as the residual water drained off the heading's lining of corrugated sheets.

'Let's leave it there, Chris,' Jim shouted. 'That's bloody close to the old workings. And we don't want to end up worse off again.'

'Okay, that makes sense to me,' Chris said. 'Let's leave the drill-string up there. We can pump cement up it and through the non-return valve in the bung, if we ever get the opportunity to try to seal it off.'

Chris told Paul to turn off the hydraulic power-pack for the drill-rig.

They had slowed the water flow from the standpipe dramatically and the flow from the ugly hole in the middle of the face was no longer torrential.

With the noise of flowing water reduced and the Dosco and the drill-rig's power-pack both switched off, the heading descended into relative calm.

'Well done, men,' Jim said. 'It's midnight – but we're winning. Let's get out of this cold and into some warm air, in the return, and take a breather.'

Down at the junction, Jim phoned the Control Room. Ken Goodall had taken over as the senior manager in the surface control for the incident.

'How are you getting on, Jim lad?' Ken said.

'We've achieved a partial seal with the rubber bung.'

'Well done, Jim! What make of water have you got now?'

'The water flow from the standpipe is about a fifth of what it's been at its worst and the flow from the hole in the face of the heading is down to about five gallons a minute.'

'What does the strata look like?'

'A mess, what we can see of it is fractured to fuck,' Jim said.

He didn't mention that he couldn't imagine how much longer it would be before it was blown away or, even, how it had lasted for so long.

'What're you planning to do now?'

'Draw breath,' Jim said. 'I'll give it some thought and phone you back when we've decided what we're going to do next.'

'Righto lad, I'll speak to yer when yer ready,' Ken said. 'Thanks Jim.'

Jim hung the phone receiver up. He turned slowly and looked at his team.

'We've got two choices,' he said.

'We need to secure that face with a bit of a wall. If we don't do that then I can't see them ever sending anyone back down this pit, it'd be too dangerous.'

'Alternatively, the partial seal might hold long enough for us to make it to the pit bottom. I can't ask any more of you, you've done enough, if you want to try to make a break for it.'

'What do you reckon, Boss?' Joey said.

Jim looked at the floor and frowned, shook his head then lifted it to look each of them in the eye.

'I think we should stay and get the wall up, otherwise we've lost the pit.'

The men all glanced at each other, except Jack who continued to look back at Jim.

'Let's get on wi' it,' he growled.

'Yeah, in ferra penny, in ferra pound,' Joey said. 'The sooner it's done the sooner we can all fuck off aht of it.'

Chris, Colin and Paul nodded.

'Thanks lads,' Jim said. 'It does have the benefit of giving us a bit of confidence about what's behind us if we do actually get to head for home.'

Jim phoned Ken and told him of their decision.

'Are you sure, Jim?' Ken said. 'No one would think any worse of you if you cut and run now.'

'I know. But the team's unanimous.'

Jim put the phone down, firm in his resolve, now they'd committed themselves.

Paul stood up and shuffled over, stiffly, to where his father's maroon, snap bag hung in the roadside.

He put his hand in, felt around and pulled out a crumpled paper bag then turned around to offer it around the group. When it was his turn, Jim accepted an aniseed ball, confident that his low blood sugar would prevent it from making him feel nauseous.

Whitacre Heath Colliery Control Room. Friday, 29 March, 1985: 00:15.
Knowing that the wives of the underground team would not be sleeping and that they would appreciate a more reassuring update, Ken Goodall called them. When he spoke to Susan Wood she told him that she would still prefer to pass the message onto Diane Beighton. Susan didn't want any confusion arising; in the event of any serious reversal, she wanted to make sure that it would be she who would be notifying her friend.

13's Head. 00:20
The first job for the team was moving a train of Readymix concrete to the junction at the end of the heading. Having manoeuvred the Dosco Roadheader as far as possible to the left side, Colin disappeared off down the road and through to the intake to switch over the conveyor belt from the heading to run in reverse.

The water flowing from the face was going to make it impossible to build the wall. Jim got Colin and Jack to run a few bags of concrete up the heading on the conveyor and Joey, Paul and Chris to form a chain from the end of the conveyor to throw the bags through to him. Jim used these to build a curved dam wall, three feet high, to form a small reservoir in front of the hole in the face from which the water still ran. He built a couple of five-inch diameter, relief pipes into this, to allow the water to discharge.

With the water flow controlled and diverted, Jim picked up the headers' sledge hammer and went to work on the large lump of coal that had rolled out of the face of the heading as the waters had first broken in, making room for the retaining wall to be erected.

The steelwork for the framework, to hold the shuttering to retain the concrete as it was placed, had been dragged up to near the face, a couple of days earlier. Paul got spanners from the fitters' tool boxes, just outside the heading, and he and Jack coupled the clamping plates onto the girders as the rest of the team erected them in a robust framework of steel uprights and horizontals.

When this was complete, Chris and Paul extended the standpipe in the strata and the relief pipes from the reservoir beyond the girder work.

Friday, 04:48

Colin operated the winch on the back of the Dosco to pull the electric concrete mixer up into position while Jim and Jack man-handled it, to guide it up the heading. Then Jim got Colin to operate the winch while the rest of them got round the narrow conveyor off the back of the Dosco to position it to take the concrete, when it was ready to be poured, from the mixer right to the top of the wall.

06:36

With a first row of corrugated sheet shuttering placed along the floor, behind the girders, they were ready to start placing the first thirty inches of concrete.

'There's one thing about it, Bosso,' Joey said. 'There's no shortage of water.'

They ran a hose from the standpipe to the mixer. At any time, two of them manned the cement mixer whilst three worked as supply men. The latter party made the short trips, loading the train of vehicles with bags of concrete from the stockpile that had been unloaded onto palettes in the main return, in readiness for building the wall, then moving it the forty yards to the entrance of the heading. There, they unloaded the bags onto the reversing conveyor to supply the two men operating the mixer. The sixth man used a wooden ramming stick to make sure the concrete filled right into the sides and corners of the wall.

Jim made sure that the extendable ten-foot-long, two-and-a-half-inch diameter threaded steel pipes were placed in the wall at various levels against the roof, floor and face of the heading. He intended that these would be used as sealing pipes to pump cement into the surrounding strata, if they ever got the opportunity to try to achieve a complete seal in the future.

To conserve their cap-lamp batteries, they relied on the headlights of the Dosco near the face of the heading and on the overhead lighting in the roadway, as much as possible.

11:51

They had been working, solidly, on the wall and it was at about a quarter of the height needed. Jim was rotating them on the jobs every hour or so but after five hours of placing concrete they were all flagging. He told Paul and Colin to wash out the cement mixer they were manning so they could all take a break.

'I'm fucking starving.' Joey said as they walked out into the return airway.

'You can forget about snap for the moment, Joey,' Jim said. 'But from now on, we'll take a break every few hours, to get our breath back.'

After half an hour of rest, Jim roused the dozing men and they got their system of concrete placing and supply going again.

13's Heading. Saturday, 30 March, 1985: 01:47

After four more lengthy, draining sessions of hard work with rest periods in between, there was a gap of only a few inches between the top of the concrete and the roof. Jim and Joey were on the platform trying to ram in more concrete.

'I reckon that's as much as we can do, Boss,' Joey said. 'It just keeps fallin' back out on us now was fast as we place it.'

'I tell you what, let's ram a few packing bags of a drier mix in, to top it off,' Jim said.

'Fuck me, Bosso,' Joey said. 'You never know when you've 'ad enough, you don't.'

They rammed the bags of concrete into place to form a better seal to the roof, completing the shuttering with lagging boards. With that done and the cement mixer washed out, the six men stood and looked at the steel frame and corrugated sheets holding back the mass of concrete they had placed. They had reduced the flow of water to a level that the pumps could handle and reinforced the failing strata of the heading with their wall.

Jim phoned Ken Goodall, who was back in charge on the surface again, and reported that they had completed placing the concrete for the first section of wall.

'Well done, Jim,' Ken said. 'Now you get yourselves out of there, sharpish.'

'Righto Ken, we're bloody ready for it,' Jim said, looking around at his exhausted team.

'Ken, can you ask Roger to get the North Side conveyors running to give us a ride out?'

'Alright, Jim,' Ken said. 'And I'll get the banksman and winding-engineman ready to wind you up the shaft, when you get there.'

'Thanks, that'd be brilliant.'

The men looked at each other. It was like a dream come true; they were actually going to try to head for home.

Jim closed his mind to the thought that the concrete still needed to cure, to give the wall the strength it needed to hold the protective barrier together. If they could just make it out of the pit, without the fractured pillar failing and the colossal mass of water behind it blasting away the retaining wall then, as long as the shafts did not end up filled with water, they would know in a few hours if the concrete was setting and the pillar and wall were holding.

In the last hour, their spirits had risen as they had started to place the last of the concrete and seen that the end was, at last, in sight. And now they were on their way; Mad Jack and Joey joking and chuckling with Jim in hoarse, cracked voices, Paul, Colin and Chris marching along behind them like maturing influences, Paul carrying his dad's delicate, maroon, snap bag, as well as his own.

As Jim led them out, it occurred to him that that pathetic, broken, thirty-foot barrier of coal had held back a billion gallons of water at over four hundred pounds per square inch for forty-eight hours since the firing of the round of shots that had weakened it, almost fatally. It was unbelievable, a miracle; his prayer had been answered.

When they reached the dip in one of the deep swillies, in the return airway's manrider road, they found that the water had receded, having drained into the lower waste of an old, sealed-off face district. Instead of being forced to swim, they were able to wade through water that was three feet deep at its worst.

Jim led the way, as the team cut through the air-doors to the intake airway, to ride out on the main belt. Roger Dawkin had started up the conveyor, remotely, from the Control Room for them.

Riding along, lying face down in his exhaustion, with the belt's motion and the hypnotic, repetitive massage as it passed over the rollers Jim was finding it a struggle to stay awake. He scrambled up onto his knees to make sure he didn't drop off to sleep, he needed to stay awake to jump off at end of the belt; it was important for him to be ready to stop the belt on a lock-out and save any of the others who might have succumbed to slumber from being hurled, at speed, into the steel chute at the transfer point at the delivery end of the conveyor.

Saturday: 02:45
Susan woke. The phone was ringing in the lounge, downstairs.

She sat up, threw off the covers and leapt out of bed. She looked at the alarm clock as she rushed out of the bedroom; she'd only been asleep for an hour.

Mercifully, the caller held on. She grabbed the handset.

'Hello, me duck, it's the pit,' Roger Dawkin said. 'Sorry to wake yer, Susan. Don't worry, there's no problem, I just needed to let yer know that Paul and the team 'ave done the job and they're on their way out the pit now. He should be back on the bank in about three quarters of an hour.'

'Is he alright?'

'He's fine.'

'Thank you, Roger. You can't imagine what a relief that is.'

'That's it, me duck. You go and get that bed warm fer 'im. 'E'll be cuddled up next to yer in it in a couple of hours'

She closed her eyes and offered up a silent prayer.

Saturday: 03:35

The half-mile walk along the Loco Road seemed interminable but, eventually, they rounded a bend and saw lights not far ahead of them. As they closed with them, the road opened out into the brightly-lit section of the pit bottom manriding station.

Their feet dragged as they walked round the curve to the air-doors through to Number 2 Pit Bottom.

'Good morning Mr Greaves, good morning, lads,' Bert Heath, the pit bottom deputy greeted them as they burst out of the third air door. 'Welcome 'ome!'

Ken Goodall had left for home as soon as he knew the team was on its way out, after sending Bert down the pit to act as onsetter for them.

'Let's get you chaps on that cage, up the pit and off 'ome to bed,' Bert said, as he waved them on to where the cage waited with the pit bottom gate slid across and the chains and rods of the cage's barrier raised in readiness.

When they were safely on the cage, Bert rang the signal for their cage to be raised. As it lifted off, Joey bellowed back down to him.

'Hey, Bertie! Tell 'em in Control to get that fuckin' kettle on!'

They stepped out of the surface airlock into the starlit night. Jim forced himself to lift his legs and walk properly, as he led his men away from the shaft. The freshness of the cold air helped, it tasted clean and full of oxygen.

Still walking, Jim shut his eyes for a couple of seconds.

"Of them which thou gavest me, have I lost none."

486

They stomped down the steps from the covered way to the pit yard.

'Let's get straight over to the Control Room and get a cup o' tea,' Jim said.

He led the way to the lamp room, where they deposited their self-rescuers and exhausted lamps. Then he waited until they were all gathered again, ready to lead them, across the pit yard car park, into the main offices and down the corridor to the Control Room.

'Good morning, Mr Greaves, hello lads,' Roger said. 'Welcome back to the land o' the livin'. We thought we'd lost you lot. Boss, the Gaffer wants you to phone him. 'Ere's 'is number.'

He handed Jim a scrap of paper.

Jack and Joey groaned as they perched on top of the table against the Control Room wall. Paul and Chris lowered themselves onto the two available chairs and Colin slumped into the Control Room attendant's chair while Roger served up the tea. The mugs warmed their hands as the sugar and heat of the tea cheered them. They grabbed biscuits from the tin that Roger had thieved for them from Angie's office and stuffed them into their mouths.

A few minutes later, Jim went round them, topping up their mugs with the teapot.

'Is everyone alright for getting home?' He said through a mouthful of crumbs.

Chris said he had his car, Joey decided to put his life in Mad Jack's hands and accepted the offer of a lift home on the mechanical engineer's infamous motorbike and side-car and Paul and Colin both said they could walk back to their homes on the Estate in the pit village.

Saturday: 04:25

Jim walked out behind them all, on his way to phone the Gaffer from Ken Goodall's office.

He paused and looked up the corridor, determined to cement a picture of the remarkable band in his memory; the three, big men at the front, the reliable and competent Chris Baxter, the reassuring, buoyant presence of his overman Joey Bondar and Jack Lloyd, as mad, maverick and dangerous as ever. Behind them, the smaller, slighter figures of Paul and Colin, the quiet, competent technicians; two lifelong, former mates, between whom the strike had opened up a gaping chasm of enmity.

As Jim stepped into Ken's office, he thought about how, despite the differences in their perspectives and, at times, their approaches, he and Paul Wood had always shared a prime motivation, that of keeping Whitacre Heath open. Having started out working to that end together, they had spent the year of the strike pursuing it through different, almost opposite means before being prepared to give everything together in their life or death struggle to contain the inrush.

He put his mug down on the desktop, shuffled round the desk and dropped into the undermanager's clean, captain's chair in his sodden pit-black.

He took another sip of tea then picked up the phone handset and dialled the number, gazing around at the reassuring civilisation of the office. The medium-sized, colliery plan on the wall still had relevance; their pit had not been consigned to history by a deluge from the waterlogged old workings shown on it. He couldn't believe they had made it back there, to the surface world, still alive.

A gruff voice answered the phone, "Ello.'

'Good morning Gaffer, it's Jim. Roger said you wanted a report.'

'Yeah. Are y'alright?'

'Yes, thanks, brilliant now, Gaffer.'

"Ow've you left it?'

Jim described how they had finally brought the rushing water under control with the rubber bung. He explained that he had not wanted to withdraw from the heading without building a minimal, reinforcing wall because he knew that, without it, there was no way that a team of men would ever be allowed to risk going back down the pit to do it, once the pit had been fully evacuated.

He outlined the next steps he had identified.

'We need to get back in there, later today, to start extending the wall,' Jim said. 'As you know, the normal design criteria for an underground dam is that it should be at least as thick as the greatest, roadway cross-sectional dimension.'

On the other end of the phone, Harry Everitt grunted and smiled to himself.

'In 13's Head that's the road's width of four metres,' Jim went on. 'So we need to add at least another three metres of concrete to our initial, securing wall.'

'Sounds sensible,' Harry Everitt growled.

'We need a limited number of volunteers to work the second half of the dayshift, transporting the girders and concrete to extend the wall,' Jim said. 'Then I'll go back down the pit with the afternoon-shift to supervise the start of the construction of the main wall then we'll need to work it on four shifts, round the clock, to make sure that the concrete's laid continuously with no separating planes of weakness. When that's done we'll be able to pump grout into the strata through the pipes I've had built into the wall to try to achieve a complete seal.'

'Sounds like a plan,' Mr Everitt said. 'A good plan.'

'Righto, Gaffer.' Jim said. 'Anything else you need from me?'

'No, that'll do,' Mr Everitt said. 'But tell the men who stayed with yer, from me, that you've all done a fine job for Whitacre Pit.'

Jim went round and into the huge area of the men's baths and relayed his Gaffer's thanks to Colin, Paul and Chris then went through to the officials' baths, where Joey and Mad Jack were showering; Mad Jack scrubbing Joey's back with his sponge as Jim passed on the commendation.

'It's all very well, these medals,' Joey said, as Jack's scrubbing pummelled at him. 'It's fine when they're pinnin' 'em on...'

'Yeah,' Jack said, pausing for a moment. 'But it don't 'alf 'urt when they rip 'em off again!'

Jim turned away with a wry smile

'Hey!' Jack shouted after him. 'Thanks for comin' an' tellin' us that, Jim. And well done yourself – Jackie Charlton would be proud of you, the old bastard.'

'Yeah, well done, Bosso,' Joey called after Jim as he hauled himself off to the undermanagers' bathroom.

Colin was impatient to get home to reassure Diane and get to bed but, as he approached the end of the pitman's shift-ending ritual of stripping, bathing, towelling himself dry and dressing, he started to take his time. Determined to avoid the embarrassment of imposing himself on his former friend on the walk home, he needed to leave Paul enough time to dress and get away on his own.

Having hung his helmet on one of the hooks in his dirty locker in the undermanagers' bathroom, Jim chucked his soaking pit boots into the bottom of the locker. He knew that, by the time the locker's heated ventilation had dried them, they'd be stiff as boards when he next went to put them on.

He stepped out of his wet overalls and hung them up, over his boots. Sitting down on the narrow bench attached to the lockers, he pulled off his sodden, coarse, woollen socks, wrung them out and hung them in his locker, next to his overalls.

It was an effort to stand up, rather than sit there and drop off to sleep but, pushing himself up like an old collier, he gathered up his towel, old, metal, soap dish, shampoo and sponge from his locker.

He had turned the shower on, before stripping, to get it running hot. He went over, stood under it and closed his eyes.

At last, he was clean but he was reluctant to get out from under the hot shower, to stand and dry himself in the chill of the bathroom. The heat from the water soaked into his aching joints and muscles but could not wash away the cold in his body, from being stuck in the chill of the heading's ventilation for countless hours in drenched clothes.

He was so tired he felt as though he could drift off to sleep on his feet. When he closed his eyes again, he started to daydream.

He thought of the men who had stayed down there with him and come together as a team to do the impossible. He pictured Paul and Colin walking home together in the dark of the morning. They had all come through.

It was a miracle. His guardian angel had saved him again.

As soon as he thought of her, he felt a presence. His eyes snapped open. He looked all around.

But there was no one there. He was alone.

He relaxed, closed his eyes again and murmured a few, simple words of thanks for their deliverance.

Paul and Susan Wood's house. Saturday: 05:23
Sensing something, Susan jumped up from the settee and rushed over to peer out through the curtains.

Paul was coming up the street, in the middle of the road, with Colin. Their heads were down but their strides were surprisingly long and light, as they stepped out, side by side.

Colin had still been taking his time getting dressed when Paul had wandered into his bay of lockers and flopped down onto the narrow, metal bench, attached to the lockers opposite, as he had after so many of their shifts in the past; to wait without a word until his friend was ready to shut his locker, so they could leave together.

The front door was already unlocked, ready for him. Susan rushed to pull it open then stood waiting, framed in the front door in her long nightie, dressing gown and fluffy, mule slippers.

She could hear their footsteps as they marched along, in step, muttering quietly to each other. looking as though they were just returning from another day trip together. As they reached the house, Paul peeled away from his friend.

'See ya, Colin,' he murmured.

Colin pulled up.

'Yeah, see you Paul.'

He moved off again.

"Morning Susan, see ya,' he said, in a stage-whisper.

'Bye Colin,' she sang out, softy.

Paul eased the garden gate shut and walked, with his serious smile, to where Susan waited.

She stepped down from their front step, to avoid towering over her husband.

He wrapped his arms around her.

'I thought I'd lost you this time,' she murmured in his ear.

'Nah. Yer know me better'n that, our Susan.'

Saturday, 05:36

His key grated into the Yale.

He couldn't believe it. He had made it home; to Gabriella. Having forced himself to accept that he had seen his last of the surface world, he had been convinced that he would never see either, ever again.

The landing light was on. He closed the door quietly behind himself and looked up to glimpse Gabriella's lissom nakedness as she appeared, running out onto the landing, throwing on her dressing gown. She flew down the stairs to him.

She put her hands on his stubble-coarsened cheeks and kissed him. Looking deep into his eyes, she warmed him with the welcome in her smile.

'Are you alright? Is everyone safe?'

'I'm fine. Just knackered. Everyone's fine.'

'Do you want anything to eat?'

'No, ta. I just want me bed. I need to be back at the pit at ten thirty.'

'What? This morning?'

'Yeah.'

'That's ridiculous. Can't they get someone else to do it?'

'No! I've got it this far. I need to see it done right, right to the end.'

Upstairs, he stripped off his freshly-bathed body, dropping his clothes onto the floor. Then he rolled into the bed, still warm from where Gabriella had been sleeping on his side. For a moment, he relished the contrast between the soft, safe, perfumed comfort of their bed and the harsh, hazardous environment that he and his team had endured for countless hours. Then he fell into the deathlike sleep of the truly exhausted, before Gabriella could join him.

No dreams, no nightmares, no visions of angels, just oblivion.

Saturday, 09:15
After what seemed like only an instant, the explosive pop, from the lounge below, of the stylus landing on the disc on the record deck woke him. It was apparent from the loud crackling of the scratched LP that Gabriella had turned up their stereo to full volume.

The distant rustle of tambourine at the start of the track gave it away; her choice of song apposite and motivating. Jim picked up the irresistible aroma of bacon frying as the song's fanfare started. The electric organ swelled and the synthesizer started to pick out the melody. At the song's first cymbal smash, he felt light-headed with the exhaustion and elation of their triumph in those interminable, perilous hours underground.

Gabriella drifted into the bedroom and drew back the curtains with a bright smile.

Jim opened one eye to see a cloudless sky as Stevie Winwood's voice rang out through the house.

"Stand up in a clear, blue morning, until you see what can be…"

'Sorry, Darling, but if you really are going to be back at the pit by ten thirty you're going to have to get up now.'

She sat down next to him on the bed and pulled the sheet back from his face.

'I've got a lovely, big breakfast for you.'

Jim felt deathly tired in every atom of his being but he still managed a weary smile for his girl.

For once, he wasn't raring to go, to get to the pit and get to work. But it would certainly be worth rousing himself for that breakfast.

While you see a chance, take it.

Psalm 23

The LORD *is* my shepherd; I shall not want.

He maketh me to lie down in green pastures: he leadeth me beside the still waters.

He restoreth my soul: he leadeth me in the paths of righteousness for his name's sake.

Yea, though I walk through the valley of the shadow of death, I will fear no evil: for thou *art* with me; thy rod and thy staff they comfort me.

Thou preparest a table before me in the presence of mine enemies: thou anointest my head with oil; my cup runneth over.

Surely goodness and mercy shall follow me all the days of my life: and I will dwell in the house of the LORD for ever.

The King James Bible

Together

Three months later: Saturday 29 June, 1985: 07:40

Somehow, he knew he'd see them.

He let his old Escort slow to a crawl. As it approached the crown of the hump-backed bridge he craned his neck. As always, it was a relief to see nothing coming at him from the other direction.

He glanced left, down into the water meadow, and smiled.

They were there and there were three of them; the old man, the little boy and, a little way behind them, the boy's father; the old man's son.

All three carried a rod and bag each. All three of them, heading for the river.

Together.

By 2015, Britain was still burning 40 million tons of imported coal every year.

Despite this, Britain's last colliery was closed down on 18 December of that year.

A personal message

If you're a miner, anywhere in the world, then you will know that we are brothers.

For those of us who are British colliers and who worked in the NCB's mines in the 1980's, even after all these years, there are some of us and some family members who still feel betrayed by those who returned to work before the strike ended. And there are some who worked who still feel bitter at the treatment they received from some of the steadfast strikers and their families. It would be wonderful if *The Enemy Within* enabled some amongst them to find some reconciliation.

As I left the industry, on 29 January, 1989, I thought at least one thing was certain; I wouldn't change, I would always be a miner. Once a pitman, always a pitman. And that has proved to be true.

If you are a British pitman, you may have never felt able, or taken the time, to grieve the loss of our amazing industry. But grieve it we should: Our comradeship, our way of life, our world. I hope that my story helps you to do this, at last, as it has helped me, and that, through it, you will have been able to celebrate and relive a little of those extraordinary, unforgettable, wonderful years we spent down our pits.

As my mate George said as we 'turned some coal' whilst reminiscing over a pint, a couple of years ago:

"Fuck me, it was a <u>hard</u> life – but I wouldn't have missed it for the world!"

God bless you,
Bob Wilson
Warwickshire pitman
October, 2016

The Enemy Within:
List of Characters

Alice Parkin	Striker's wife
Angela Barker	General Manager's Secretary, Whitacre Heath Colliery
Archie	Picket and striking miner from Barnsley Area
Angus Dewar	Colliery Mechanical Engineer
Arnie Campion	NUM Regional Delegate, Midlands
Arthur Lambert	Faceman
Arthur Scargill	President of the National Union of Mineworkers
Barbara ('Barbie) Whittaker	Assistant Secretary, Canley Colliery
Bernie Priest	Senior Overman, reporting to Ken Goodall
Bert Heath	Pit Bottom deputy
Bev	Woman Police Constable, Neil Bradford's friend
Bill Metcalfe	Services Manager (Deputy Manager, Services)
Billy Thompson	Deputy on 22's
Brett Redfern	Apprentice fitter
Brian Kettle	Faceman (a ripper), Paul Wood's mate
Carl Bessemer	Police constable
Chris Baxter	Drilling Team member
Chris Turnbull	Colliery General Manager, Railsmoor Colliery

Cyril	Prisoner, Ian Flint's cellmate
Clive Hamilton	Shaftsman
Colin Beighton	Electrician, Paul's friend
Dan Crowley	Header, 13's Heading (chargehand)
Dan Wainwright	Striking collier
Dave Coton	Header, 13's
Davy Tonks	Shaftsman
Des Proctor	Shearer driver and chargehand on 85's Face then 15's Face
Derrick Hill	Colliery Surveyor, Whitacre Heath Colliery
Diane Beighton	Colin's wife
Dick Flavel	NUM Lodge Secretary, Whitacre Heath Colliery
Doreen	Diane Beighton's mother
Emily Beighton	Colin's & Diane's older daughter
Gabriella	Jim Greaves' girlfriend
Gary Hedges	Header, 13's
Gary Wallender	Deputy on 22's (afternoon shift)
Gerald Chipman	Materials Supply Officer
Graham Dent	Deputy Electrical Engineer
Harry Everitt	Colliery General Manager, Whitacre Heath Colliery
Horace Wood	Faceman (a chocker) and Paul's father
Howard Freer	Conveyor 'button lad'
Ian Flint	Miner and pickets' ringleader from Barnsley Area
Ian MacGregor	Chairman of the National Coal Board
Irene Mundy	Proprietor of Whitacre Heath VG store
Ivan Wykes	Shuey McFadden supporter
Jack Charlton	Colliery General Manager, Canley Colliery
Jackie (Jacqueline) Flint	Teacher, Ian Flint's wife
Jasper Ford	Header
Jed Davenport	Shaftsman (chargehand)
Jenny	Paul's & Susan's daughter

Jerry Malin	Control Room Attendant (nights)
Jim Greaves	Non-Statutory Undermanager
Joey Bondar	One of Jim Greaves' overmen
Joss Kemp	Deputy on 22's District
Julie Keighley	Nurse, Neil Bradford's fiancé
Kathy	Member of Women Against Closures
Kath Taylor	Landlady of the Horse & Jockey at Bentley
Keith Deeming	NUM Lodge President, Whitacre Heath Colliery
Ken Goodall	Statutory Undermanager
Kevin Clark	Electrician
Laurence ('Lol') Collier	Production Manager, Warwickshire Coalfield
Les Parker	Colliery Overman, Senior Overman, reporting to Jim Greaves
Linda	Jackie Flint's sister
Lorna Kettle	Brian's wife
Lynne	Woman Police Constable, Neil Bradford's friend
Mad Jack Lloyd	Assistant Mechanical Engineer
Malcolm Kennedy	Shuey McFadden's henchman
Margaret Thatcher	Prime Minister
Martin Wood	Paul's & Susan's son
Matthew Flint`	Ian's and Jacqueline's son
Maureen	Member of Women Against Closures
Maurice Huntley	NCB's National Bore-Master
Mike Taylor	Afternoon Shift Senior Overman
Norman Bramble	Faceman (chock-fitter)
Neil Bradford	Police constable
Pat Wood	Horace's wife, Paul's mother.
Paul Wood	Senior Fitter
Peter Linton	Control Room Attendant (afternoons)
Phil Grant	Colliery Electrical Engineer
Ralph Rawlinson	Area Director, South Midlands Area

Ray Bentinck	Personnel Manager
Roger Dawkin	Control Room Attendant (days)
Roy Gently	Union of Democratic Mineworkers' Delegate, Midlands
Sally	Horace and Pat Wood's Highland Terrier
Sammy Crowther	Miner, Ian Flint's mate
Sheila	General Manager's Secretary, Canley Colliery
Sheryl Skinner	Striker's wife
Shirley	Canteen Manageress, Dick Flavel's partner
Shuey McFadden	NUM representative
Sidney	Lamp room attendant (nights)
Sophie Beighton	Colin's & Diane's younger daughter
Stan Townsend	Deputy on 85's then 15's District
Stephen Wheatcroft	Chief Mining Engineer, South Midlands Area
Steve Bircher	Striking collier
Susan Wood	Paul's wife
Ted Carter	NUM Lodge Committee member, Whitacre Heath Colliery
Terry Bond	Colliery Mechanical Engineer, Canley Colliery
Terry Harrison	Deputy in charge of 13's Heading
Tim Hope	Operations Manager (Deputy Manager, Operations)
Toby Sweet	Faceman
Todd Crisp	Leader of South Midlands Area Drilling Team
Tom Gibbons	Grade II deputy
Vanessa Crowther	Sammy's wife
Veronica	Receptionist
Vic 'Warbler' Worrell	Deputy, occasionally acting overman
Wes Hands	Striking collier
Wilf Albrighton	Assistant Surveyor
Willie Stokoe	Night Shift Senior Overman

Glossary of Terms

Words and expressions used in collieries in 1980's Warwickshire. These could vary, not just in relation to the other areas and regions of the country but between individual collieries within the same coalfield or county.

Advance Face – a face that extracts the coal by extending its gate roads and advancing away from the main roads. The opposite to a retreat face. *(See Note 5)*

AFC – the abbreviation for Armoured Flexible Conveyor/Armoured Face Conveyor. Synonymous with Panzer. *(See Note 1.2)*

A-frame – the powered support at the end of the face, in the gate road. *(See Note 19)*

Air-doors – the opening that enabled miners to pass through the barrier between intake to return, without allowing the air to short-circuit the intended ventilation paths.

Anchorage – *See Note 7.*

Arches – roadway supports, formed of an "I"-section steel bow supported by "I"-section steel legs, set on wooden foot-blocks. The legs and bows were coupled together with arch-plates (also sometimes erroneously referred to as fishplates) and nuts and bolts. For face gate-roads they were typically 10 feet wide by 8 feet high or 12 feet wide by 10 feet high. An arch was also referred to as "a ring". Consecutive rings were linked by steel struts, to provide the continuous, supportive structure for the road.

Auxiliary fan – a fan used, usually in conjunction with air-ducting, to increase the ventilation to a section of the mine, e.g. a development heading or a face heading. An auxiliary fan either blew the air in or (less frequently) drew the air out (which was replaced by fresher air).

Back-rip – an operation that enlarged a crushed roadway by opening out the road above the old arches before setting new supports then removing the deformed steel.

Back-ripper – a man employed on a back rip. Often working in a team of three but sometimes working in a pair.

Backshifts – the afternoon and nightshifts at a colliery.

Bait – the Geordie word for snap. A word that was well-used in Warwickshire, due to the numbers of miners who had relocated to that county's mines from Northumberland.

Bank – the word used by colliers for the surface. Its strict meaning was a colliery's surface premises but miners used it to refer to the whole, wider, surface world, when underground.

Banksman – the man responsible for loading and unloading men and (separately) materials onto the cage and signalling to the winding-engineman and onsetter for movements of the cage to be undertaken.

Bedframe – the massive chassis that provided the backbone to a shearer (a coal-cutting machine). *(See Notes 1.1 and 17.2)*

Belt – A conveyor belt (an endless moving belt) used to transport coal and stone. Some belts were also authorised and safe for manriding.

BJD, British Jeffrey Diamond – a manufacturer of notoriously unreliable shearers.

Blackleg – a worker who fails to join fellow union members in industrial action.

Bonus – the incentive bonus. *(See Note 13)*

Borer – a handheld drilling machine.

Boss – a term of address that might be applied to a member of the colliery management, not applied automatically, it tended to indicate a degree of respect.

Bottom chain – on a panzer or face conveyor. *(See Note 16)*

Brattice – sheeting used to deflect air into particular areas to improve ventilation and dilute flammable or noxious gases.

Bull's head – a handheld electrically powered boring machine, used to drill holes in coal or stone.

Button – a conveyor's on / off switch.

Button-man or button-lad – a conveyor attendant who made sure that he stopped the inbye conveyor when the outbye conveyor stopped, ensuring no pile-ups occurred and who cleaned up and loaded any of spillages of coal, back onto the belt.

Can of powder – a lockable, heavy, thick cylinder, made of stiff canvas, used to carry sticks of explosive ("powder") into the mine.

Canaries – the small birds, kept at the colliery in order to comply with the Regulations relating to the Mines & Quarries Act. They were used in the event of an underground fire because small birds and animals are more susceptible to carbon monoxide poisoning than humans. An important, mining, cultural totem.

Cap (shaft cap) – the covering installed over a disused shaft, usually made out of concrete.

Cap-lamp – a rechargeable, battery-operated light carried by a miner and usually attached to the miner's safety helmet.

Capel – the attachment connecting the winding rope to the chains that supported the cage. *(See Note 2)*

Car – also known as a tub or a mine car; a wheeled rail-vehicle for transporting stone or coal or materials (supplies) underground.

Carbon monoxide (CO) – a potentially-lethal, poisonous gas, arising from the incomplete combustion of carbon.

Carbon monoxide poisoning – *see Note 14.*

Caving – occurred behind a longwall face when the powered roof supports were drawn in to the face conveyor, after the taking of a cut.

Chew – a piece ("plug") of tobacco of the right size for chewing.

Chock – a powered roof support (unless used to refer to a wooden chock – *see Note 22*).

Chock block – a wooden block that was thirty inches long with an eight by eight-inch square cross section that was used to build a wooden chock. *(See Note 22)*

Chock test – *see Note 11.*

Chockers – facemen who drew in the powered roof supports after each cut. *(See Note 1.5)*

CO (carbon monoxide) – see carbon monoxide, above.

Coal preparation plant – the large establishment on the surface of the mine where coal was sized and separated from stone and shale, largely by flotation methods, and prepared for sale.

Coalface-training instructor – workman appointed under the National Coal Board Scheme of Training for Coalface Work to supervise and instruct a trainee at the face.

Colliery – an underground mine that extracted coal.

Colliery General Manager – a statutory position held by a suitably qualified mining engineer. *(See Note 6)*

Colliery Overman – a senior overman in charge of a shift, reporting directly to an Undermanager.

Contraband – illegal objects and materials (e.g. cigarettes) that could ignite any explosive accumulation of methane and air underground in a colliery. *(See Note 3)*

Control Room – the surface room in which information on the progress of operations and the status of underground workings was collated. Manned, round the clock, by a control room attendant (or two or more at large collieries). Frequently referred to by colliers, staff and management as "Control".

Cut – the depth of coal taken with each pass of the shearer (coal-cutting machine). 30 inches in width through the whole length of a face.

Deployment centre – the room where the preferred deployment of a shift was plotted with spring-loaded name plates on its wall. The room from where the miner drew his tallies each shift. *(See Note 3.1)*

Deputy – a colliery official. *(See Note 1.3)*

Deputy Manager, Operations – *(see Note 6)*. In "The Enemy Within" referred to as the "Operations Manager".

Deputy Manager, Services – *(see Note 6)*. In "The Enemy Within" referred to as the "Services Manager".

Detonator – a small, thin, cylindrical, brass-cased explosive device which caused a stick of explosive to explode when inserted into it and detonated by the passage of a small electrical current from a shot-firing battery.

"Dets and powder" – colliers' slang for detonators and sticks of explosive.

Development – roadway drivages, created to prepare a new coalface or create access roads (intake and return) to a new area of a mine.

Dint – excavation of the floor, underground, usually to address roadway closure due to the floor heaving up.

Dintheader – see Note 20.

Direct haulage – see Note 10

District – an underground production unit, consisting of the longwall coalface the two gate-roads (tunnels) that served the face (maingate and tailgate) and the connecting road with its air-doors at the outbye end of the two gate-roads. Often known by a geographical location and a sequential number, e.g. "South West 5's".

Dosco -large, tank-like tunnelling machines with movable booms to profile out an underground drivage to form a road underground (tunnel).

Downcast shaft – the main shaft of the colliery up which coal was wound. The fresh air for the ventilation of the mine was drawn down this shaft.

Dowty – British manufacturer of excellent powered roof supports. Also a term used for a Dowty prop; a manually-operated hydraulic prop.

Drager – a handheld monitoring device for measuring the level of carbon monoxide in the air underground. Described in detail in the text of "The Enemy Within": The Drager pump was like a small rubber concertina between two curved pieces of grey plastic which fitted comfortably in the palm of a man's hand. The user took a Drager tube, a long thin phial of crystals, broke the ends off and shoved it into the rubber housing on the hand-pump. This had to be fully depressed so that it would take a full, calibrated draw through the tube of crystals. Drager was a manufacturer of breathalysers and this test worked in a similar way to those only, instead of testing for alcohol on a driver's breath, it measured the level of carbon monoxide in the air.

Drill-string – a number of drill-rods connected to a drill-bit.

Drivage – a new road (tunnel) being driven (i.e. excavated) in solid ground.

Electrical panel – *see Note 4.*

Endless Haulage – *see Note 10.*

Face deputy – an official in charge of a face district. *(See Note 1)*

Face start-line – the face-line driven out on which the face equipment was installed. The location of the face before it started extracting coal.

Face team – *see Note 1.5.*

Fall – a mass of roof, rock or coal, which collapses underground.

Fan – the main fan ventilated the whole of a colliery's underground workings. Auxiliary fans provided additional, more localised and directed airflows underground. Both types were axial fans, in design, by the 1980's.

Flame lamp – also known as a Davy lamp, after its inventor. A hand-held methane and black-damp (oxygen-deficiency) detector. All officials carried a flame with a re-lighter and were required to be proficient at reading varying percentage levels of methane concentration. The Mines & Quarries Act and its Regulations defined the percentage numbers of flame lamps that needed to be carried by men working underground.

Flights – steel bars which spanned the pans of an armoured face conveyor and were pulled along by the chains on either side of the pans to haul the coal of the face.

Floor plate – *see Note 7.*

Gaffer – respectful term of address used when talking to or about the Colliery General Manager in a Warwickshire colliery.

Gate or gate-road – a roadway leading to a face. Virtually all of England's coalfields (except Warwickshire) were in the old Danelaw, the regions taken and held by the Vikings. Gate, in this instance, was derived from the Norse word for a road or street.

Gearhead – A gearhead on a shearer *(see Note 1.1)* incorporated the ranging arm that raised and lowered the cutting disc. Alternatively a gearhead on a panzer was an extended pan that housed the face-end gearbox and motor *(see Note 15)*.

General Manager – (See Colliery General Manager *(and Note 6.)*.

Goaf – another word for the waste, the area behind the coal face where the supports had been drawn and the roof allowed to cave / fall in.

Gob – another word for the waste or goaf.

Grade I and II deputies – *see Note 1.3.*

Half-shift – *see Note 12.*

Haulages – *see Note 10.*

Head or heading – a development drivage (excavating a tunnel) underground in coal or stone.

Header – a man who worked driving tunnels underground.

Headgear – the huge, steel, gantry-like, framework construction erected over the top of a shaft which incorporated the headstocks and winding wheels. The winding ropes which lifted and lowered the cages through the shaft went over the headgear's winding wheels.

Heading – see "Head", above.

Headstocks - the huge framework holding the winding wheels over the shaft.

Heating – an underground fire. *(See Note 8.)*

HMI – Her Majesty's Mines Inspectorate with statutory powers to hold the statutory managers and officials (i.e. general managers, statutory undermanagers and deputies) accountable for operating mines safely and efficiently. (See Inspector, below).

Incendive spark – any hot particle that could ignite an explosive mixture of methane and air in the same way that a lighter works on a gas cooker.

Incentive – *see Note 13.*

Inrush – a body of water or material which can flow when wet and which floods into underground mine workings.

Inset – an access to a seam part way down the shaft.

507

Inspector – a person (Her Majesty's Inspector) appointed by the Government to ensure good working practices are employed and compliance with relevant Legislation, particularly the Health & Safety at Work Act and the Mines & Quarries Act and the Regulations pertaining thereto. (See HMI, above).

Installation – the work involved in transporting equipment to site underground and positioning and commissioning it.

Intake – fresh air roadway leading to the working part of a mine.

Jim Crow – a heavy duty, manual device for bending rails, used to create curves when laying tracks. *(See Note 9.)*

Keps – devices for supporting a cage and holding it in positon at the top of a shaft.

Lagging – boards or metal sheets placed between the flanges of steel arches. (The boards were known at some collieries as "slobbing").

Lamp cabin – the large room on the 'dirty side' of the pit where miners stored and drew their cap-lamps, self-rescuers and flame safety lamps. The racks in which the miners left their lamps which had an integral charging system to recharge the cap-lamps' batteries.

Leg – the lower part of steel arch road support. (A pair of legs and a crown formed an arch) (see "arch").

Loader – a large, mechanised coal bunker into which a main conveyor belt from a part of a mine emptied and from which the loader operative could regulate the flow into a train of empty mine cars, in pits where the coal was transported to the pit bottom by trains.

Locker – pit head baths container for clothes. Each miner had a clean locker for his home clothes and, on its mirror image the dirty side of the pit head baths, a dirty locker for his pit-black.

Loco road – this tunnel was the road along which all of the coal travelled in a mine in which trains were used to transport the coal to the pit bottom. It was always in the main intake airway for such a mine.

Lodge – the branch of the National Union of Mine Workers or NACODS at a colliery.

Longwall mining – method of mining in which coal is extracted over a long face, typically at least 200 metres in length, with a tunnel at each end to provide the ventilation circuit and to enable the mineral to be transported out, supplies to be transported in and men to access and egress the working.

Machine cable man – *see Note 1.5.*

Machine driver – *see Note 1.5.*

Main fan – a large, electrically-powered axial fan installed at the surface to draw air out of the upcast shaft to induce airflow throughout the mine.

Manrider – any conveyor or train that was legal and suitable to use as a means to convey people underground.

Mine – synonymous with colliery and pit. "Mine" was the word used in the Mines & Quarries Act for underground workings for the extraction of coal.

Mines & Quarries Act – the Legislation governing workings of mines and quarries in the United Kingdom. It is supported by relevant Regulations that provide more detail.

NACODS – the National Association of Colliery Overmen, Deputies and Shotfirers, the officials' union.

Non-Statutory Undermanager – *see Note 6.*

NUM – the National Union of Mineworkers.

Officials – *see Note 1.5.*

Onsetter – a man employed to allow men off and onto the cages at the pit bottom and to load and unload vehicles on and off the cages. He was responsible for signalling to the banksmen and winding-engineman to cause cage movements to occur. (See banksman).

Operations Manager – a simplification used in this book for the Deputy Manager, Operations – scc Note 6

Overman – an official, "superior to a deputy and inferior to an Undermanager". *(See Note 1.3)*

Overwind – an incident in which a cage is wound too far at the pit top and is held in the headgear by a safety device, (e.g. King's Patent, Ormerod hook or Bennett's catchgear).

Pack – a wall constructed at the end of the face to bound the waste area that caved behind the face. Its purpose was to protect the arched supports of the roadway and provide a seal to prevent air leakage through the waste.

Packer – a person who builds a pack (a two-man operation). *(See Note 1.5)*

Packing bag – a robust, brown-paper sack used to pack cavities underground, when filled with dirt or crushed stone, or to build a face-end pack.

Pan – part of an armoured face conveyor or stageloader through which the chain was guided. Numerous pans were coupled together to make up the required length. *(See Note 1.2)*

Panel – a type of electricity supply equipment. *(See Note 4)*

Panzer – a chain conveyor, armoured flexible conveyor, AFC. *(See Note 1.2)*

Panzer driver – *see Note 1.5.*

Part (of a mine) – an area of a mine that fell under the legal jurisdiction of a Statutory Undermanager and bore his name.

Pickrose – *see Note 10.*

Pit – synonymous with "colliery" and "mine". Pitmen always talked about their place of work as "the pit".

Pit bottom – the area underground, in the vicinity of the shaft bottom. The marshalling area for men and materials at the upcast / supply / Number 2 Shaft and for handling the loading of coal in the downcast / Number 1 Shaft.

Pit-black – the workwear clothes, worn by miners for underground work or accessing their place of work. (In many instances, men stripped off and worked virtually or completely naked, particularly on hot districts or headings.) By the mid-1980's, each NCB miner had three sets of orange workwear shirt and overalls, provided and laundered under a contract between the NCB and Sketchleys. Prior to this, miners just brought old clothes from home to work in. These got particularly black, inside and out, because miners would often wear them for years, whilst never having them washed.

Pit top – the area on surface, in the vicinity of the top of the shafts, performing the corresponding purpose to the pit bottom.

Powered roof support – also referred to as a chock. The hydraulically-powered roof supports on a mechanised longwall coalface. Each had three pairs of hydraulic legs, mounted on a heavy, sledge-like steel base, to support the big, rectangular, steel beam which pressed up to the roof. A chock could weigh from two tonnes on the lowest faces to twelve or even fifteen tonnes on high faces. It supported four feet (1.2m) of the face length so there would be over 160 of these units on a typical 200m longwall face.

Pre-shift deputy – *see Note 1.3.*

Pre-shift inspection – a statutory inspection undertaken by the pre-shift deputy who tested for gas, and examined a district to make sure it was safe before the next shift of men entered the district to work.

Ram – a pneumatic or hydraulic piston.

Ram's head – a handheld, hydraulically-powered boring machine, used to drill holes in coal. (Smaller and less powerful than an electrically-powered 'bull's head' boring machine).

Retreat Face – a longwall face in which the roads are driven out to their full extent before the face is driven out. The face then retreats back along the gate roads. Where it was possible to use it, retreat working had advantages over an advancing face. *(See Note 5.)*

Return air – air in a mine that has passed through the workings and is flowing towards the upcast shaft.

Return or return airway – roadway (tunnel), along which the air travels from the face and out of the mine.

Ride – ascend or descend the shaft in the cage, or travel to and from work, underground, in manriding cars or on belt conveyors.

Riders – men congregating to ride, at the top or bottom of the shaft.

Ring – an arch, (See "arch").

Rip – the working area at a face-end in which the gate road was formed and its supports erected by a team of three rippers.

Ripper – one of the team employed to extract the gate profile material and set gate supports. *(See Note 1.5)*

Road or roadway – a tunnel.

Roof – the upper surface of an underground excavation.

Rope capping – *see Note 2.*

Rope-man – a man employed to maintain and extend, splice or install haulage ropes.

Royalty – an amount of money received by the owner of the coal lease, usually a specified amount per ton extracted.

Salvage – recovering all of the plant and equipment from a worked-out district or part of a colliery to be transferred, overhauled or sold for scrap.

Scab – a worker who works through a strike. A strike-breaker.

Screw – a piece of pigtail chewing tobacco.

Self-rescuer – *see Note 14.*

Services Manager – a simplification used in this book for the Deputy Manager, Services. *(See Note 6.)*

Shaft – vertical opening which connects the surface with the underground workings. Fresh air enters the mine by the downcast shaft, circulates the workings and is exhausted via the upcast shaft.

Shaftsman – a member of the team which inspected and maintained the shaft, including its lining, winding ropes and headstocks. Shaftsmen also slung large and awkward loads under the cages to transport them down or out of the pit and under-took rope-capping. *(See Note 2.)*

Shearer – coal-cutting machine on a coal face. *(See Note 1.1)*

Shot – a borehole primed with explosive and a detonator, enclosed with tamping which can then be fired to break strata.

Snake – the act of pushing over the armoured face conveyor after the shearer (coal-cutting machine) has taken a cut.

Snaker – a man employed to operate the chocks to push the AFC over to the coal face after the machine has taken a cut. *(See Note 1.5)*

Snap – sandwiches etc. taken to eat part way through the shift. Sometimes carried in a special 'snap tin'.

Snap-time – twenty-minute meal break and rest period, four hours into a weekday shift. Usually all machines and conveyors stopped. Sometimes that period was used for maintenance and setting steel at the rips at the ends of the face.

Splice - joining of an endless rope or a conveyor belt.

Spontaneous combustion – also referred to by the slang expression spon' comb'. *(See Note 8.) (See also "heating")*

Stable – *see Note 18.*

Stageloader – the short, chain conveyor connecting a face conveyor with the gate road belt conveyor.

Start-line – see "face start-line".

Statutory Undermanager – *see Note 6.*

Stone dust – crushed limestone (calcium carbonate). The ignition of naturally occurring methane gas was serious enough, but if this propagated a coal dust explosion then the consequences could be devastating throughout the mine. To help reduce the risk of this happening, stone dust was introduced to provide a concentration of suspended non-flammable dust particles in the path of the flame of an explosion, it was hoped that this would reduce temperatures and arrest an explosion. Stone dusting became mandatory on 1 January 1921.

Stopping – (also known as a "stank") *see Note 21.*

Strata bolts – *see Note 7.*

Strike – industrial action involving the withdrawal of labour in order to exert pressure on management.

Stuffing-box – device enclosing a drill-rod, at a borehole, to prevent an inrush when boring into an area liable to contain either water or gas.

Sump – a container at the bottom of a shaft, or any other place in a mine, that is used as a collecting point for drainage water.

Swilley – steep-sided dip or basin shape affecting a seam or roadway.

Sylvester – a manually-operated, pulling device. *(See Note 17)*

Tailgate – the return airway down which the supplies were transported to the face. Also known as the return or supply gate. Most longwall faces tended to have some degree of gradient running through their length (due to the inclination of the coal seam) and, for this reason, the tailgate would be the higher of the two gates so that the gradient facilitated the transporting of the coal down the face, thereby reducing the loads on the face conveyor motors.

Take – the area contained within the boundaries or limits of a mine, from which coal was to be extracted.

Tally and tallies – *see Note 3.*

Thirl or thurl – a joining of roadways, usually from opposite directions. "To thirl" was the act of breaking through to make a connection between two roadways.

Transfer point – a point in the conveyor belt transportation system where coal or stone is transferred from one belt conveyor to another.

Tree – any prop for supporting the roof. A Dowty hydraulic prop could be referred to as "a tree" as well as any piece of timber with a circular cross-section that was suitable for use as a roof support.

Troughing – steel ducting or transportable, flexible ducting used to deliver air from an underground auxiliary fan to the inbye end of a heading ('dead-end'). Particularly useful in roadway drivages (headings).

Tub – a steel container with four wheels to hold and transport coal or stone or other materials. (Also known as a mine car).

Turning coal – producing coal.

UDM – Union of Democratic Mineworkers, formed in 1984 as an alternative to the NUM.

Undermanager – *(See Note 6.)*

Upcast Shaft – vertical opening through which air returned to the surface after ventilating the mine workings. This shaft was used for winding men and materials (supplies).

Waste – The area which caves behind a coal face as the coal is extracted. Also known as the goaf or gob.

Winding rope – connected the cage to the winding-engine.

Winding sheaves – the drums, powered by a winding-engine, that wind the winding ropes.

Wooden chock – *see Note 22.*

Working – an underground area in which the coal is being mined or extracted.

Guidance Notes

Note 1. Shearers, the face conveyor and manpower deployment for a longwall face district:

1.1 Shearers: Longwall faces had two shearers, massive machines which spanned the armoured face conveyor to move up and down the face, cutting the coal with vertically spinning discs.

In a low seam of thirty inches, the discs would have a diameter of approximately two feet. In thick seams where the extraction height could be up to fourteen feet or more they could be as much as eight feet in diameter.

Each disc had approximately thirty steel-carbide tipped picks, each the size of a man's hand, installed on the three, thick, steel vanes that corkscrewed round them.

The distance that the discs cut into the face was known as the web-depth and enabled the machines to remove a cut that was two and a half feet (75cm) deep. The discs rotated at an amazing rate of two revolutions per second.

Although facemen would normally be at least a yard away from these fearsome components, there were accidents in which a miner was unfortunate enough to have got caught by a disc, usually when replacing worn, shearer disc picks. In such terrible cases, the unfortunate faceman would be reduced to mince-meat in a fraction of a second.

The maingate machine was usually a "double-ended ranging-drum shearer" (DERDS) used to cut the majority of the face, taking the higher coal and forming the roof with the leading disc whilst the trailing disc took the lower coal to form the floor at the same time. When you looked up at the roof you could see the see the straight lines of the score marks, about half an inch deep, made by the picks as they sheared the coal.

The tailgate machine was only required to do the cutting necessary to keep the tailgate end of the face conveyor moving over and this was a single-ended ranging-drum shearer (SERDS) with only one disc. This enabled the three rippers, who erected the steel arches to form the tailgate face-end road and the two packers, who built the pack wall that sealed the waste from the road and provided support to protect the arches at that end, to keep working.

The length of the maingate shearer's second gearhead made that machine about 40% longer than the tailgate machine.

The shearers consisted of several, heavy, steel-boxed units, mounted on a bedframe

The bedframe was a long, heavy-duty steel chassis that provided the machines' backbone, supporting the heavy, steel-boxed machine sections and straddling the panzer with an arch shape that left a gap of about eighteen to thirty inches over the conveyor (depending on the height of the face).

The shearer's three units were: the gearheads which carried the coal-cutting discs on their ranging arms which enabled the machine driver to raise and lower the level at which the disc cut through the face; the power-pack, containing the electrically-powered pump which provided the hydraulic power to move the machine and operate the ranging arms; and the haulage section which provided the force to move the shearer through the face.

[See Note 17.2 Shearer Bedframes]

1.2 The face conveyor: The four terms AFC, Armoured Flexible Conveyor, Armoured Face Conveyor and "Panzer" were synonymous for the conveyor on the coalface in the 1980's.

All pitmen understood that the panzer had made its German inventor a fortune, as the winner of a competition launched in the 1930's by Hitler, himself, to find a design of face conveyor that was flexible and easy to advance at the same time as being durable. Hitler wanted this in order to boost Germany's pre-war production to fuel his nation's re-armament.

The face conveyor was made up of steel pans, each being five feet long and thirty inches wide. The pans were linked by 'dog-bone' connectors, distinctively-shaped steel links the size of a small dumb-bell, their two heads slotted into cups on the ends of adjacent pans, holding the pans together whilst providing the required flexibility.

Steel flights at metre spacings, spanning the pans and drawn by heavy steel chains on either side of the conveyor, dragged the coal off the face. The chains and the ends of the flights ran in 'races' on either side of the pans. These were vertical channels, three inches high, built onto the top and bottom of pans which kept the flights down in position to scrape the coal off the face on the pans.

The pan-sides were steel plates. They bolted onto the pans that made up the armoured face conveyor, immediately in front of the powered roof supports, to keep the coal on the panzer and to provide an anchorage for the shearer, the coal cutting machine, as it travelled through the face.

Toe plates were installed on the face-side to dig under loose coal and keep the pans on the floor that had been cut by the shearers.

[See Note 15. Drive sections or gearheads, and Note 16. Installing panzer chain]

1.3 Officials:

District Overman – responsible, on one shift, for overseeing production, coal clearance. material supplies, supply and maintenance of services (extending rope haulages serving an advancing face, moving electrical panels to keep them up with the face etc.). Managed all of the deputies, supporting them in maintaining and enforcing discipline within the workforce.

Face Deputy: (Grade I deputy). Reported to the overman and managed and led the face team. Statutorily responsible for coal production and all work carried out on the face and within the district, the safety of the district and the health and safety of the face team. Maintained and enforced discipline within the workforce.

Pre-shift Deputy: (Grade 2 deputy role). Undertook the statutory pre-shift inspection of the whole district to ensure the district was safe to enter and work.

Tailgate Deputy: (Grade 2). Managed the tailgate road. Managed 'the supply lads' and ensured timely delivery of the materials required at the face. (Timber, steel arch bows and legs, corrugated sheets, cement for pump-packing etc.).

Maingate Deputy: (Grade 2). Responsible for the coal clearance system of belt conveyors and their operators.

The Tailgate and Maingate deputies would manage any back-rips or dents working to enlarge their roads to combat roadway closure (see below)

1.4 Face-team:

Maingate shearer ("Number 1 Machine) driver	1
Maingate shearer's No.2 driver	1
Tailgate shearer ("Number 2 Machine) driver	1
Machine cable men	2
Snaker	1
Chockers	3
Rippers (maingate and tailgate)	6
Packers (maingate and tailgate)	4
Chock Maintenance Fitter	1
Fitters	2
Electricians	2
Face-trainee	1
Panzer driver	1
	26

1.5 Face-team roles:

Maingate shearer ("Number 1 Machine) driver: drove the shearer that cut the majority of the face-line (typically, a length of about 200 metres in total), controlling the shearer's speed of transit and the position/height of the leading arm and disc in the direction being cut (taking the upper level of the seam extracted and forming the roof of the face). Also cut out into the maingate road at the face end and 'ranged-out' the road's arched profile to form the maingate road.

Maingate shearer's second driver: controlled the position of the trailing disc/ranging arm to take the lower level of the seam extracted and form the floor for the face.

Tailgate shearer driver: drove the shearer that cut the last thirty metres of the face at the tailgate and cut out into the tailgate road at the face end and 'ranged-out' to form the tailgate's arched profile. This was to keep the tailgate packers and rippers operating productively and to keep the tailgate advancing to enable more cuts of the whole face-line to be taken.

Machine cable men: moved through the face with the loop of the big, rubber sheathed cables which powered the machine to prevent them becoming caught and damaged. The tailgate cable man typically snaked the panzer over at the tailgate end. This role was eventually made redundant by more reliable, cable management systems.

Snaker: used the snaking rams integral to the powered roof supports' (chocks') bases to ram over the armoured face conveyor (AFC/'panzer') to the face after each cut by the Number 1 Machine. The man responsible for ensuring that the panzer and, therefore, the face-line remained straight.

Chockers: advanced the hydraulically-powered roof supports to the panzer, after each snake of the panzer, by dropping each support, individually, and drawing it in, using the snaking ram, before pumping it back up to the roof.

Rippers (maingate and tailgate): three men at either end of the face who erected the arches (composed of three, curved, heavy steel RSJ girders (one bow and two legs)) which supported the gate road at yard intervals. The rippers coupled up consecutive arches with steel, tubular struts and lined the arches with corrugated steel sheets. They packed the space between these sheets and the roof and sides, opened up by the shearer, with brown paper packing bags filled with stone they had shovelled in to them.

Packers (maingate and tailgate): two men at either end of the face who erected the pack at the face-end, on the face-side of the road, to fill the gap for about two metres, left when the coal was cut out by the shearer. This provided some support to protect the steel arched supports of the gate road and to provide some seal against air short-circuiting through the waste. The pack, at its most elementary, was formed by packing bags filled with stone or, at its more sophisticated, by hanging a sturdy, canvas, waterproof bag (like the bedroom of a frame tent) and pumping a mixture of cement and water into it.

Chock Maintenance Fitter: maintained the chocks (powered roof supports). In particular, replaced damaged hydraulic hoses which powered the legs and snaking rams on the chocks.

Fitters: maintained and repaired all mechanical equipment on the face, principally, the two shearers.

Electricians: maintained and repaired all electrical equipment on the face, including the two shearers, the two drive motors on the panzer and its lockout system and the face tannoy (communication system).

Face-trainee: was undergoing the training, required by Law, before a man could be employed on face-work.

Panzer driver: was not 'face-trained'. Stood out in the maingate to operate the controls which provided power to the armoured face conveyor (AFC/panzer), which ran through the face-line, and stageloader which carried the coal from the face conveyor to load it onto the maingate belt conveyor. Provided the communications link between the face and the surface control room, keeping the control room attendant (and, through him, the senior management) informed of the progress of the shearers through the face and of any problems affecting production etc.

Note 2. Rope-capping:

It was a legal requirement under the Mines and Quarries Act and its Regulations that six feet be cut off the end of each winding rope, every six months. This rope sample was examined, tested and measured to ensure that the amount of strain in the rope (the amount the rope had stretched) was not excessive and that the rope was safe and within the specified safety limits.

The cage would be supported, to enable it to be uncoupled from the winding rope during this operation, by running a trolley onto the top deck of the cage loaded with two, ten inch, webbed girders that were long enough to extend wider than the shaft on both sides of the cage. When the winding-engineman received the signal from the banksman to lower the cage steadily, the girders came to rest on the floor of the pit top so that, when the underside of the roof of the cage made contact with the girders, the cage was supported, releasing the winding rope of its load.

On the end of each rope there was a capel, a casting which connected the winding rope to the four, massive chains from which the cage hung. The capel was secured to the rope by splaying the thick wires which made up the winding rope and fusing them into the structure of the capel by filling it with molten, white metal. This was known as "capping the rope".

Taking the six-foot sample of rope involved removing and replacing the capel, so rope-capping was another of the essential, highly skilled and hazardous duties of a colliery's shaftsmen.

Note 3. Tallies and contraband:

3.1 Tallies: On arrival at the mine, a miner's first stop was at the opening between the main entrance and the deployment centre, known to colliers as the tally office window. Here, the deployment centre clerk would issue him with his pair of tallies from the rack prepared for men supposed to be on the forthcoming shift.

Each man had a "silver tally", a thin, square, hexagonal, oval or circular, grey metal plate, about an inch and a quarter wide, and a similar shaped, brass tally. Both tallies had the colliery's name and "NCB" embossed on them and the miner's identification number stamped on them. There was a hole in each to enable them to be clipped, with a dog clip, on to the miner's belt.

No miner would be allowed to board the cage without unclipping and handing his silver tally to the banksman at the pit top. The banksman placed the tallies for each cage-load, or "run", of men into a metal cylinder which he placed in the compressed air tube system which blew the tally container across the colliery surface, back to the deployment centre.

On returning to the surface, the miner was required to hand his brass tally to the banksman.

The presence of a miner's silver tally and absence of the brass his brass tally indicated that he was still underground. This was a way of accounting for the safe return of the miner to the surface. This had been found to be essential in the event of men being trapped or lost underground due to explosions, inrushes of water and roof-falls.

3.2 Contraband: The banksman was notified of the number of miners he was required to search, each shift, for contraband. This involved a quick search of their clothing, "observing the proprieties", in the words of the Mines & Quarries Act, i.e. not groping their private parts (often ignored).

Contraband was anything that might cause a naked flame or incendive spark underground and risk igniting a methane explosion. This included cigarettes, lighters and matches

Miners' collective horror of explosions was so great that even the most addicted smoker never thought about cigarettes whilst underground, particularly with the nicotine hits that many obtained from taking snuff or chewing tobacco.

Methane explosions still occurred, infrequently, in British coalmines in the 1980's. In the mid-nineteenth century these had caused massive disasters at numerous pits. In many cases, they had wiped out entire shifts of several hundred men working underground at the time.

These all-enveloping explosions are a particular mining phenomenon and their extent and power are in their very nature.

Three things are required to cause them; an explosive mixture of methane and air, a flame or incendive spark and coal dust.

The sequence of events starts with an accumulation of methane and air, usually in a layer near the roof because of the relative lightness of the methane. A spark, perhaps from a pick striking 'fool's gold' (iron pyrites) ignites this first explosive mixture.

The blast from the methane explosion is concentrated by the confined nature of mine workings and raises the fine, coal dust lying on the floor and in the flanges of the arch supports into the air in advance of the flame of the burning methane, creating another, perfect, potent, explosive mixture. The coal dust and air mixture requires considerable heat to ignite it and the methane explosion provides this. Whilst methane is easier to ignite, once it combusts, its fierce heat detonates the coal dust explosion. This creates the last and most catastrophic phase of the conflagration so that, when the pocket of methane mixture is exhausted, the coal dust explosion continues and the blast in front of it, raising more and more coal dust creating a self-sustaining beast; the terrifying, self-propagating, coal dust explosion.

The first preventative measure was to provide adequate ventilation to dilute and flush out the methane and to sample the air to make sure dangerous levels were not allowed to accumulate.

The next prevention was to ban the use of anything with a bare flame or which could result in a spark. This was the reason for the abhorrence that was apparent, and which trainees were imbued with, for the pitman who would betray his mates and their dependants by bringing in the potential to ignite a methane explosion. It was also why heavy equipment had to be intrinsically safe, with switchgear sealed in heavy, airtight steel containers so that no spark could escape in the event of arcing from an electrical short.

The mine management and officials were required by law to prevent the build-up of coal dust and to dilute it by scattering inert stone dust amongst it.

Finally, as a last line of defence, the law required that stone dust barriers should be maintained within specified distances of the face. If all else had failed, the face and all the men on it would be blown up but, when the blast of the explosion hit the stone-dust barriers, suspended from the steel arches in the gate road, their shelves and boards would be blown down, cascading the white powder to dilute the coal dust kicked up by the blast to make it non-explosive, thereby saving the rest of the mine and its workforce.

Note 4. Electrical switchgear; "panels":

The electrical power to the coalface was provided through electrical panels. These large, steel, flameproof and airtight cabinets were section switches, electrical switchgear designed to enable an electrical connection to be made safely to the high voltage underground supply (typically 550kV or 110kV at the face). At the outbye end of a bank of panels was a circuit breaker to enable power to be disconnected safely from the panels.

The bank of panels in the tailgate were usually mounted on trolleys and moved forward with the face as it advanced or retreated on a short track, laid alongside the tailgate's supplies haulage track.

The coalgate's electrical panels, were slung on a platform from monorail over the belt conveyor. These powered the maingate shearer, the panzer's motors and the stageloader (the chain conveyor that carried the coal from the face and onto the gate belt). This was where the panzer driver stood to operate the panzer and stageloader and act as the communications link between the face and the surface.

Note 5. The benefits of retreat faces over advancing faces were:

Driving the roads to the full extent of the workings 'proved' the coal seam so the management could have more confidence in their being no poor roof or faults or 'wash-outs' where stone replaced the seam of coal;

The rippers job was easier because the road was already formed and the steel already set;

There was no need to build packs at the face-ends to protect the gate roads because these could be allowed to close behind the face. This saved time and the cost of packing materials and removed a limitation on the rate at which the face end could move over; and

With 'packers' not being needed to build the packs, there was a 20% saving in manpower.

Note 6. Statutory management and officials:

"The Enemy Within" contains references to a number of statutory and non-statutory roles.

The Mines and Quarries Act defined three statutory roles that were required in the management and supervisory structure of every mine:

"Manager", or Colliery General Manager

"Undermanager", known as a "Statutory Undermanager"

"Deputy" *(see Note 1.3)*

The role of "overman" was also mentioned in the M&Q Act but no title was ascribed to it. It was just described as "an official superior to a deputy and inferior to an undermanager". *(See Note 1.3)*

The Colliery General Manager was required to hold a First Class Certificate of Competency in Mine Management.

The Mines and Quarries Act required everywhere in a British coal mine to be under the control of a competent, Statutory Undermanager, holding either a First Class or Second Class Certificate of Competency in Mine Management. This accountability involved being in charge of a significant part of the mine, twenty-four hours a day, seven days a week.

The Statutory Undermanager's appointment and the extent of their part of the mine had to be clearly indicated on a plan and both had to be notified and acceptable to the Her Majesty's Inspector of Mines. From this statutory perspective, in *The Enemy Within*, Ken Goodall was in charge of all parts of the mine and Jim, as the 'Non-Stat", was nominally acting under his authority.

The Non-Statutory Undermanager position gave management experience opportunities to mining engineers. The role almost invariably involved working regularly on one of the two 'back-shifts' (afternoons or nights) and taking charge of the mine on that shift on behalf of the Colliery General Manager. It was highly unusual for them to be given round-the-clock responsibility for a significant part of a mine and for them to have their working life based more around the dayshift working pattern, although it did happen on occasions.

For the sake of simplicity, in *The Enemy Within*, I have modified the other two main titles, changing "Deputy Manager, Services" to "Services Manager" and "Deputy Manager, Operations" to "Operations Manager". The Deputy Manager, Operations was, effectively, the Number 2 man at any pit. The Deputy Manager, Services was, effectively, a senior Undermanager with no designated part of a mine under his responsibility. Holders of both deputy manager positions would be required to hold a First Class Certificate of Competency in Mine Management because they were expected to be bound for higher things.

In reality, there was not much need for the Deputy Manager, Operations. He was, hierarchically, the Undermanagers' superior. However, the fact that the former's role was not statutory meant that there were limitations on what he could instruct the Statutory Undermanager to do.

The Deputy Manager, Services might have been a help to the Statutory Undermanager in developing the access to the reserves in his part of the mine. However, too often, the people in these roles failed to make much impression on this, due to their lack of statutory authority.

Note 7. Anchorages and floor plates:
Strata bolts would be used to secure plates to the floor underground to act as an anchorage. Strata bolts were steel rods that were five or six feet long and could be secured with ampoules of resin in holes drilled into the floor.

Note 8. "Heatings":

Heatings involved the spontaneous combustion of coal in the waste behind the face. This increased the possibility of a naked flame coming into contact with an explosive mixture of methane and oxygen and gave off carbon monoxide, an irreversibly poisonous gas, which could rise to lethal levels in the underground confines. They had the potential to blow up a pit or kill a production district or the men in it.

Heatings could either be shallow, in which case the flammable material on fire was close to the gate road, or deep-seated, when the fire took hold deeper into the waste.

The spontaneous combustion that resulted in heatings was caused by a number of factors; the heat coming from the ground due to the geothermal gradient, the exothermic oxidation of materials like iron in the coal and rock in the waste, and the heat generated by friction as the slow consolidation of the waste crushed the coal and carboniferous material that had caved into the waste.

The other essential factor in causing heatings was oxygen. This came from the air passing through the waste between the imperfect seals of the maingate and tailgate roadsides.

As well as the important preventative measure of sealing the roadways with the roadside packs that were built as the face advanced and by shuttering the arches with lagging boards and pumping in a grout mix of water and gypsum behind them, the most effective defence was to keep the face moving. As the face advanced and the roof caved behind the chocks as they were drawn in, the waste further behind the face became consolidated, causing the passage of air to become progressively restricted, the further back the waste was from the face.

When a face was static for any time, like pit holidays or, as 22's was in *The Enemy Within*, during salvage or equipment transfer, the airflow could continue for longer periods at more consistent flow rates through the waste. When that rate was sufficient to accelerate the rate of oxidation whilst being too low to cool the heating strata, there was a danger that some pocket of flammable material, typically coal and shales from the roof, would start to combust.

The high calorific value of Warwickshire's coal and the depth of its seams helped to make it the coalfield most afflicted by these fires. Statistically, there were more fires in the Warwickshire Coalfield than the rest of the country put together, by a factor of 19 times.

[See Note 14 on Spontaneous Combustion and Self Rescuers]

Note 9. A Jim Crow:

A Jim Crow was an implement to bend rails to form curves in rail track. It consisted of a big, curved steel yoke, a few inches thick, with a large claw on either end to hook onto the rail to be bent. In the mid-point of the curve was a large gauge bolt which would be wound through a heavily-threaded, six-inch diameter hole to apply pressure to the side of the rail to be bent to a curve. By applying pressure gradually over a number of points, it was possible, with skill, strength and patience, to bend a rail to form a curve of the requisite radius.

Note 10. Pickrose haulage motors and types of rope haulage:

Pickrose haulage motors powered the rope haulages used underground to transport supplies. Pickrose was the name of the company that manufactured them.

Like all of the plant and equipment the National Coal Board purchased for underground use, Pickrose haulages were painted in the NCB's operational colour of matt white. They consisted of a big cable drum, driven by an electric motor, secured to a steel base-plate. The company's evocative logo of a pick and rose was always embossed on their sides.

They were normally held in position by girders, laid across the motors' steel base-plates. These girders would be held down by plates anchored by six-foot strata bolts, secured into the floor with ampoules of resin.

An **endless haulage** was used to move supplies and equipment along a rail track. The front and the back of the train of vehicles was clipped onto the continuous, steel cable which ran along in the middle of the two rails of the track and round a return wheel at either end of the haulage run and round the drum of the haulage motor, off to one side at the outbye end of the haulage.

A **direct haulage** dragged a vehicle or small train up an inclined track (or lowered it down, where there was sufficient gradient to propel the vehicle or train) using a single rope with a pommel on its end which dropped into the coupler of the front vehicle to be towed.

Note 11. 'Chock tests' – Load-testing of powered roof supports:

Normally, when a face ended its life, the aim was to salvage all of the equipment with residual life and transport it out of the mine and then transport it to the NCB's Equipment Overhaul Workshops at Swadlincote in South Derbyshire.

With face-to-face transfers, the equipment would be staying

underground to be reused, going straight from one face to the next. In these cases, tests had been introduced using specially designed test-rigs to test the load-bearing capacity of each chock (powered roof support). The test would check that no leaking seals or weak relief valves were letting the chock drop before it had taken its designed load. This was to check that the right level of support was going to be provided to prevent the roof from breaking down and causing falls of ground on the coalface.

In terms of Jim Greaves' view that the fitters in this instance might have been tampering with machinery unnecessarily; a fundamental tenet of Reliability Centred Maintenance is that unnecessary maintenance interventions should be avoided to prevent equipment failures from being induced.

Note 12. Half-shift:
Although this overtime period was only an extra couple of hours in duration, a half-shift was so-called because, being paid at a rate of "time and a third", its pay was roughly equivalent to half that of a normal shift.

Note 13. The Incentive Bonus System:
The Incentive Bonsu System had been introduced as part of the National Coal Board's national pay settlement with the NUM in 1983. It was calculated on the basis of tonnes produced per week against the colliery's specific, weekly target. All mineworkers received bonus payments with face-workers receiving the full bonus and other workers receiving lesser percentages, dependent upon their grade.

Note 14. Spontaneous combustion and self-rescuers:
It was a legal requirement for every man underground in British coal mines to carry a self-rescuer. The rescuer was threaded onto a narrower, buckled strap, attached to the pitman's belt. It consisted of a shiny, metal container, about the size of a half-brick, which opened up to yield the mouthpiece and converter inside. Breathing through this device turned poisonous carbon monoxide into carbon dioxide.

This potential life-saver was only designed and intended to provide the wearer with sufficient time to escape, making his way out of the dangerously contaminated air, it did not enable them to continue to work in high levels of carbon monoxide.

Carbon monoxide's lethality arises from the reaction it causes in the blood. In healthy blood, red blood cells absorb oxygen from the lungs as the haemoglobin they contain is converted into oxy-haemoglobin which carries the oxygen and delivers it to tissue around the body. When carbon monoxide is breathed in and absorbed from the lungs into the blood it reacts with the haemoglobin to create carboxy-haemoglobin. This reaction is irreversible, which means that the haemoglobin is irreparably damaged and the blood's ability to absorb and carry oxygen around the body is reduced. The amount of reduction is dependent on the concentration of carbon monoxide in the air and the length of the person's exposure to it.

Repeated shifts working on a district with a fire progressively increased the amount of haemoglobin rendered useless. A short term exposure of a few minutes could have the same effect, if the concentration of carbon monoxide was sufficient. In both cases, this could be debilitating and even deadly. The antidote to this was the natural replacement of red blood corpuscles over time, away from the high carbon monoxide levels.

Carbon monoxide poisoning by exposure to a vehicle's exhaust fumes is well known as a means of committing suicide. When its level rises sufficiently in the event of a fire underground, it could be similarly effective at quickly rendering its victims' unconscious and delivering swift deaths, thereafter.

Note 15. Drive sections or "gearheads" on panzers:
Drive sections on panzers consisted of a pan that was over three times the length of the normal pans and which rose up to form the robust unit to house the gearbox and electric motor to drive the face conveyor and to raise the shearers to enable them to range out the full height of their gate roads.

The panzer's maingate drive, at the delivery end, dragged the chains and flights and their load of coal through the face. The tailgate drive, at the other end of the face, was needed to throw slack chain onto the top of conveyor and drag the weight of the bottom chain through the enclosed underside of the pans to take that load off its counterpart at the maingate end of the face.

Note 16. Installing panzer chain:
The first half of the chain, on installation, was used to form the "bottom chain", in the space between the surface of the pan and the base plate underneath.

Every fifth pan in the conveyor was fitted with an inspection cover, the steel plate of the cover could be unclipped and lifted out so that the men could access the bottom chain to work on it.

Typically, drain rods would be used to pull the chain down into and through the hidden cavity. The complexity of getting the chain into the covered underside of the pans and coupling it up was more time consuming than installing the top chain.

Note 17. Sylvesters and bedframes:

17.1 The Sylvester: was a hazardous implement that provided pulling force by the action of a steel lever operating to move a small, steel box along a steel ratchet bar with one hook and chain attached to one end of the ratchet bar and another to the steel, levered box.

Occasionally, during the operation of a Sylvester, the lock on the ratchet box failed and the tension in the implement would be released by the lever bar flying forward under enormous force. This frequently resulted in some injury to the operator.

17.2 A shearer bedframe was fabricated as a single, long, massive steel casting. This robust chassis was the shearer's backbone, giving it enormous strength and good stability. It supported the heavy, steel-boxed machine sections and straddled the panzer with an arch shape that left a gap about eighteen inches over the conveyor. Being too long to fit in the cages, when they were transported down the shaft into the pit they had to be slung under the cage by the colliery's shaftsmen to lower them down the shaft. Getting them into shaft at the top of that journey and out of the shaft in the pit bottom involved highly skilled manoeuvring by the shaftsmen.

Note 18. Stable:

On a working face, a stable was an area in the seam, opened out in the coal-face and supported safely, to enable the men to work on the face-side of the shearers. This was to remove the risk of them being crushed or buried by coal 'hading off' (i.e. breaking off and falling away from) the coalface.

Note 19. A-frames:

A-frames were powered supports which sat at the gate-end of the face, within the roadway itself, when the face was in production, one in the maingate and another in the tailgate. They were used to push over the heavy gearhead of the armoured face conveyor or panzer.

An A-frame had a bulky, steel sledge-like base, two long hydraulic legs (one front, one rear) that provided the support to the long narrow beam that pressurised up to the top, middle section of the arches, known as bows, nearest to the face.

In high roads, at the end of thicker seam faces, the A-frame also provided the platform for the rippers who set steel, i.e. erected the three-piece steel arches of a bow and two arch-legs on an advancing face. On a retreating face, the A-frame would support the bows when the face-side leg of an arch was taken to allow the panzer or armoured face conveyor to pass as it was snaked over in the maingate, where it delivered onto the stageloader which took the coal away from the face in that road.

Note 20. Dint-headers:

Dint-headers crawled on heavy duty caterpillar tracks like roadheaders, but lacked the latter's boom which was used to cut out an arch profile. They had a long, powered steel drum, the width of the machine, mounted on the front that drove a short, heavy-duty steel conveyor which loaded out the coal that had been loosened by shot-firing. The front conveyor could drop to the floor or raise to the roof. A metal linked conveyor ran through the machine and this delivered the coal onto a narrow conveyor belt that would be extended through the face-line as the machine advanced.

Note 21. A stopping:

These barriers were used to seal off worked-out face districts. They consisted of two walls, four metres apart, built with strong, brown-paper sacks, called packing bags, filled with dirt. The gypsum-based, pinkish, plaster-type material known as hardstop was mixed with water and pumped in to fill the void between the walls and create a seal on the road.

Note 22. A wooden chock:

A wooden chock was a support that consisted of wooden chock blocks that were two feet long with a six by six-inch square cross-section. Each chock was built by placing two of the blocks parallel to each other, one foot apart. The next layer of two chocks was laid across the ends of the first pair of chock blocks. The chocks were built up, in this way, until they reached the height needed.

Note 23
NCB
National Coal Board
Hobart House, Grosvenor Place, London SW1X 7AE
CHAIRMAN
Ian MacGregor

June, 1984
Dear Colleague,

YOUR FUTURE IN DANGER

I am taking the unusual step of writing to you at home because I want every man and woman who has a stake in the coal industry to realise clearly the damage which will be done if this disastrous strike goes on a long time.

The leaders of the NUM have talked of it continuing into the winter. Now that our talks with them have broken down this is a real possibility. It could go on until December or even longer. In which case the consequences for everybody will be very grave.

Your President talks continually of keeping the strike going indefinitely until he achieves "victory".

I would like to tell you, not provocatively or as a threat, why that will not happen however long the strike lasts.

What this strike is really about is that the NUM leadership is preventing the development of an efficient industry. We have repeatedly explained that we are seeking to create a higher volume, lower cost industry which will be profitable, well able to provide superior levels of earnings while still being able to compete with foreign coal. To achieve this, huge sums of money are being invested in new equipment; last year it was close to £800 million and we expect to continue a similarly high rate of investment in the years ahead. Our proposals mean, short term, cutting out some of the uneconomic pits and looking for about 20, 000 voluntary redundancies – the same as last year. The redundancy payments are now more generous than ever before for those who decide not to take alternative jobs offered in the industry.

However long the strike goes on I can assure you that we will end up, through our normal consultative procedures, with the same production plans as we agreed with your representatives on 6th March last.

But the second reason why continuing the strike will not bring the NUM "victory" is this: in the end nobody will win. Everybody will lose – and lose disastrously.

Many of you have already lost more than £2,000 in earnings and have seen your savings disappear. If the strike goes on until December it will take many years to recover financially and also more jobs may be lost – and all for nothing.

I have been accused of planning to butcher the industry. I have no such intention or desire. I want to build up the industry into one we can all be proud of.

But if we cannot return to reality and get back to work then the industry may well be butchered. But the butchers will not be the Coal Board.

You are all aware that mines which are not constantly maintained and worked deteriorate in terms on safety and workability.

AT PRESENT THERE ARE BETWEEEN 20 AND 30 PITS WHICH ARE VIABLE WHICH WILL BE IN DANGER OF NEVER RE-OPENING IF WE HAVE A LENGTHY STRIKE.

This is the strike which should never have happened. It is based on very serious misrepresentation and distortion of facts. At great financial cost miners have supported this strike for fourteen weeks because your leaders have told you this . . . . .

That the Coal Board is out to butcher the coal industry.
That we plan to do away with 70, 000 jobs.
That we plan to close down around 86 pits, leaving only 100 working collieries,

IF THESE THINGS WERE TRUE I WOULD NOT BLAME MINERS FOR GETTING ANGRY OR FOR BEING DEEPLY WORRIED. BUT THESE THINGS ARE ABSOLUTELY UNTRUE. I STATE CATEGORICALLY AND SOLEMNLY. YOU HAVE BEEN MISLED.

The NUM, which called the strike, will end it only when you decide it should be ended.

I would like you to consider carefully, so we can get away from the tragic violence and pressures of the mass pickets, whether this strike is really in your interest.

I ask you to join associates who have already returned to work so that we can start repairing the damage and building up a good future.

Sincerely,

Ian MacGregor